Geothermal
Systems

Geothermal Systems

 John Elder

The University of Manchester
Department of Geology

1981

Academic Press

A subsidiary of Harcourt Brace Jovanovich, Publishers

London · New York · Toronto · Sydney · San Francisco

ACADEMIC PRESS INC. (LONDON) LTD.
24/28 Oval Road
London NW1

United States Edition published by
ACADEMIC PRESS INC.
111 Fifth Avenue
New York, New York 10003

British Library Cataloguing in Publication Data
Elder, J
  Geothermal systems.
  1. Geothermal engineering
  I. Title
  621.4   TJ280.7   80-49663
  ISBN 0-12-236450-3

Printed in Great Britain by
Alden Press, Oxford, London and Northampton

Preface

The substance of this book is an attempt to bridge the gap between large-scale volcanic and small-scale hydrothermal systems. From this point of view the book should be of interest to geologists, geochemists and geophysicists who are interested in mechanisms which involve the earth's heat.

More specifically, this book is for scientists and engineers who are now, or in the future will be, involved in projects designed to exploit the internal thermal energy of our planet. This branch of planetary engineering is very new, our techniques are crude and we have a lot to learn. As with other projects which set out directly to exploit nature — for example, hydroelectric power generation — most of the total system is not under our control. But in the case of geothermal power the situation is even worse since we do not understand, except in part, how the natural system works. Consider a typical geothermal power project. Many of the tasks involved can be handled by more or less standard engineering techniques: pilot holes can be drilled in an area delineated by a variety of geological, geochemical and geophysical measurements; aquifer performance and discharge characteristics measured; financial and logistical matters considered; production holes drilled; a power station designed and built. All this is fine until the engineer in charge begins to realize that the natural system has a life of its own: ground levels are falling; discharge is falling and the enthalpy rising; land slips increase; a small phreatic eruption destroys the main pipe line. It is my belief and experience that it is vital for all concerned to have a broad and detailed knowledge of the widest possible range of processes that can occur in hydrothermal systems. They are much more complex and dynamic in their nature than an oil deposit or a river.

In writing this book, I have tried to avoid what I call the "bumble-bee syndrome". This refers to the apocryphal story of the applied mathematician who is alleged to have deduced that the bumble bee could not fly. Thus: the observation of the rapid and persistent rearrangement of the earth's crust requires some global-scale convective system within the mantle; the observation of extrusive basaltic magmas and the results of laboratory experimental petrology require partial melting within the mantle at depths of order 10^2 km; the observations of the persistent discharge of water at the surface in geothermal areas requires a convecting body of water to be maintained in the crust. Insufficient observational data, however, prevents this ideal approach in many instances. Nevertheless, from this point of view the sets of mathematical relations presented here are not so much deductions from the relations of Newtonian mechanics and thermodynamics but phenomenological scenarios which provide quantitative and precise models of the system of interest. Of course, they are simplifications, perhaps over-simplifications, but they encapsulate the essential factors which describe the individual processes operating within the system.

This book was started officially in 1973, but in reality in 1954. Early in that year I was fortunate, as a graduate student, to have a temporary research job at the Applied Mathematics Laboratory, DSIR, Wellington, for work related to the New Zealand geothermal project. That study envisaged the possibility of the use of various laboratory-scale models. In 1956, as a graduate student at Cambridge, I built what I believe was the first laboratory model of a hydrothermal system and showed that at a Rayleigh number of order 10^3 it was compatible with the field data from Wairakei. Also at about that time, R. A. Wooding had begun his work with numerical models. Laboratory and numerical model work has subsequently proved a fruitful method of investigation and generator of ideas about hydrothermal and lithothermal systems.

The outstanding problem in getting this book together was the great gap between descriptions of strictly volcanic systems and purely hydrothermal systems. For a time my own progress was nil. Then I saw that two distinct pieces of work when put together provided a link. These are my own studies of the development of the lower Tertiary lava plateau of central West Greenland and a simple model of hydrothermal ore genesis. This material is the linch pin of the work, given in Part II. There is, however, an aspect of these only sketchily described here. At the moment studies of this type are either strictly physical, as here, or strictly petrogenetic. Little has yet been done to put these two aspects explicitly together. Some work is in progress and I expect that a book like this written ten years hence will have this marriage of ideas as its central theme.

I have unashamedly plagiarized my own work (Elder, 1966). That material, where possible brought up to date, forms the kernel of this work. Around it I have wrapped a shell of physical studies of volcanic processes. Since all these systems derive their energy and rock-substance from the mantle, they need to be seen in the perspective of the global heat engine. Nevertheless I do not wish to give the impression that this book is a study of the hot earth. Not at all! I wish to focus attention not so much on the origin, nature and location of the reservoirs of heat but rather on the question of how the heat gets out. This is a study of heat and mass *transfer* in lithothermal and hydrothermal systems deriving their energy from a vigorously convecting mantle, with emphasis on the mechanisms by which the energy is delivered to the surface.

The book deals mainly with an aspect of volcanology. Volcanic systems are treated as if they were localized independent systems. Clearly this is a simplification — we look at the earth as it were through a high-pass filter, set to a scale of 100 km or so. Where it is necessary to consider the role of processes which operate on a longer scale — the information not seen directly through our high-pass filter — the system inputs are modified by means of simple kinematic representations of the larger-scale processes. The procedure works because the local processes are relatively rapid, with time-scales of order 1 Myr, while global processes have time-scales of order $10-10^2$ Myr. In any event, an author has to draw the line somewhere, much as an artist uses a finite piece of canvas.

The material is arranged according to two criteria: size of system, large to small; and proximity of the working substance to the surface.

The studies described in Parts I and II do not pretend to be complete in themselves. They deal only with certain aspects of the larger-scale geothermal systems. They do however provide an essential conceptual path to the studies of hydrothermal systems. These lithothermal systems act as large heat and matter reservoirs and exchangers which derive their energy ultimately from the mantle. Although many aspects of the behaviour of hydrothermal systems can be described as if they were closed and isolated, an understanding of their overall behaviour relies on a knowledge of the systems in which they are embedded.

A further important restriction of the material here should be noted. The focus of our interest is in systems operating within the earth as it now is. In other words I restrict the discussion to systems which had their origin in the current phase of geological history — the past few hundred Myr. Matters concerning the early geological history of our planet are only referred to in so far as they have a direct bearing on the current situation.

Considerable emphasis is placed on the use of models based on the notions of reservoirs interconnected with pipes. This frequently provides a compact sketch of a system in terms of the capacitances and resistances of the system elements, the masses of material and their residence times within the system elements. This type of description represents the structure and behaviour of a system by means of an equivalent circuit.

Much of the vast literature on hydrothermal systems is to be found in internal reports of various agencies. For the systems described in Parts III and IV there are two useful compilations: for Wairakei (Pritchett *et al.*, 1978) and Ohaki—Broadlands (NZMWD Report, 1977).

Finally, I presume that readers will be familiar with the book "Chemistry and Geothermal Systems" by A. J. Ellis and W. A. J. Mahon (Academic Press, 1977). It is an excellent book in its own right and for readers of this book provides a wealth of descriptive and background material.

Acknowledgements

My own experience in geothermal and volcanic studies has dominated the writing of this book and I wish to express my gratitude to the many people who have helped me, especially to my friends and colleagues in the Department of Geology at Manchester University, the New Zealand DSIR, the Geological Survey of Japan, the Societa Lardarello and the Geological Survey of Greenland.

A major part of the text was written in 1978 while I was on sabbatical leave at the Department of Geology, Auckland University, New Zealand, and supported as Energy Research Fellow by the NZ Energy Research and Development Committee. I wish to note my particular gratitude to Frank Studt (then NZ DSIR geothermal coordinator) and Nick Brothers (Professor of Geology, Auckland University) for making this possible.

The manuscript was typed by Elaine Lock, Joan Norcott, Ada O'Brien and others. All the drawings were drafted by Philip Stubley. The arduous task of reading and refereeing the text was done by Dr. Arnold Watson (Engineering School, Manchester University; now of KRTA, Geothermal Power Consultants, Auckland). I am most grateful for all this help.

J. W. E.
Geology Department
Manchester University

Contents

PROLOGUE

PLATE. Steam plumes rising above Wairakei, North Island, New Zealand. The two big plumes are about 1.5 km apart and rise to a height of 300 m. Their combined discharge of about 0.2 ton/second is almost entirely from the borefield. View from a height of 4 km at $175°41'E$, $38°42'S$ about 50 km distant from the plumes, looking to the East from the western side of Lake Taupo across the northern bays of the lake — the town of Taupo is situated on the extreme right of the lake coast. In the distance there is low cloud and on the horizon, 100 km distant, the Huiarau Range. (Photograph taken January 1978.)

1. Prologue

The earth is a heat engine.

The components of this heat engine, the matter and energy involved in its operation, occur in a variety of forms. The matter which contains the heat engine is rock-substance. The matter which acts as the agent for transporting the energy within the heat engine can be rock-substance, solid or partially molten, derived often from the substance of the heat engine itself, or water-substance. The energy which allows the heat engine to perform work, or rather to rearrange its matter, is largely the internal thermal energy of the matter of the heat engine itself. The heat engine of the earth can be considered as an ensemble of interconnected partially isolated heat engines. The focus of this book is on those heat engines in which the transport of water-substance is an essential feature — hydrothermal systems. They are considered in detail but the early part of the book sketches how they fit into the ensemble of systems which run the earth.

Water is a more or less ubiquitous constituent of the crust of the earth. It is a vital ingredient in many geological processes. Of these, two broad classes can be distinguished. First, the role of water may be confined to local processes involving pre-existing water often at the scale of the granularity of the rock. Many metamorphic processes are of this type. So are many phenomena of soil and rock mechanics. Such processes are not the concern of this book. Second, water itself or as a constituent of a magma may be transported to, through, and out of, the system of interest. Hydrothermal and some lithothermal systems are of this type. These are the processes which concern this book. Our interest focuses on systems in which transportable water plays an active role in the dynamical behaviour of geological systems, particularly those which carry matter and energy to or near to the solid surface of the earth.

The role of temperature is pre-eminent in global dynamics. Suppose it were otherwise. Consider the dynamics of an isothermal earth. Global motions would be possible owing to pre-existing unbalanced density variations. These motions would however be one-shot — heavily damped and one way. Newton is alleged to have observed that apples fall from trees: what is often overlooked is that a particular apple falls but once. A motion of this type is not what is observed in the earth (in any event the bulk of the effect would last a time of order ν/ga, a few 10^3 yr only; a similar pattern recurs on the global scale every few 10^2 Myr). Clearly the earth is not like a clock wound up 5000 Myr ago and then left. Somehow the earth has to repeatedly rewind its own clock. A highly damped isothermal earth cannot work. Temperature is the quantity which unlocks the internal energy of the earth by turning the global system into a highly unstable one. This arises because thermal energy can be transported rapidly by mass transfer and only slowly by thermal diffusion.

Studies of the earth's heat are very ancient. It is of particular interest here that one of the earliest references to geothermal phenomena is to a warm spring that fed a stream near Troy (Homer, circa 800 BC). Systems like this which rely on the combined action of the transport of matter and energy are the subject of this book. The point of view centres around modern studies of thermal convection. Although many phenomena in which convection plays a role have been known and described qualitatively since time immemorial, the term "convection" was coined in 1834 by William Prout to refer to the method of transfer of heat by fluid motion identified about 1797 by Count Rumford. Numerous observations by many people, for example Marangoni, Thompson and, notably, Benard, culminated in the first study in the modern idiom by Rayleigh (1916). He analysed the stability and form of weak convection in a thin horizontal layer of fluid heated from below. He clearly identified the role of the buoyancy forces and the stabilizing effect of the combined role of viscosity and heat conduction. Since that time, what is now a distinct branch of continuum mechanics has been established in which convection plays a central role in the description of the dynamics of stars, the Earth's atmosphere—ocean system and the interior of the earth, as well as a vast field of application to engineering. Convective phenomena exist on a very wide range of scales from the evaporative, surface-tension driven effects in a thin layer of paint to the sunspots found in the thick outer layers of convective stars. Many of the convective phenomena now recognized do not occur within the Earth since particular phenomena are determined by the scale of the system and the properties of the working substance. Convection in a permeable medium provides the key element in the description of the mechanics of hydrothermal systems and is a central theme of this book.

TYPES OF MODEL

The analysis of a particular geological situation will involve many elements. Those that involve thermal convection are usually only a part. Further, the situation is generally so complex and poorly understood that the best model that can be developed must of necessity be greatly simplified and constructed out of idealized elements.

The first features to be considered are as follows:

(i) *Scale.* Systems of interest range from global-, core-, mantle-, crystal-scale to very small-scale systems in the upper crust. The scale has the most profound effect on the behaviour of the system. The earth as a whole can be treated as a vigorously convecting system of solid rock-substance, whereas a slab of the same material 1 km thick is mechanically inert.

(ii) *Closed or open system or sub-system.* If there is a net transport of matter through a system it is necessary to consider the discharge and recharge of matter at the boundaries of the system. Examples of systems of this type are geothermal areas with mass discharge in steaming ground, hot springs, fumaroles and phreatic eruptions; volcanic systems both in orogenic and mid-oceanic regions; the mantle itself seen as open with respect to the core; the upper mantle seen as open to the lower mantle.

(iii) *Steady, quasi-steady, unsteady or pulsatory system.* Often the key feature of the system of interest is its temporal behaviour. For example the most characteristic feature of the surface zone of a volcanic system is its pulsatory behaviour so that it would be pointless to develop models which could only manifest a steady state.

(iv) *Changes of state.* If a fluid element in its path around or through the system experiences one or more changes, or a continuous change of state, quite different system behaviour will be produced compared to that of a similar, homogeneous system. There are many possibilities: a simple change of state at a distinct interface, as in steaming ground or at the surface of an igneous intrusion; a simple but continual change of state, as in the fluid in a fumarole or erupting volcano; gross differentiation by crystal fractionation; progressive change as a fluid particle moves through regions of different pressure and temperature, for example during mantle convection.

(v) *Changes of system geometry.* A convective system may be capable of rearranging itself to such a degree that gross changes of the system geometry result. A simple example is the role of the high-level reservoirs in a volcanic system, which behave like a heavy bag pumped up by the mantle source, prior to each eruption (Elder, 1976, chapter 14). A more extreme example is the effect of the lateral displacement of a portion of the so-called lithosphere because of its interaction with the mantle below (Elder, 1976).

The next feature of importance is the effect of the state of matter in the system on the mechanics.

(i) *Wholly "solid"*. This system can be represented for example as a viscous fluid with a large kinematic viscosity of order 10^{16} m^2 s^{-1}, the dominant overall parameter is then the Rayleigh number A.

(ii) *Wholly "fluid"*. This is a similar situation except that the kinematic viscosity is very much smaller, for example of order 0.1 m^2 s^{-1} for a basaltic magma.

(iii) *Distinct parts of the system solid and other parts fluid*. In this case each part would be analysed separately with the joint behaviour treated by appropriate interactions between the parts.

(iv) *Partially fluid*. Here two distinct cases arise:
(a) lithothermal systems, in which melt permeates an otherwise solid rock matrix;
(b) hydrothermal systems, in which fluid water-substance permeates a solid matrix.
Such systems are treated by means of the Darcy approximation for flow in a so-called porous medium. The dominant parameter in a convective system is then A_m.

Convection will be very weak or negligible for $A \lesssim 10^3$ and $A_m \lesssim 40$. The gross ability to transfer heat and matter in vigorous convection is proportional to $A^{1/3}$ for a viscous fluid and to A_m in a porous medium. The only situation known to me in which convection, if it occurs, is not vigorous, is the suggestion that weak, low Rayleigh number convection occurs in the body of the ice-caps of Antarctica and Greenland (Hughes, 1972).

OBJECT OF THIS BOOK

The net surface heat flux through the earth's surface, over more than 99% of its area, is of the order of 50 mW m^{-2}. There are numerous localized thermal areas, however, where the heat flux is of the order of 100 times normal; hot springs and steaming ground discharging some 100 kg sec^{-1} km^{-2} of hot water and steam. It is the purpose of this book to discuss those physical mechanisms which produce thermal areas and the phenomena found in them.

Not only are these phenomena of interest in themselves, or because their energy can be used to produce power, but a study of them provides a thermal window into the depths of the earth. However, while studies of the thermal history of the earth encompass the whole earth and studies of volcanism provide information to depths of order 100 km, the hydrothermal systems responsible for thermal areas penetrate only to depths of the order of 10 km.

The discussion will centre around the hydrothermal system found in the Taupo district of New Zealand, because these are the most extensively studied.

The emphasis is on the behaviour which arises from the thermo-dynamic nature of water and from the permeability of the ground. Field data are combined with theoretical ideas and data from laboratory scale models to give a general view of a hydrothermal system as a dendritic structure, relatively simple at depth but increasingly complex near the surface.

In spite of the great diversity of the phenomena and the small body of quantitative observations, restricted both in space and time, many features can be described in terms of a few simple physical principles. This description proceeds without recourse to anything but the simplest results of geological or chemical investigations. The geological structure is regarded as a passive element merely defining the boundaries of each part of the heat and mass transfer system and establishing the conductivity and the permeability of the medium within the boundaries of the system, while chemical processes are considered not to have a major role in controlling any of the heat and mass transfer mechanisms, but rather, these mechanisms provide the physical environment in which chemical processes occur between the hydrothermal fluids and the rock. (Our interest is in the world of Neptune, rather than that of Vulcan.)

The outstanding and distinctive features of geothermal systems arise not only from the variety of phase changes of the working sub-stance but from the possibility of several phases being simultaneously present in various mixtures. For water-substance there is evaporation, simply boiling, flashing to produce a steam—water mixture. For rock-substance there is total melting, partial melting within an immobile solid matrix, partial melting within an otherwise fluidized granular solid, a multitude of partially molten differentiates in which a variety of crystallizing phases may be retained by the ambient fluid or otherwise deposited. Even for entirely solid mobilized rock-substance we must allow for phase changes especially in global and meso-scale systems. For wet rock-substance there are even more possible phase changes.

Geothermal systems could be approached from the point of view that the essential controls are the phase changes. In this view the sub-systems would be those parts within which the working substance is in a uniform phase state separated from other sub-systems by phase change interfaces.

In order to put the energy of various systems into perspective we need a suitable reference level. One day, from a global point of view, the fossil fuels will be exhausted. There remains the energy from the sun or from the earth's interior. Hydrosphere systems provide hydro-electric and some tidal and wind power. The most widespread trap for solar energy is, however, plants. The net productivity of chemical energy in plants ranges from $0.2{-}8 \, MW \, km^{-2}$, tundra to tropical rain-forest, corresponding to a net efficiency of $0.1{-}4\%$. Taking a value of efficiency of 1% as probably as good as could be achieved by farming, for example sugar cane, and assuming optimistically that only half the chemical energy is lost in gathering, processing and distributing the product the net production is $1 \, MW \, km^{-2}$. This is a fair reference flux. System with energy fluxes of $0.1 \, MW \, km^{-2}$ will be a poor proposition while those with $10 \, MW \, km^{-2}$ are valuable. The latter figure is typical of present geothermal power projects.

EARLY WORK ON HYDROTHERMAL SYSTEMS

The nature and origin of geothermal waters is part of a larger question to do with the proportion of volatiles which are "juvenile", originating in the deep interior of the earth to contribute to the growth of the atmosphere and ocean, or are merely re-cycled meteoric water.

Pioneer work on gases from volcanic and hot spring areas was done by Allen and Day (1935), Day (1939), Jagger (1940) and others. While the role of water-substance was early recognized as vital, there was much confusion about its origin and role. Various physical and chemical studies suggested that the bulk of geothermal water was meteoric (see, for example, White, 1957). Nevertheless the key to the problem was to be found in isotopic geochemistry. Numerous studies of individual isotopes were made but the major step was H. Craig's combined measurements of both hydrogen and oxygen isotopes which revealed the signature of meteoric water — a fundamental result (see, for example, Craig et al., 1956, Craig, 1963).

Of the numerous speculations (reviewed by Banwell, 1957) about the nature of thermal areas, the work of Einarsson (1942) on the weak hot springs of West Iceland was the first quantitative and comprehensive exposition. He contends that these springs are not physically different from ordinary cold springs except that due to the greater depth of penetration of water, it has become heated, and that this heat does not necessarily come directly from volcanism but comes simply from the normal heat flux through the ground. His evidence is:

(1) It is unnecessary to invoke an additional heat source when the total heat transmitted through the area including the springs does not differ from the local average of about $200 \, \mathrm{mW \, m^{-2}}$.

(2) The springs are strongly aligned (in the manner of cold springs) along basaltic dykes, particularly where they pierce strata dipping at $4°$ or more, and especially at lower levels not more than 100 m above sea level. The distribution of the springs shows no connection with volcanism. Whereas the acid springs of the southern quaternary volcanic area are strongly grouped, these springs are scattered over the western tertiary plateau.

(3) The alkaline spring water is similar to normal ground water, allowance being made for the higher temperature.

(4) Extensive post-glacial erosion has left moraine, bound by sinter, at spring outlets. This would not be possible if the springs did not exist till after the post glacial erosion and thereby dates some springs at least 10^5 years old. While changes are noticed in spring outputs probably due to local ground movements, this continuity of discharge is difficult to support merely by the transient effects of local volcanism (of which there is none) but is supported without difficulty by the normal heat flux.

(5) Such evidence is suggestive, but Einarsson now takes the vital step of attempting a quantitative exposition of a hot spring. His calculations agree moderately well with the quantitative observations; in particular that the discharge temperature is proportional to the discharge, and with the results of his physical manipulation of actual springs.

Bodvarsson (1948, 1949, 1950, 1954, 1961) has elaborated Einarsson's work and attempted to apply it to the intense areas of Central Iceland. The 1961 paper is an excellent summary. He notices in these areas the high thermal gradient near the surface and shows how this could arise from a slowly moving vertical current of hot water which is cooled near the surface by conduction of heat to the surface. At this point the features of a spring-type hydrothermal system are established. He draws attention to the limiting case of a column of water everywhere at its boiling point — the so-called BPD relation exploited by Banwell (1957). However, in spite of his contention to the contrary, he has not produced evidence to require some other kind of hydrothermal system for the intense areas. But he has shown that the high regional heat flux in Iceland is produced by the regional volcanism, and has attempted to discuss the intense areas by local volcanism using a conductive model.

The earliest large-scale investigation and exploitation of geothermal energy was in the Tuscan thermal area near Larderello (Penta, 1954). Initially interest was directed to the wet surface zone of boric water, but later drilling revealed a deeper zone of steam. During the period of intense exploitation this steam has become superheated. Unfortunately detailed physical investigations have been lacking till quite recently (Burgassi, 1961; Burgassi *et al.*, 1961) and little attempt has been made to produce a physical model.

A model presented here (Elder, 1966) considers the Tuscan hydrothermal systems as not fundamentally different from those in New Zealand but in which the water table is at great depth (2 km) and the discharge mechanism is similar to that of steaming ground.

The successful utilization of geothermal power at Larderello, where the ground apparently contained super-heated steam, together with concepts borrowed from oil prospecting and an excessive preoccupation with magmatic steam as the sole heat source, strongly influenced early ideas about the Taupo systems in New Zealand. Henderson (1950) has calculated the heat transferred by steam which rises from magma at depth through a permeable overburden and also for the case where the steam collects in a reservoir before passing to the surface up a bore hole. The work of Elder and Kerr (1954), while correcting the calculations of Henderson, pointed out that his work would perhaps be of value for Larderello but could not be expected to apply to the water-saturated Taupo systems. They began to formulate the mathematical problem of the flow in a porous medium (taken up by Donaldson, 1958), advocated model experiments to study the interaction between a bore and the surrounding medium, and began a study of the flow characteristics of bores.

Model work was begun in 1957 (Elder, 1966). For the intense thermal areas, the key step, which had barely eluded Bodvarsson, was to notice that the high surface thermal gradients were typical of boundary layer phenomena, such as those of a free convective system. But could such a free convective system exist in the porous material of the Taupo area, and if so could it transport sufficient heat? By means of simple two-dimensional and three-dimensional scale models, using 2 mm diameter sand in pots up to 20 cm deep, the first details of the heat transfer, temperature and velocity distributions for a free convective system in a porous medium were obtained. The models in the first instance were conceived as representing the flow in the whole of the debris-filled Taupo depression and though the measurements were crude, it was clearly shown that such a free convective system could transfer sufficient energy to operate the Taupo thermal areas. The viewpoint here was that the hydrothermal system was based on the continuous closed circulation of water and that the discharge at the surface was only a small "leak" which had little effect on the system as a whole. There was then, indeed, a second type of hydrothermal system — a convector — to be found, complementary to the single-pass spring-type system of Einarsson and Bodvarsson.

Parallel with this work has been a series of investigations relating to weakly convecting systems, extremely important, not because of its direct bearing on the geophysical problem where the convection is strong, but because of the clarification and understanding it gives to convection in a porous medium. Lapwood's (1948) work considered the conditions under which convection could occur at all and was (not very successfully) checked experimentally by Rogers *et al.* (1951). Wooding (1957, 1958, 1959, 1960a, 1960b, 1962, 1963) has enormously extended and generalized the work begun by Lapwood.

While this book deals with hydrothermal systems in general, it arose from a physical study of the Taupo hydrothermal systems. Studt (1957, 1958a, b, 1959) has principally contributed to this, studying both the systems themselves and, together with Modriniak (Modriniak and Studt, 1959), their geological and geophysical environment. However, very little quantitative attention has been given to different modes of discharge of hot fluid at the surface, in spite of the fact that these problems are most amenable to physical and theoretical investigation. Benseman (1959, 1965) has begun studies of geysers and steaming ground and Banwell (1957, 1961), has made detailed studies of the behaviour of bores.

OUTLINE OF THE ARGUMENT OF THIS BOOK

At each stage we consider a conceptual model based either on field observations or a laboratory model or analogue. The model does not necessarily correspond in detail to conditions in the ground, but rather it is an idealization of a possible system suitable for analysis and discussion. The creation of a conceptual or laboratory model which would correspond in detail to an actual system would be difficult, since much field data is lacking, and premature, when the behaviour of the simpler systems is not fully understood. The argument proceeds by discussing a sequence of models of increasing complexity.

The centre of the stage, in this book, is occupied by hydrothermal systems. Their short-term behaviour can be described in isolation. Nevertheless, when we are concerned with their long-term behaviour, in particular, their origin and local persistence, it is necessary to consider the larger-scale systems in which they are embedded. Since the essential working substance is water, derived as rainfall or directly from sea water, the key question about conditions at depth is the mechanism of the supply of thermal energy. An outline of the thermal structure of the outermost part of the solid earth allows identification of the thermal lithosphere. The oceanic lithosphere is treated in some detail to highlight the important role of oceanic hydrothermal systems in controlling the heat transfer over upwelling regions of the oceanic mantle. A similar process occurs in continental volcanic zones but its effect on the continental lithosphere is more localized and of negligible global significance.

Systems in which the bulk of the mantle discharge occurs as extrusive volcanism do not produce large intense hydrothermal zones. The bulk of the energy is lost with the lavas. In order to produce intense hydrothermal zones magma must in part be trapped below the surface. The circumstances in which trapping can occur are treated in detail.

The mechanism by which the energy of the trapped lithothermal system is actually handed over to the hydrothermal system is revealed in the study of hydrothermal ore deposits. In many ways this aspect is the linchpin of our understanding of the gross features of the origin of hydrothermal systems.

From this point on, the argument treats hydrothermal systems as small-scale, more or less self-contained high-level systems provided with inputs and outputs controlled by ambient conditions in the lithothermal system at depth or the local surface hydrology.

The simplest and crudest model of a hydrothermal system is that of a reservoir provided with inlets and outlets for energy and fluid, but in which details within the reservoir are completely ignored. This is a zero-dimensional model. While the capacity of the reservoir is important, both energy and chemical content are assumed to be uniformly dispersed throughout the reservoir volume, so that apart from the flows at the inlets and outlets the only independent variable is time. It is found that the natural Taupo systems, for example, were in a steady state with respect to conditions at the inlets and outlets so that there must exist a source of water, energy and chemicals from outside the reservoir.

The reservoir is elaborated by adding to it a pipe-like structure through which water and energy can flow; attention is concentrated on conditions along the pipe rather than over its cross-section. This is a one-dimensional model. A relation between the flow and its enthalpy can be deduced; in particular a distinction can be made between systems which have a "dry" discharge (superheated steam) or a "wet" discharge (in which steam production by flashing at depth or evaporation near the surface is permitted).

If the medium in which the hydrothermal system occurs is now considered to contain many "pipes" and "reservoirs" the medium will have similar behaviour to that of a homogeneous porous medium. A model is considered in which the flow of energy and water through the system is produced by free convection in a porous medium. This is a three-dimensional model. The pipe model and the free convection model are extreme approximations to all possible hydrothermal systems.

Except for the "dry" systems, it is found possible to describe all the known physical phenomena of thermal areas by requiring at depth a body of hot water only — it is not necessary to invoke water which originates at depth — so called "juvenile" water. Steam is produced by flashing or evaporation.

Exploitation can have a profound effect on the hydrothermal system. This is a very diverse field with a voluminous literature. I have decided to concentrate attention on the one vital aspect of such exploitation, the long term effects of exploitation. I do this by describing the exploitation of a hypothetical system, modelled somewhat on Wairakei.

Part I

GLOBAL SCALE PROCESSES

The cascade of energy from large to small scales of motion dominates the thermodynamic evolution of the interior of the earth. This energy flow is derived from a single closed reservoir, the entire body of the earth, which was filled with matter and energy about 5000 Myr ago. A sketch of the heat and mass transfer processes which operate on the global scale and have a bearing on the transport of matter and energy to smaller scale sub-systems embedded in the outermost zone of the solid earth is presented.

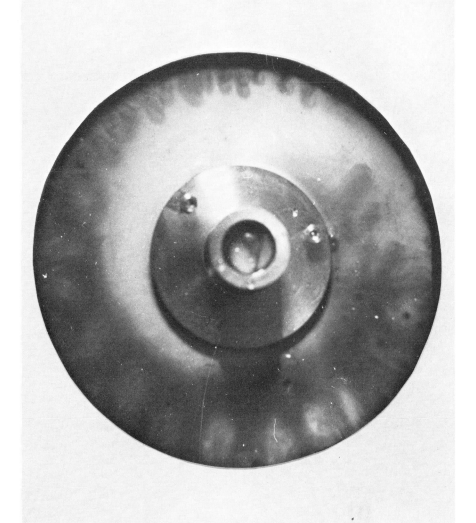

PLATE. Laboratory model of mantle convection: plan view, looking vertically downward, into a thin horizontal annular layer of fluid supported above and below by sheets of perspex, from a camera attached to the annulus. The inner rim of the annulus is cold, the outer rim is hot and the entire apparatus is rapidly spinning about the vertical axis of symmetry. This produces a radial acceleration field analogous to the radial gravity field in the earth. Irregular but discrete eddying motions, revealed by dye in the fluid, can be seen in the "sub-layer" region near the outer rim which represents the earth's surface. The equivalent Rayleigh number in the model is about 10^6, so that in the actual earth, with a value of 10^9, the sublayer will be relatively thinner in the ratio about 0.1.

2. Thermal Structure

The relative vigour of the heat and mass transfer subsystems within the earth is revealed by consideration of the global energy budget. This provides the constraints and setting in which models of the various subsystems can be developed. The measurable data are described here and the important deduction is made that the dominant contribution to the global heat loss is through surface conduction and not by mass transfer. The thermal structure of the near surface zone of the earth is described, but is discussed in detail in the following Chapters 3 and 4.

2.1. SURFACE TEMPERATURE

From the point of view of life on our planet, by far the most important geothermal system is that of the atmosphere-hydrosphere with its working substances air and water (including ice). The gross time-scale of this system obtained by noting the intervals of fluctuations during ice ages is probably of order 10^4 y with individual subsystems being much more rapid. Our interest here in this system, however, is restricted to two things. The temperature of the solid surface is determined by this system alone and provides a powerful external constraint on the interior by restraining the surface temperature to be nearly uniform and constant in time with respect to the much higher temperatures of the interior.

The temperature distribution near the atmosphere—solid earth interface

The mean temperature as a function of elevation above the earth's solid surface is shown in Fig. 2.1. The striking feature of this profile is a zone of about 100 km in width within which the temperature is low.

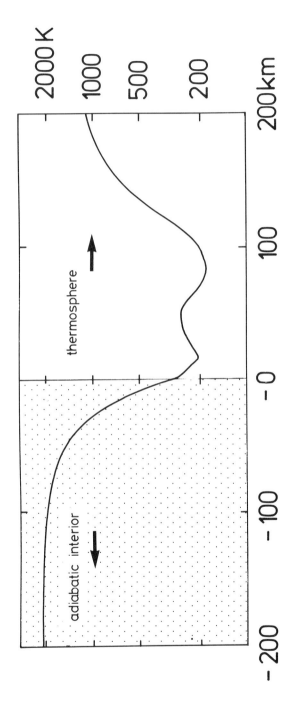

Fig. 2.1. Mean temperature profile as a function of height above mean sea-level.

The interior temperature distribution (see, for example, Elder, 1976) has two zones: (i) a deep vigorously mixed nearly adiabatic interior, with (ii) a less mobile surface zone in which, as the surface is approached, molecular processes predominate and across which is a nearly uniform outward vertical heat flux of about $50\,mW\,m^{-2}$.

The atmosphere also has two distinct zones (Jacchia, 1967); (i) an outer "thermosphere" or "heterosphere" in which photodissociation is important, mixing is weak, and radiation dominates and varies with the 11-year sunspot cycle, with temperatures being uniform above about 300 km and ranging from 650–2000 K from sunspot minimum to maximum — the entire thermosphere contracting and expanding in proportion; (ii) an inner "homosphere" in which mixing is sufficiently vigorous to maintain nearly uniform composition, apart from the local temperature high near 50 km owing to preferential radiation absorption by ozone.

Atmosphere energy transport time-scales

Whereas the time-scale of radiative processes in the atmosphere are effectively zero, other less rapid processes are also important (data taken from Sheppard (1967) and Saunders and Sheppard (1967)).

(i) The next most important process in the lower atmosphere is water vapour transport. The mass of water vapour in a vertical column, of cross sectional area $1\,cm^2$, the "precipitable water", is about 2 gm (precipitated, it would give a 2 cm thick layer) which compared to the total mass of air of about 1 kg in the column is a proportion of 0.2%. Taking the global average rate of precipitation as about $1\,m\,yr^{-1}$, the mean life time of water vapour in the troposphere is (amount/rate) ≈ 7.2 day. The corresponding life-time above the troposphere is about 1 yr.

(ii) Turbulent transport processes expressed as eddy diffusivities are: $1\,m^2\,s^{-1}$ in the near surface (0–100 m) constant stress layer; 10–$10^2\,m^2\,s^{-1}$ in the first few kilometres; $10^7\,m^2\,s^{-1}$ for the large scale motions of the troposphere.

These processes are sufficiently slow compared to that of radiation that they provided a means of temporarily trapping energy within the atmosphere. Assuming therefore that the atmosphere is in energy balance and the proportion of the incident flux that is trapped is small we can anticipate that a radiative model of the atmosphere will be a good one.

Simple model of the earth's near surface temperature

Consider a rapidly rotating perfectly conducting sphere placed in a uniform beam of radiation of flux f. Then the energy input per unit area

$$\tilde{f} = f/4$$

Taking the mean solar flux $f = 1.4\,\text{kW m}^{-2}$ then $\tilde{f} = 0.35\,\text{kW m}^2$. If the sphere is also a black body radiator its temperature, $\tilde{T}$ will be determined by Stefan's relationship

$$\tilde{f} = \sigma T^4$$

where $\sigma = 5.673 \times 10^{-8}\,\text{W m}^{-2}\,\text{K}^{-4}$, Stefan's constant. Hence $\tilde{T} = 280\,\text{K}$. Given that the mean surface temperature, T of the earth is 287.4 K and that the temperature gradient in the lower atmosphere is about $-4\,\text{K km}^{-1}$, the temperature $\tilde{T}$ corresponds to that at an altitude of a little less than 2 km — a typical cloud height. This result is almost too good to be true, but taken at face value it tells us that the overall control of the atmosphere temperature is radiative transfer. Fig. 2.2 shows in more detail the latitudinal variation of T and

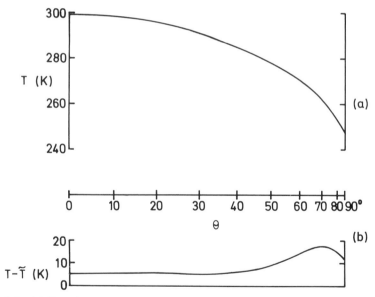

Fig. 2.2. (a) Zonal average temperature, T, as a function of latitude, θ, (north or south). Compiled in 0 (10) 90° and plotted with abscissa proportional to cos θ, the zonal area proportion. (b) Difference, $T-\tilde{T}$ between T and temperature obtained from average insolation flux $\tilde{f}(\theta) = \sigma \tilde{T}^4$. Based on f in Wm^{-2} for $\theta = $ 0 (10) 90°: 426, 421, 403, 374, 337, 292, 243, 202, 183, 176.

$(T - \tilde{T})$. The difference, $(T - \tilde{T})$ is below its mean value near the equator and above it near the poles. This relatively small contribution to the overall variation arises from the net polarward heat transport within the meridional circulation of the atmosphere and ocean.

This is an extremely important result. Near surface temperatures are determined by radiation. Since our sun is relatively youthful and its output has not varied appreciably throughout geological time the temperature of the earth's surface has been approximately constant throughout geological time. In the light of the evidence of paleontology and sedimentology the result is expected. The argument is perhaps more logically put in reverse: given the paleontological evidence we deduce that a radiation dominated model is appropriate.

2.2. SURFACE HEAT FLOW

Conductive surface flux

The simplest way to determine the heat transferred at the surface is to drill a hole in the crust, measure the temperatures down the hole, and measure the thermal conductivity of a sample of the rock. There is a surface zone, extending down to about 100 m, in which temperatures are variable because of diurnal heating and movement of water and water vapour. But below this zone almost everywhere there is a steady increase of temperature with depth, typically 25 K km^{-1} in a region of thermal conductivity K of, say, 2 W m^{-1} K^{-1}. The heat flux f and the temperature gradient $\mathrm{d}T/\mathrm{d}z$ for heat transferred by thermal conduction are related by

$$f = K\mathrm{d}T/\mathrm{d}z.$$

This in effect is the definition of K. Thus for the above figures $f = 50\,\mathrm{mW\,m^{-2}}$.

Two distinct regions can be recognized.

(1)   Normal areas, covering more than 99% of the earth's surface, are those in which the heat flux lies typically in the range 0—200 mW m^{-2}, with an average of about 50 mW m^{-2}. A recent compilation gives a value of 52 mW m^{-2} ± 50% s.d. (Jessop *et al.*, 1975). The vertical temperature gradient is nearly constant to depths exceeding 1 km, and the surface heat flux variations are gradual, usually being negligible over distances of the order of 1 km.

(2)   Thermal areas are those in which the heat flux can reach $1 \, kW \, m^{-2}$, though it is generally much smaller, with possible average values of the order of $1 \, W \, m^{-2}$ over areas of the order of $10^3 \, km^2$. Large horizontal variations of heat flux and vertical variations of vertical temperature gradient are possible over distances of 1 m.

Two extremely important results emerge from a global study of the surface heat transfer.

(a)   There is everywhere over the earth's surface a *net* outward flow of heat. Heat is being continually brought to the earth's surface from its interior and lost.

(b)   Apart from certain narrow areas, to be discussed later, the heat flux is nearly uniform over the surface. In particular it is roughly the same over continental platforms as over oceanic basins.

A schematic representation of the global surface distribution of conductive heat flux is shown in Fig. 2.3. This is from a recent compilation (Chapman and Pollack, 1975) following the first comprehensive collection by W. H. K. Lee and a number of successors (Lee and MacDonald, 1963; Lee and Uyeda, 1965; Horai and Simons, 1969; Lee, 1970; Jessop *et al.*, 1975). Unfortunately, even today the data is very unevenly distributed: even on the $5° \times 5°$ scale (2592 of them cover the globe) about 60% of the earth's surface is as yet unmeasured. If the global fine structure of the heat flux distribution is determined on the scale of the upper mantle, representations like that shown are only of value for the grossest of statements, as given above. In particular it is pointless to quibble about the precise value of the global mean heat flux; the value of $50 \, mW \, m^{-2}$ is satisfactory.

Regional heat flow data are shown in Figures 2.4—2.6. for three distinct regions: continent (Europe); continental margin (East Asia); ocean (Indian). The range of variation of magnitude of heat flux is quite small, being least on continents. The picture from oceanic regions is rather scrappy owing to lack of data but a striking feature is the occurrence of zero heat flux predominantly in the youngest oceanic crust. (Zero heat flux values are also found in geothermal areas on land but are generally too localized to show up on the 1000 km scale).

Clearly the patterns on this scale reveal the thermal "weather" of the mantle muted through the low-pass filter of the upper mantle and crust. Here I am mostly interested in the qualitative evidence that the mantle is not thermally homogeneous and that statistically

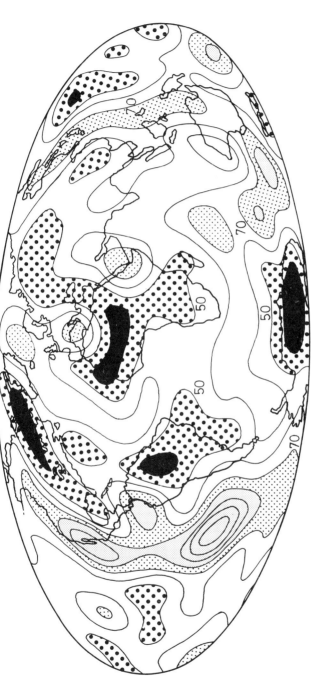

Fig. 2.3. Surface conductive heat flow distribution. Spherical harmonic representation of order 12 based on actual data averaged over $5° \times 5°$ elements supplemented with estimated values in empty elements (68% of the total number) by means of a presumed relation of heat flow and "age" (from Chapman and Pollack, 1975). Contour values in $mW\,m^{-2}$.

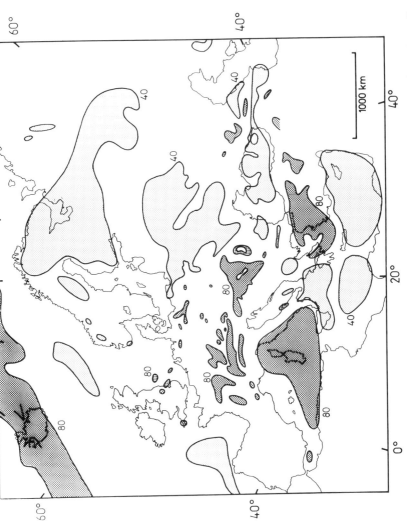

Fig. 2.4. Surface conductive heat flow distribution of Europe (2605 data points). Adapted from heat flow map of Europe 1:5000 000. V. Čermák and E. Hurtig (1977) IASPEI International Heat Flow Commission. Contour values in mW m^{-2} .

Table 2.1. *Apparent relation of heat flow, in mW m^{-2}, with age of a tectonic province, measured from the most recent major event. From a compilation by Chapman and Pollack (1975).*

Continents

Tectonic province	N. America	Australia	Europe	Asia	Typical continent
Archean shield	41	43		36	38
Proterozoic shield	55	73	38	45	44
Precambrian platform	49			45	
Phanerozoic non-orogenic	53	61	72		46
Caledonian orogeny	48	85	65	56	52
Hercynian orogeny	62	58	67	73	59
Mesozoic orogeny	80				
Cenozoic intermontane trough					41
folding	75	80	78	63	73
volcanism					92

Oceans

Tectonic province	Age (Myr)	N. Pacific	S. Atlantic	Indian	Typical ocean
Jurassic	136	54			49
Early Cretaceous	100–136	49	48	53	58
Mid Cretaceous	76–100	58	59		59
Late Cretaceous	63–76	60	59	55	56
Anomaly 13–25	38–63	60	53		65
Anomaly 6–13	20–38	67	29	64	83
Anomaly 5–6	10–20	93	73		
Anomaly 0–5	0–10	118	90	100	103

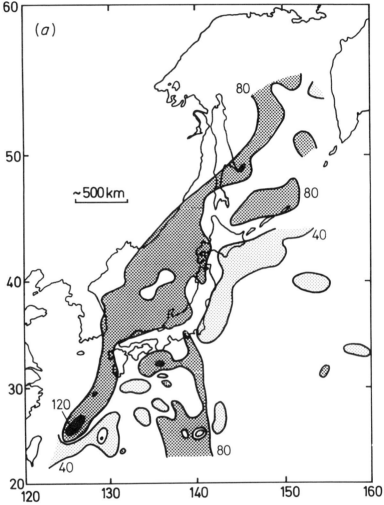

Fig. 2.5. (a) (above) Surface conductive heat flow distribution in the Western Pacific near Japan. Contour values in mW m^{-2}. (b) (facing page) Estimated temperature distribution at 10 km depth below sea level near Japan, obtained by linear extrapolation. After Watanabe *et al.*, (1970), and Uyeda, Nomura and Watanabe (1976). Contour values in $^{\circ}$C.

the spatial heat flow variation is globally minor except near the discharge and recharge zones of the upper mantle.

If the surface data is extrapolated downwards we can obtain a picture of the temperature at a particular depth. In the case of data from near Japan this is shown in Fig. 2.5. Over most of the area temperatures at 10 km depth lie in the 200—400°C range. There are

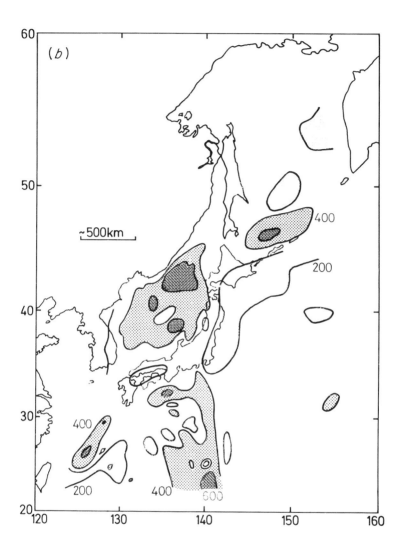

also extensive hot patches with temperatures above 600°C. In this most thermally disturbed region we presumably have an indication of the range of possible temperatures high in the mantle and crust.

Review of regional surface heat flow data

The relationship of the heat flow pattern to that of the crust is weak. There is however a rough apparent correlation of heat flow

GEOTHERMAL SYSTEMS

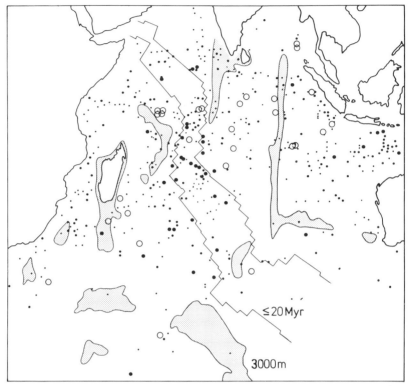

Fig. 2.6. Indian Ocean heat flux distribution. The band shows ocean floor younger than 20 Myr. (Adapted from a compilation by Anderson and Langseth, 1977). Circular symbol values in mW m^{-2}: •, < 20; •, 20–25; ·, 50–100; ○, > 100.

with tectonic age as indicated in Table 2.1 and Fig. 2.7. For oceanic regions with mass discharge and subsequent lateral transport from up-welling sites, the data indicate typical heat flux of about 110 mW m^{-2} in the source region, diminishing with lateral distance (and time) to a constant value about 55 mW m^{-2} in 40 Myr — an exponential time-scale of about 30 Myr. The thickness of the region which cools by conduction is this time-scale, τ is about $2(\kappa\tau)^{1/2}$, in this case about 60 km. This suggests that a substantial part of the laterally transported upper mantle is excessively heated above normal as it leaves the source region. For continental regions with plutonism in orogenic systems, the heat flow, initially about 80 mW m^{-2}, diminishes with time to a constant value about 40 mW m^{-2} in 500 Myr — an exponential time-scale of about 350 Myr. This time-scale is sufficiently close to the time-scale of the overall removal of a dead orogenic system

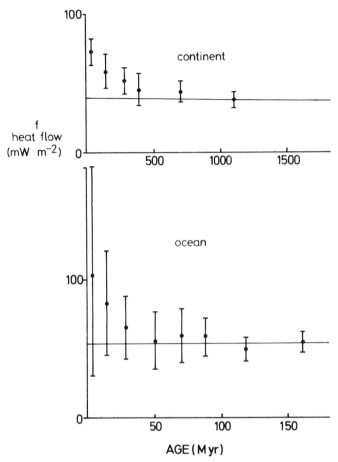

Fig. 2.7. Surface conductive heat flux as a function of most recent tectonic age. Standard deviation range indicated. From Chapman and Pollack (1975).

by erosion (Elder, 1976) to suggest that the uplift of deeper material owing to the isostatic response to removal of the upper material is the main control in the post tectonic phase.

2.3. GLOBAL ENERGY AND MASS TRANSFER SUMMARY

Since in this work all the systems of interest are embedded in the crust and upper mantle it is simplest to consider the global system on two scales (i) core—mantle—upper mantle and crust; (ii) upper mantle and crust.

The estimates which follow are little better than order of magnitude estimates, but are made to give a perspective on the heat and mass transfer rates in lithothermal and hydrothermal systems. These data are collected in Table 2.2.

Upper mantle—mantle—core system

There are three interfaces across which transfer can occur.

Mantle—core interface. Estimates of conditions here are speculative. Taking the model of Elder (1976) we have that the core radius satisfies

$$dr/dt = -fr/\rho\kappa a$$

where $a = 6370$ km, the radius of the earth, $\kappa = 1700$ kJ kg^{-1}, enthalpy exchange, and f is the crustal surface heat flux. With $f = 50$ mW m^{-2} we obtain a mass flux of 1.6×10^{-8} kg s^{-1} m^{-2} and a heat flux of 27 mW m^{-2}. The total mass flow rate out of the core is about 2.5×10^6 kg s^{-1}. For the purpose of comparison it is convenient to convert these quantities to equivalent values at the crustal surface by multiplying by $(r/a)^2$ to give an equivalent mass flux, 0.5×10^{-8} kg s^{-1} m^{-2} and an equivalent heat flux, 8 mW m^{-2}. These latter quantities represent the maximum possible contribution of the core material to the upper systems. They are sufficiently small to be largely ignored in discussions of the present upper mantle.

Upper mantle—lower mantle interface. Assume that the entire global lithosphere over a depth of 100 km is recharged every 150 Myr. Taking the mean density as 3.3 g cm^{-3} this requires a mass flux of 7×10^{-8} kg s^{-1} m^{-2}. The mean temperature of the recharged layer is perhaps 1000–1500°C, say, 1250°. Let the recharge be at temperature $(1250 + y)$°C, the discharge at $(1250 - y)$°C. Then for a specific heat of 1 kJ kg^{-1} K^{-1} we find that, if the difference of flux of the recharge—discharge system is equal to the surface heat loss of 50 mW m^{-2}, then $y = 357$°C, say 350°C. Thus the temperatures of the recharge and discharge are about 1600°C and 900°C, and the corresponding heat fluxes are about 110 and 60 mW m^{-2}.

Global surface conduction. The global average conductive surface heat flux of about 50 mW m^{-2}, already discussed, gives over the global surface of 5.1×10^8 km^2 a total power of 25.5×10^{12} W.

Crust and upper mantle systems

There are several subsystems.

Crustal kinematics. If the areal rearrangement of the crust occurs with typical velocity u, the rate of working of the viscous stress is of order $\mu u^2 /h$ where μ is the mean viscosity of the lithosphere and h is its thickness. Taking $u = 3\,\mathrm{cm\,yr^{-1}}$, $h = 10^2\,\mathrm{km}$, $\mu = 10^{21}\,\mathrm{P}$ the energy flux is about $1\,\mathrm{mW\,m^{-2}}$.

New oceanic basaltic crust. Take the basaltic layer as being 5 km deep, $3.6 \times 10^8\,\mathrm{km^2}$ in area, and of mean age 150 Myr. The enthalpy lost, largely to heating sea water at a nominal $0°\mathrm{C}$ in the discharge area, for a magma temperature of $1100°\mathrm{C}$, specific heat $1\,\mathrm{kJ\,kg^{-1}\,K^{-1}}$ and latent heat $400\,\mathrm{kJ\,kg^{-1}}$ is $1500\,\mathrm{kJ\,kg^{-1}}$. The total power is then $1.72 \times 10^{12}\,\mathrm{W}$. This estimate assumes that all the magmatic energy is dissipated directly to the ocean. If material is intruded as well as extruded throughout the basaltic layer of mean ambient temperature say $150°\mathrm{C}$ (assuming an ambient thermal gradient of about twice normal), the above estimate will be $1.5 \times 10^{12}\,\mathrm{W}$, an insignificant alteration.

Land volcanism. The bulk of this occurs in young fold belts as predominantly quartz-rich magmas. Thus performing the estimate as if the material were all a granitic magma of latent heat $270\,\mathrm{kJ\,kg^{-1}}$ at input temperature of about $750°\mathrm{C}$ the enthalpy is about $1000\,\mathrm{kJ\,kg^{-1}}$. The excess thickness of the upper crust in young belts averages 35 km, corresponding to an elevation of 2.8 km. If a third of the excess (a guess) over a reference sea level crust is of magmatic origin, the average thickness of volcanics (intrude or extruded) is about 5 km. Allowing for the proportion of the globe covered by young fold mountains, about 8% with an assumed life of 200 Myr, requires a global average heat flux of $0.25\,\mathrm{mW\,m^{-2}}$.

Orogenesis. Additional to the energy of crustal rearrangement, in orogenic belts there is accumulation of gravitational energy. Relative to the reference sea level crust the energy density is

$$\tfrac{1}{2} g(H^2 \Delta\rho + h^2 \rho_c)$$

where ρ_c, ρ_m are the densities of the crust and mantle and $\Delta\rho = \rho_c - \rho_m$. In isostatic equilibrium

$$h \rho_c = H \Delta\rho.$$

For a mean elevation of 3 km and 8% of the surface covered by young fold belts for a life-time of 200 Myr (see following), the energy flux is less than $0.1 \, \mathrm{mW \, m^{-2}}$.

The energy required to laterally compress the geosynclinal sedimentary deposits in order to confine them to narrow belts can be estimated as follows.

Metamorphism. Suppose all the sediment is in a closed cycle and the energy of chemical transformation is about $400 \, \mathrm{kJ \, kg^{-1}}$. For a mass flow of $2 \times 10^{13} \, \mathrm{kg \, yr^{-1}}$ this corresponds to about $0.8 \, \mathrm{mW \, m^{-2}}$.

Erosion. An estimate of the energy lost owing to erosion of fold belts can be obtained from the global average vertical uplift rate, w. Two different estimates are possible.

(i) Direct measurement of the mass transported to the ocean gives about $1.6 \times 10^{13} \, \mathrm{kg \, yr^{-1}}$. This is equivalent to $w = 3 \times 10^{-13} \, \mathrm{m \, s^{-1}}$, and is probably an underestimate.

(ii) From considerations of the isostatic response of a crustal block to erosion (Elder, 1976) if the thickness removed is ΔH for a net loss of elevation Δh then

$$\Delta H = (\rho_c / \rho) \Delta h,$$

where ρ_c is the density of the removed upper crustal material and ρ that of the mantle. The time-scale of this process,

$$\tau = (\rho / \Delta \rho) T,$$

where $\Delta \rho = \rho - \rho_c$ and T is the erosional time-scale, about 30 Myr. Thus the local uplift rate is $\Delta H / \tau$ which for $h = 5$ km, say, is about $5 \times 10^{-12} \, \mathrm{m \, s^{-1}}$. This is probably an overestimate since the time-scale τ is smaller than that obtained from studies of the geological column. To obtain the global value this needs to be multiplied by the proportion of the earth's surface covered by young fold belts, about 8%. Thus $w \approx 4 \times 10^{-13} \, \mathrm{m \, s^{-1}}$. Taking this latter figure and a temperature at depth of typically $\theta = 1500°C$, the corresponding heat flux $f = \rho_c w \theta$ is about $2.6 \, \mathrm{mW \, m^{-2}}$.

The gravitation energy, estimated above, is also lost.

Radioactivity. Direct measurements are extremely non-uniform. Here we wish to find an upper bound. Noting that the continental crust heat flow differs by no more than 20% from that for oceanic

Table 2.2. *Global upper mantle heat balance.*

	Power	Mass rate	Enthalpy
	(10^{12} W)	(10^6 kg s^{-1})	(kJ kg^{-1})
INPUT—OUTPUT			
(i) Input:			
Supply from mantle	57	36	1600
(ii) Output:			
Discharge to mantle	32	36	900
Surface conduction	25	0	0
New oceanic basaltic			
crust	1.6	1.1	1500
Land volcanism	0.1	0.1	1000
INTERNAL REARRANGEMENT			
Crustal kinematics	0.5	53	0
Orogenesis:			
(a) Construction	0.1	0.8	0
(b) Metamorphism	0.4	0.6	400
(c) Erosion	1.3	0.6	(1500)
Earthquakes	0.01	0	0
INTERNAL POWER GENERATION			
Radioactivity	$\lesssim 1$	0	0

Notes:
  (1) A flux of 2 mW m^{-2} give a global power of 10^{12} W.
  (2) A volumetric rate of $1 \text{ km}^3 \text{ yr}^{-1}$ (at density 3 g cm^{-3}) is about 10^5 kg s^{-1}.
  (3) Enthalpy relative to nominal surface value at $0°\text{C}$.

crust; and the continental crust covers an area of only 30% of the whole surface, then at most 6% of the global crustal heat flow is from radioactive decay.

Earthquakes. From a compilation of large earthquakes recorded since 1904, the annual energy release is estimated to be about $1.2 \times 10^{10} \text{ W}$.

Global geothermal power. An amusing, but rather suspect, estimate (upper bound) of the global geothermal power can be made as follows. Considerations of the chloride content of sea-water suggest that a recharge rate of 10^7 ton yr^{-1} is required (Rubey, 1951). Let us assume that all this chloride comes from geothermal water at nominal temperature $250°\text{C}$ and chloride concentration 1000 PPM, values typical of New Zealand hydrothermal systems, so that $1\,g\,Cl$ corresponds to about $1\,MJ$ of energy. The global chloride recharge corresponds to about $3 \times 10^{11}\,W$, or 60 times the geothermal output of the North Island of New Zealand, or 1% of the heat output from normal areas of the globe of about $2.5 \times 10^{13}\,W$.

Global surface mass transfer heat flux

The vital question that interests us here is whether or not there is one or more upper mantle systems, in particular that responsible for magma transport, which is more vigorous than that which drives the upper mantle. There is no such system. The most vigorous is indeed magma transport but even this requires only about 3% of the energy supply to the upper mantle. From this point of view all crustal systems are minor superficial aspects of the global energy balance.

The global surface mass transfer heat flux is much smaller than the conductive heat flux. This is quite different from the situation in geothermal areas where the heat flux is dominated by mass transfer.

3. The Thermal Sublayer

The uppermost zone of the earth, with the systems embedded in it, is dominated by the recharge—discharge system of the upper mantle. Some understanding of the mechanism of this dynamically most important zone is essential in understanding the systems in it. An outline of an appropriate model is presented here; a fuller, though elementary, discussion has been given elsewhere (Elder, 1976).

The mantle is represented as a vigorously convecting system for which the thermal structure is dominated by a relatively passive surface zone — a thermal boundary layer, referred to as the thermal sublayer. This sublayer, modified by lateral rearrangement, is more usually referred to as the lithosphere.

This chapter considers only global thermal average aspects of the thermal sublayer. Also, the role of surface discharge is ignored: its role in the modification of the lithosphere on the sub-global scale is described in the following chapter.

3.1. CONVECTION

Studies of the gross shape of the Earth show, to an accuracy of about 1 in 10^5, that the Earth behaves like a self-gravitating mass of inviscid, slightly compressible fluid in uniform rotation. Departures from this state of uniform rotation can arise from imbalanced density variations, considered here to be produced by variations of temperature which arise because the Earth is not in thermodynamic equilibrium. A first approximation to a description of such effects by means of a dynamical theory is to consider the fluid viscous. This approximation is a good one, provided the stresses in the fluid are sufficiently small and will therefore apply nearly everywhere throughout the Earth, except possibly in earthquake and vigorous tectonic regions.

Let us consider for a moment the dimensional analysis of the convective motion of a homogeneous fluid sphere of mass M, radius a, spinning at the angular rate Ω, with initial available energy per unit volume Q. Let the fluid have density ρ, specific heat c, kinematic viscosity ν, thermal diffusivity κ, and co-efficient of cubical expansion γ. If we make the Boussinesq (1903) approximation and note that the acceleration due to buoyancy has the form $\gamma\rho\theta g$ with the acceleration of gravity $g \approx GM/a^2$, where G is the gravitational constant, and the temperature scale $\theta \sim Qa^2/\rho c \kappa$, the problem is defined by

$$\nu, \kappa, \gamma\rho\theta g, \Omega, a.$$

These five quantities involve only length and time, so that three dimensionless quantities are required to specify the system. A convenient choice is the following:

$A = \gamma\rho\theta ga^3/\kappa\nu$ Rayleigh number,

$\sigma = \nu/\kappa$ Prandtl number,

$R = 2\Omega a^2/\nu$ Rotation parameter
(square root of the Taylor number).

Inserting typical values:

$$\gamma = 10^{-5}\,\mathrm{K^{-1}};\ g = 10\,\mathrm{m\,s^{-2}};\ \theta = 10^3\,\mathrm{K};\ a = 6370\,\mathrm{km};$$

$$\kappa = 10^{-6}\,\mathrm{m^2\,s^{-1}};\ \Omega = 7.10^{-5}\,\mathrm{rad\,s^{-1}};$$

we find:

$$A \sim 10^9;\ \sigma \sim 10^{22};\ R \sim 10^{-2}.$$

These values indicate the following: convection will be important; convective inertial effects will be negligible; and rotation effects will be very small (Chandrasekhar, 1961).

It is important to note that systems of this kind do *not* have an imposed velocity scale. Whenever motions occur within such systems the velocities are internally determined, often through a balance between viscous and buoyancy forces. For vigorously convecting systems, it is found that the global velocity scale, namely that obtained by determining the average over the flow volume of the root mean square velocity at a point $\tilde{q}$ is given empirically by

$$\tilde{p} \equiv a\tilde{q}/\kappa \approx 3A^{1/3},$$

where the coefficient 3 is determined from laboratory experiments (Elder, 1976). At $A \sim 10^9$, we have $\tilde{p} \sim 10^3$. The quantity $\tilde{p}$ is a Peclet number, a measure of the relative importance of the rate of heat transfer by the flow to that carried by thermal conduction. Within the earth Peclet numbers are large and the role of mass transport dominates in all large-scale convective systems. The corresponding Reynolds number

$$\tilde{R} \equiv a\tilde{q}/\nu = \tilde{p}/\sigma$$

is very small so that all motions are dominated by viscous forces, the role of inertial forces being negligible — so-called Stokesian flow.

Within the mantle of the Earth *all* effects of the inertial accelerations are negligible compared to the others that are present. It is worth noting that this applies to all quantities which are in effect derived from the inertial acceleration, such as kinetic energy, and that such quantities are not only to be ignored but will produce nonsensical results if incorporated as an essential feature in a dynamical model. In this connection a great source of confusion has been through estimates of the Rossby number, a measure of the ratio (all other inertial forces/coriolis force), which is close to zero and seems to imply dominance of rotation — as in the earth's atmosphere — but all this overlooks the fact that the entire inertial forces are negligible.

There is a much more important point. Since the inertial acceleration is negligible the equation of motion does not explicitly retain a time-dependent part. Thus, whereas in a fluid at high Reynolds numbers most of the temporal variation arises from the flow itself, here that is not the case. Temporal variation must arise in other ways. In other words the flow adjusts itself everywhere immediately to the viscous constraints placed on it. This arises because in effect the rate of diffusion of vorticity is rapid: effectively infinitely rapid for the mantle.

The temperature scale in this model may be determined by a variety of factors. The initial temperature scale may be determined not internally but by external processes. If the Earth formed by the gravitational collapse of a diffuse cloud of matter and a proportion ξ of the energy were trapped then the initial temperature was

$$\theta_0 \sim \tfrac{4}{3}\pi\xi G\rho a^2/c,$$

where c is the specific heat of rock-substance. If the origin was by cold accretion so that $\xi \approx 0$, the temperature of the interior is set solely by the internal energy. On the other hand, even if only a small portion of the energy is trapped (for example with $\xi = 0.08$, $\theta_0 = 3000$ K), the early phase of a planet's thermal history will be dominated by θ_0, but in any event the long-term behaviour will be determined by $Qa^2/\rho c\kappa$.

3.2. MECHANISM OF THE SUBLAYER

An archetypal system of very great interest and wide application is the interface which develops in the thin surface zone of a wide, deep layer of fluid which is vigorously convecting owing to heating from below or cooling from above. This is a system in a state of thermal turbulence. The motion is not at all steady. The layer as a whole is filled with an ever-changing collection of vigorous eddying motions so that the layer is very well mixed and (in the laboratory) the mean temperature is nearly the same throughout the layer. There is, however, a thin surface region, the sublayer, in which the vertical gradient of the mean temperature is large and in which the motions are rather coherent. Instabilities develop in this surface interface and when they reach finite amplitude localized portions of the interface fall into the interior flow. While so doing they entrain fluid from the ambient sublayer thereby inducing the ascent of deep hot fluid into the region of the sublayer which is being evacuated of cold dense fluid. Subsequently this hot fluid is cooled by conduction to the surface until that portion or a nearby portion of the sublayer again becomes unstable. This process occurs more or less at random over the areal extent of the sublayer. The thickness of the sublayer δ is given by

$$\gamma g \theta \delta^3 / \kappa \nu = A_c \approx 10^3,$$

where θ is the temperature difference across it.

This zone at the top of the mantle is the major active zone of the entire earth. It acts as a buffer between the well mixed interior and the crustal systems.

It is important to emphasize that within the sublayer molecular processes — both thermal conductivity and viscosity — are important, whereas in the interior, molecular processes play a negligible role in the global mechanics. In particular, therefore, the values of κ and ν appropriate to evaluation of this model are those of the material in the sublayer and *not* those of the interior. The single most characteristic parameter of this model is the length-scale $(\kappa \nu / \gamma g \theta)^{1/3}$ evaluated for the uppermost layer of the system: it determines the fundamental scales of length, time and velocity for all macrogeological systems.

It is convenient to express the mean heat flux f in terms of the properties of the sublayer, where θ is the temperature drop across the sublayer, so that

$$f = \rho c \theta / \delta$$
$$\delta = \zeta \theta^{-1/3}$$
$$\zeta \approx 9 (\kappa \nu / \gamma g)^{1/3}.$$

The coefficient 9 is derived from the sublayer critical Rayleigh number $\approx 9^3 \approx 730$ (see below). Note that the temperature gradient at the top of the sublayer is about θ/δ, but the actual temperature reaches θ at a depth of about 2δ. Below this depth, the mean temperature is constant in the laboratory, but in the earth it will lie close to the adiabatic temperature. Other properties of the sublayer are given in Table 3.1.

Table 3.1. *Properties of thermal turbulence of a cooling layer. Comparative values for an earth at* $A = 10^9$, $\Delta T = 2770$ K.

Surface heat flux, mean N	$N = 0.1 A^{1/3}$	~ 100
Sublayer thickness, δ	$\dfrac{\gamma g \Delta T \delta^3}{\kappa \nu} \approx 10^3$	$\sim 10^2$ km
RMS temperature fluctuation, $\tilde{\theta}$	$\tilde{\theta}/\Delta T \approx 2 A^{-1/6}$	$0.063: \sim 430$ K
RMS velocity, $\tilde{q}$	$a\tilde{q}/\kappa \approx 3 A^{1/3}$	0.015 m yr^{-1}
Mean velocity, $\langle q \rangle$	0	0
Global temperature, $\tilde{T}$	$\sim \Delta T$	3000 K
Spectral cut-off frequency	κ/δ^2	$\sim 10^{-8}$ cycle yr^{-1}

Note: $f = NK\Delta T/a$ defines N.

A mechanistic model of the sublayer

The characteristic feature of the thermal sublayer is its local sporadic ejection of cold discrete blobs into the interior. A mechanistic model can be developed from this observation (Howard, 1964). It should be noted that most studies of this topic refer to a layer of fluid not only cooled from above but also heated from below — an arrangement which permits a statistically steady state to be maintained and which has two similar sublayers, one on each of the two confining horizontal planes. The situation described here is for a fluid layer cooled from above only.

Consider the idealized situation in which the layer of fluid, $z \geqslant 0$ is at rest and at a uniform temperature ΔT. At time $t = 0$ the upper boundary, $z = 0$, is set at relative temperature zero. Initially thermal conduction will be the only process that can operate. The solution of the heat conduction equation for the temperature field near the upper boundary is:

$$T(x, y, z, t) = T(z, t) = \Delta T \operatorname{erf} u; \qquad u = z/2(\kappa t)^{1/2}.$$

This is a temperature field uniform over horizontal planes in which a surface zone of cooled fluid progressively grows downward.

For an interval of time $0 \leqslant t \leqslant \tau$, say, the cooled layer, although statically unstable, will be dynamically stable due to the combined stabilizing role of thermal diffusion and viscosity. If, however, the Rayleigh number for the whole layer is sufficiently large, at some time τ (considerably shorter than that to cool the entire layer, of order a^2/κ), instabilities will grow in the cooled layer. It is observed in fluids of high Prandtl number that the consequent growth and ejection of cold blobs occurs in a time short compared to that of the conductive phase, τ. Thus we envisage a period of gestation in which the cooled layer grows by conduction followed by a short interval in which the cooled fluid is ejected out of the cooled region to be replaced by deep hot fluid, thereby more or less restoring the original conditions near the upper boundary. Another gestation—ejection phase ensues. This process is observed to occur more or less at random over the upper surface. Since the temperature is primarily determined by the conduction phase the temporal mean temperature profile is

$$\bar{T}(z) = \frac{1}{\tau} \int_0^\tau T(z, t)\, dt,$$

so that

$$\bar{T}(z)/\Delta T = (1 + 2\xi^2)\, \mathrm{erf}\xi - 2\xi^2 + \frac{2}{\pi^{1/2}} \xi \exp(-\xi^2)$$

where

$$\xi = u(\tau) = z/2(\kappa\tau)^{1/2}.$$

The form of this function, together with the temperature profiles throughout the conductive phase are shown in Fig. 3.1.

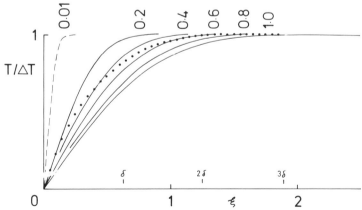

Fig. 3.1. Temperature profiles as a function of depth coordinate, ξ. (i) Full lines: $T(\xi, t)$ for $t/\tau = 0.2, 0.4, 0.6, 0.8, 1.0$; the profile for $t/\tau = 0.01$ is shown, dashed. (ii) Dotted curve: temporal mean $\bar{T}(\xi)$.

It is important to notice that this model does not involve ν except through τ. The mean temperature profile is determined during the gestation phase solely by κ, but the termination of its growth and disruption by both κ and ν. In any event the form of $\bar{T}(z)$ is determined by κ and only the vertical scaling requires a knowledge of ν as well.

The identification of the time-scale τ is complicated by the non-uniform time-dependent vertical gradient of $\bar{T}(z)$. In the simplest case of convective instability, a uniform constant gradient is considered and the onset of convection is determined by the Rayleigh number being equal to a critical value. The more general case is still a matter of some controversy and it is necessary to be rather heavy handed in treating it. The simplest approach is to define a length-scale, δ, of the mean temperature profile such that the temporal mean surface heat flux $\tilde{f} \equiv K\Delta T/\delta$, and thereafter refer to the sublayer Rayleigh number

$$A_\delta = \gamma g \Delta T \delta^3 / \kappa \nu.$$

The temporal mean surface heat flux is given by:

$$f = \frac{K}{\tau} \int_0^\tau \left(\frac{\partial T(z, t)}{\partial z} \right) \mathrm{d}t \equiv K\Delta T/\delta,$$

so that

$$\delta = \tfrac{1}{2} (\pi \kappa \tau)^{1/2}.$$

Note that f is the same as $K(\partial \bar{T}/\partial z)_0$. The depth $z = \delta$ is reached on the mean profile at $\xi = \xi^1 = \delta/2(\kappa\tau)^{1/2} = (\pi/8)^{1/2} = 0.627$, where $T/\Delta T \simeq 0.80$.

The instability of the layer will commence when the sublayer scale thickness, δ, has increased sufficiently for the sublayer Rayleigh number, A_δ to reach a critical value, $A_{\delta c}$ which on the basis of the value for a uniform profile will be of order 10^3. Comparison of $\bar{T}$ with measured profiles indicates a value in the range 700—750. The experimental data shown in Fig. 3.2 fits the computed mean profile for a sublayer Rayleigh number of 700. Further confirmation of this value has been obtained from direct measurements of τ as shown in Fig. 3.3.

The root mean square temperature fluctuations,

$$\theta'(z) = \left[\frac{1}{\tau} \int_0^\tau (T - \bar{T})^2 \mathrm{d}t \right]^{1/2},$$

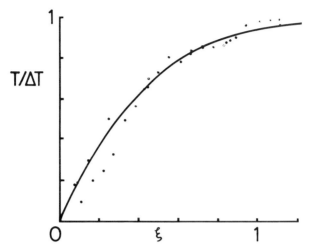

Fig. 3.2. Mean temperature profile experimental data, together with the model profile. Adapted from Howard (1964) and drawn to correspond to a layer of fluid cooling from above.

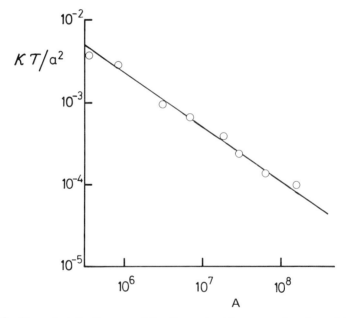

Fig. 3.3. Characteristic time, τ, of the thermal sublayer as a function of Rayleigh number, A for a layer of fluid cooled on its upper surface. Adapted from data of Somerscales and Gazda (1969), obtained by counting mean level crossings in the zone of maximum temperature fluctuations. The drawn line corresponds to a sublayer critical Rayleigh number of 700.

can also be computed. This is rather stretching the model to its limits since the role of the blobs is ignored. The computed profile of θ' has a peak within the sublayer of magnitude $0.18\,\Delta T$, close to values $(0.16 \pm 0.02)\,\Delta T$ measured in the laboratory (Somerscales and Gazda, 1969).

The above description is more than sufficient for our present purpose. It is necessary, however, to point out that the details of the mechanics are complex and poorly understood (see for example Elder, 1968, 1969).

Modification of the simple model

The simple sublayer model considers recharge occurring throughout the upper zone right up to the surface. Clearly this does not occur in the earth. In the terms of the model this must arise because of some extra resistance of the surface layers to the recharge. Assuming, therefore, that broadly the viscosity of the material decreases with depth mainly through increasing temperature, let us restrict recharge to a zone below a depth, h, where the surface layers shallower than this depth are considered immobile. In effect we are representing the viscosity variation by a step function in depth.

The model of the thermal sublayer with a stiff upper zone is now specified by:

(i) $z = 0$, $\quad T = 0$

(ii) $0 < z < h$, initially $T = 0$, an arbitrary choice

(iii) $z \geqslant h$,

    (a) at $t = 0$ set $T = \Delta T$;

    (b) $0 < t \leqslant \tau$, $\partial T/\partial t = \partial(\kappa\theta_z)/\partial z$;

    (c) $t > \tau$, reset $t = 0$ and repeat from (a).

Notice the essential feature here that after an interval τ the deep zone is recharged with deep hot material, but the *upper zone* is left alone. This is repeated until the system is in statistical equilibrium, usually after about 5τ, depending on h. The initial temperature distribution is arbitrary since its effect is progressively lost.

It is straightforward to solve this system numerically. The explicit solution for $h = 0$ provides a useful check.

Mean temperature profiles are shown in Fig. 3.4 for various values of $\zeta = h/2\sqrt{\kappa\tau}$. Larger values of ζ lead to a somewhat fatter layer and lower surface gradient, as shown in Fig. 3.5. The effect is pronounced for ζ about unity.

The role of the immobile surface zone is best appreciated by considering the surface gradient $(\partial T(z, t)/\partial t)_0$ as a function of time as

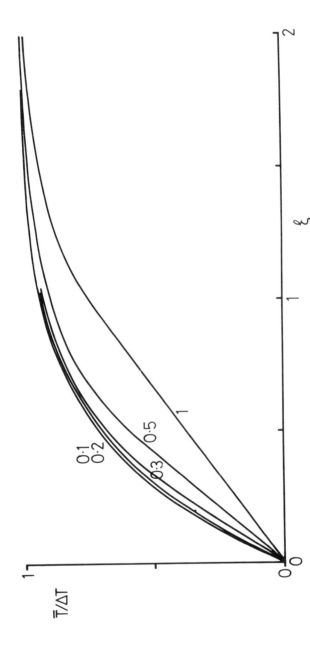

Fig. 3.4. Temporal mean temperature profiles as a function of the depth parameter, ξ, for various values of the permanent layer thickness parameter, ζ.

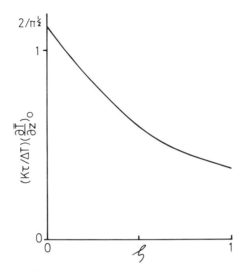

Fig. 3.5. Mean surface gradient in dimensionless form as a function of ζ.

shown in Fig. 3.6. With $\zeta = 0$, the simple case without an immobile surface layer, the surface gradient is a maximum at $t = 0$ and thereafter falls monotonically until $t = \tau$, whereupon the recharge cycle is restarted. For non-zero values of ζ the surface gradient maximum is increasingly delayed and muted at larger ζ until beyond about unity the temporal variation of the surface gradient is negligible. The upper layer is acting as a low-pass filter to the step-function-like input at its base: its role is that of a thermal capacitance which tends to retain its thermal content.

Constraints on the upper zone thickness

What limits can be placed on the thickness of the upper zone? Two ideas present themselves.

Clearly the continental crust is not being recharged in the manner described here. Thus taking its thickness as 35 km, and $\delta = 100$ km, say, corresponding to $\xi = 0.63$, gives $\zeta = 0.22$ as a minimum. For an oceanic crust of 5 km, we have $\zeta = 0.03$. All we can say here is that a value of 0.2 or more is compatible with the persistence of the crust.

The global variation of surface heat flux, ignoring localized geothermal areas, has a range of typically $\pm 50\%$ about its global average, and extreme values are unknown. If in the model we take $\zeta \approx 0$ as indicated in Fig. 3.6 there would be much greater variation in heat flux and high values would *not* be uncommon, they would be scattered

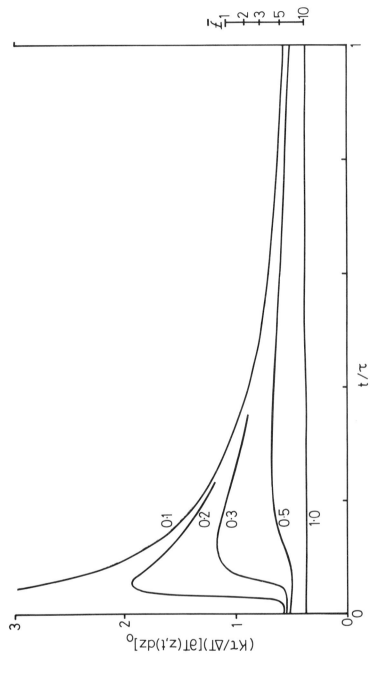

Fig. 3.6. Surface temperature gradient as a function of time t/τ for various values of ζ.

about with a spacing of order δ. We do want some variation, however, and values in the range 0.1–0.4 seem suitable.

We can partially combine these two ideas. Under the continents in any event we must have $\zeta \gtrsim 0.2$, so that heat flow and heat flow variations are muted, whereas under the oceans perhaps $\zeta \lesssim 0.1$ so that the overall heat flow is somewhat above the global average and the heat flow variation is much spottier.

3.3. PARTIAL MELTING

Convection within the mantle, seen on the global scale, produces an outer layer, the lithosphere, which is relatively passive and which is supplied with matter and energy from the mantle interior. As a conceptual picture averaged over time intervals of order 10^2 Myr, this is fine.

The geothermal systems of interest in this book are, however, seen as features of lithothermal systems, intense localized systems imbedded in a relatively passive ambient and relying for their operation on the transport of magma. Our knowledge of the mechanism of the production of magma within the mantle is inadequate. A few general remarks are all that is necessary here, although that would not be the case if this book were preoccupied with lithothermal systems rather than hydrothermal systems.

In effect I take the view that in some zone typically of depth 10^2 km the mantle is globally capable of producing a partial melt but the subsequent behaviour of that melt is dominated by its passive ambient. In other words lithothermal systems are not controlled by the productive capacity of their sources but rather by the structures in which they are imbedded.

Global layer model

Consider a system in which solid convection provides heat and matter to an upper partially melted zone. The global mean temperature distribution will be as sketched in Fig. 3.7. Below the interface, defined by the melting temperature T_m, a thermal sublayer buffers the well mixed interior. Above the interface we have a free convection system in a porous medium, with temperature close to T_m except in a thin surface zone. The details in this surface zone are ignored.

The solid convective system can be described by the relations

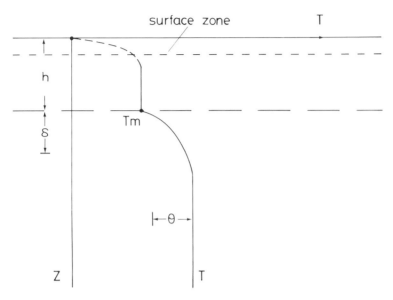

Fig. 3.7. Schematic mean temperature profile for a model with a global partially melted layer.

above but with

$$\theta = T - T_m \quad \text{for} \quad T \geqslant T_m \quad \text{and} \quad \theta = T \quad \text{for} \quad T < T_m,$$

since then no partial melting is possible. In particular the heat flux $f = \rho c \kappa \theta / \delta$.

The upper convective lithothermal system can be described by the relations for a porous medium. In particular the heat flux

$$f = \rho c w T_m,$$

where w is the mean vertical volumetric flux. From Darcy's relation

$$w = kg(\Delta\rho/\rho)/\nu,$$

where $\Delta\rho$ is the density difference between the magma and the ambient solid and ν is the kinematic viscosity of the melt.

The requirement that at the interface the temperature is T_m (effectively a constant since the variation with pressure (depth) is relatively small) and that the heat flux is continuous, forces the conclusion that the upper layer is self-adjusting through its permeability. Thus equating the two expressions for the heat flux

$$k = \kappa\nu\theta/g\delta T_m \, (\Delta\rho/\rho).$$

We are saying the upper layer is thermally transparent and that the partially melted zone determines its own permeability in order to handle the power supplied. To obtain an estimate of k take $T = 2700$ K, $T_m = 1800$ K, $\delta = 100$ km, $\nu = 0.1\,\mathrm{m^2\,s^{-1}}$, $\kappa = 10^{-6}\,\mathrm{m^2\,s^{-1}}$, $\Delta\rho/\rho = 0.1$. We have $k = 5 \times 10^{-11}\,\mathrm{m^2}$, about 0.5 darcy. This value is not unreasonable but the model is really quite *ad hoc*. We search (rather desperately) for some other clues to the nature and existence of the partial melt state of the upper mantle.

Theoretical sketch: global partial melting

The development of a zone of partial melting in the upper mantle can be sketched with the following model. It is necessary to remark at the outset that the model is very crude, but it does highlight the features of the zone of interest here in a simple manner.

Consider the local volumetric fraction, e, of rock in the liquid state. For a given planet e will be a function of position and time and we presume that it is extremely random both in space and time. On the global scale, however, in the manner of global thermal histories, we consider as a first step only values of e, at a given time, averaged over surfaces of uniform depth. Thus we consider $e = e(z, t)$. This approach does not imply that partial melting necessarily occurs beneath each point of the surface.

The melt fraction is presumed to be small and, furthermore, in the light of the discussion above, movement of melt contributes negligibly to the gross energy budget. This suggests the following:

(i)   the overall thermal behaviour, in particular the temperature, $\theta(z, t)$, can be described by a non-partial melting model, except for regions in which the melting point, T_m is less than θ;

(ii) in those regions sufficient melt is produced to maintain the temperature at T_m.

For the purpose of this illustration we use a simple global thermal history (Elder, 1976). The melting point can be expected to be a rather elaborate function of pressure but here we take

$$T_m = T_0 + \beta z,$$

where for example $T_0 = 1000-1500°\mathrm{C}$, $\beta = 5-10\,\mathrm{K\,km^{-1}}$. The figures shown below are for $T_0 = 1200°\mathrm{C}$, $\beta = 7\,\mathrm{K\,km^{-1}}$.

In the region, $\theta > T_m$ conservation of energy requires

$$\rho * c * \theta = (1 - e)\rho * c * T + e\rho(L + cT),$$

where $\rho*$, $c*$ are the density and specific heats of the solid rock, ρ, c those of the liquid, and L is the latent heat of melting. It is convenient to write.

$$\rho = (1 - \xi)\rho*,$$

where ξ is the relative density change on melting and is usually about 0.1. Then, on rearrangement,

$$e = (\theta - T_m)/[(1 - \xi)(L/c) - \xi T_m], \qquad \theta > T_m$$

$$\approx c_*(\theta - T_m)/L.$$

Otherwise $e = 0$, $\theta \leqslant T_m$.

In a model of this crudity it is sufficient to take $\rho \approx \rho_*$, $c \approx c_*$. For typical values of $c_* = 1\,\mathrm{k\,J\,kg^{-1}\,K^{-1}}$, $L = 500\,\mathrm{kJ\,kg^{-1}\,K^{-1}}$ with $(\theta - T_m) = 100\,\mathrm{K}$ give $e = 0.2$. The values of e suggested by this model are probably much too high, perhaps by an order of magnitude.

The temperature profile is calculated from

$$\frac{\theta_s}{\theta_0} = W^{-3}; \qquad W = \left(1 + \left[\frac{A_0}{A_c}\right]^{1/3} \frac{\kappa t}{a^2}\right)$$

$$y = z/\delta; \qquad u = y + 0.6y^2; \qquad \theta = \theta_s(1 - e^{-u}).$$

Temperature profiles for a number of cases are shown in Fig. 3.8. The profiles $\theta(z, t)$ for no melting are characterized by a sublayer thickness which increases with time and a deep temperature which diminishes with time. The time-independent melting curves are thereby intersected by a zone of melting which has the features:

(i) vertical extent diminishes with time;
(ii) amplitude $(\theta - T_m)$ diminishes with time;
(iii) depth to partial melting increases with time;
(iv) no partial melting at all after a finite time.

The corresponding profiles of volumetric melt fraction are shown in Fig. 3.9.

The most important feature for this work is that the depth to the partial melt zone, to be regarded as the source depth for higher level magmatic processes, increases with time. Two pieces of evidence give some support to this picture.

Deductions from the heights of volcanic systems. The discussion below on the lithostatics of magmatic systems suggests that the ultimate height, h, of sufficiently voluminous extrusive structures, such as lava plateaux, is more or less proportional to source depth, H;

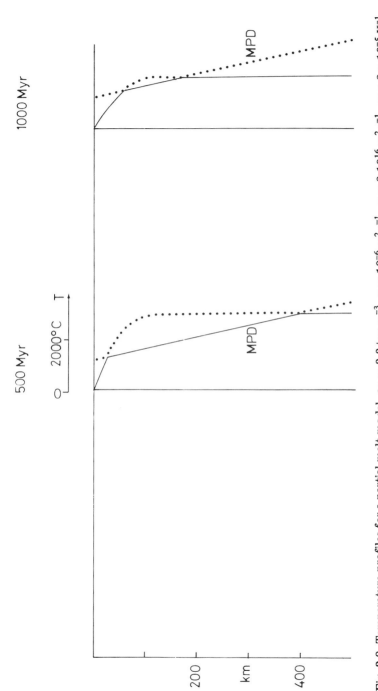

Fig. 3.8. Temperature profiles for a partial melt model. $\rho* = 3.3\,\mathrm{ton\,m^{-3}}$, $\kappa = 10^{-6}\,\mathrm{m^2\,s^{-1}}$, $\nu = 3.10^{16}\,\mathrm{m^2\,s^{-1}}$, $\gamma = 2 \times 10^{-5}\,\mathrm{K^{-1}}$, $c_* = 1\,\mathrm{kJ\,kg^{-1}\,K^{-1}}$, $\theta_0 = 5000°\mathrm{C}$, $a = 1738\,\mathrm{km}$ (moon).

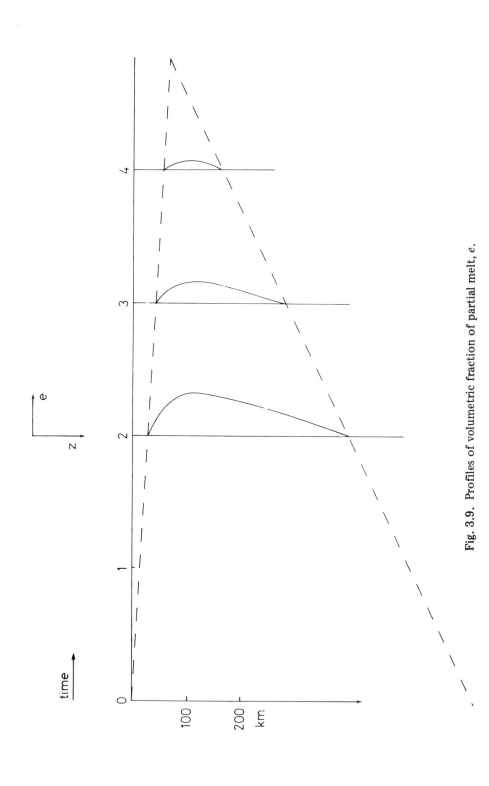

Fig. 3.9. Profiles of volumetric fraction of partial melt, e.

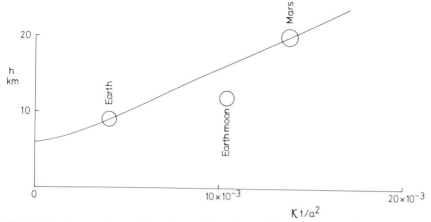

Fig. 3.10. Comparison of simple partial melt model results on maximum heights of planetary volcanic systems as a function of active planetary age with data from Earth, Mars, and the Earth Moon. For the Earth and Mars, $t = 5000\,\text{Myr}$, both active; for the moon the active life is plotted for $t = 1000\,\text{Myr}$. All with $\kappa = 10^{-6}\,\text{m}^2\,\text{s}^{-1}$.

namely $h \sim H(\Delta\rho/\rho)$. Taking data from the planets Earth, Mars, Earth Moon for the greatest heights of volcanic systems together, with estimates of their respective age as measured by kt/a^2, we obtain Fig. 3.10. The correlation is not too bad.

Estimates of melt fraction from surface measurements. On the global scale the electrical conductivity profile, obtained from natural geomagnetic soundings, as shown in the compilation of Fig. 3.11 (Alldredge, 1977), has two more or less distinct regions.

(i) Below about 500 km the (sparse) conductivity data can be fitted to $\sigma/\sigma_0 = (r/r_c)^{-\gamma}$, where r is the radial position, $r = r_c$ at the core—mantle boundary; $\gamma \approx 25$, a constant; and the conductivity at the core—mantle boundary $\sigma_0 \approx 10^5\,\text{mho m}^{-1}$.[†] This behaviour is presumably the effect of pressure on the electronic distribution.

(ii) In the zone above 500 km depth the conductivity is low and there appears to be considerable spatial variation. Conductivity is typically $10^{-2}\,\text{mho m}^{-1}$. These normal areas have the conductivity of olivine at temperatures of $1500°\text{C}$ and 300—400 km depth. There are however anomalous regions in which the conductivity exceeds $10^{-1}\,\text{mho m}^{-1}$.

[†]Note: $1\,\text{S m}^{-1} = 1\,\text{siemen m}^{-1} = 1\,\text{mho m}^{-1} = 10^{-2}\,(\text{ohm cm})^{-1}$.

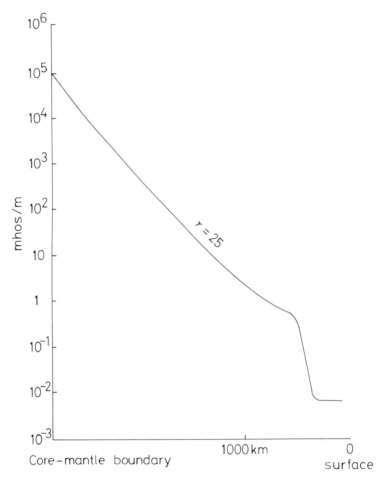

Fig. 3.11. Conductivity profile in the mantle (after Alldredge, 1977).

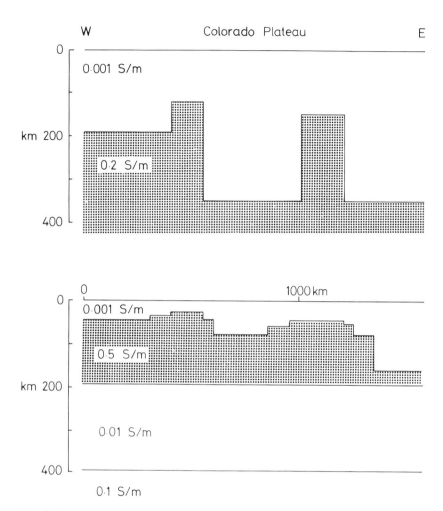

Fig. 3.12. Possible distributions of partial melting. Alternative, two-dimensional conductivity models for section across western USA at 38°N (adapted from Shankland and Waff, 1977).

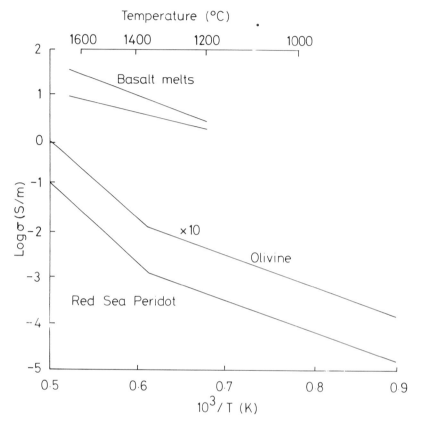

Fig. 3.13. Laboratory values of electrical conductivity for solid olivine and basalt melts (taken from a compilation by Shankland and Waff, 1977).

For example, Fig. 3.12 shows data from a detailed study in the western USA (Shankland and Waff, 1977). These conductivities in the range $0.1-1$ mho m^{-1} suggest a connected melt fraction perhaps as large as 0.1 (see also Murase *et al.*, 1977), obtained from

$$\sigma = [\sigma_m + (\sigma_s - \sigma_m)(1 - \tfrac{2}{3}e)]/[1 + \tfrac{1}{3}e(\sigma_s/\sigma_m - 1)],$$

where σ_m, σ_s are the conductivities of the liquid and solid, the values of which for possible materials are shown in Fig. 3.13.

Studies of the reduced seismic velocities in the "low-velocity-zone" suggest that, averaged over regions of 10^3 km extent, mean values of melt fraction are $0.01-0.03$ (see for example, Stocker and Gordon, 1975).

4. The Oceanic Lithosphere and Oceanic Hydrothermal Systems

Lithothermal systems are imbedded in the lithosphere. They can be thought of as pieces of magmatic plumbing installed in a passive stiff container. The nature of the container is therefore of interest. Indeed, as will be discussed in Part II, its role is vital. Information about the lithosphere can be obtained from a variety of measurements, but those of surface heat flux are the least equivocal. They show that the mature lithosphere is thermally a passive barrier about 10^2 km thick.

In the global picture just presented the production and destruction of lithosphere is seen as a random process in space and time. Local details are ignored. If however we look not on the global scale but on the sub-global scale of oceans and continents it is possible to identify some of the local processes, in particular the role of lateral displacement of portions of the lithosphere. Furthermore the time-scales of these can be identified and are found to be much longer than those of lithothermal systems. Thus we can assume not only that the sources of lithothermal systems are potentially available at all times in the mantle, but that discharge toward the surface will be in an ambient which changes only slowly and is not determined by the lithothermal system.

4.1. THERMAL IDENTIFICATION OF THE LITHOSPHERE

The global heat flow data, especially that from the ocean basins allows identification of the "lithosphere" (see for example, Sclater and Francheteau, 1970). This outermost zone of the mantle is considered to be stiffer and more coherent than the deeper interior. In the simple kinematic models to be described below a description of the heat flow data is possible with a surface layer in which heat is transferred solely by conduction. This model originates from the crude correlation of heat flow with age, the heat flux, f, for the oceans is given by

$$f \approx 500(t/\text{Myr})^{-1/2} \, \text{mW m}^{-2}.$$

An equally good model would be possible with a fluid-like system in which the "lithosphere" was regarded as a thermal boundary layer.

In this book, however, the main interest in this question is to provide a background to the setting of oceanic hydrothermal systems so we shall not quibble about detailed aspects of these lithosphere models.

Homogeneous half-space model

Consider a uniform region $z \geqslant 0$ below a horizontal surface $z = 0$, flowing outwards from $x = 0$ with uniform velocity u, and with temperature $T = T_1$ on the vertical surface $x = 0$, cooled to $T = T_0$ on $z = 0$. This presumes a narrow source region, and that the depth of the flow is greater than that of the cooled surface zone. Then if

$$x = ut,$$

we have

$$T = T_0 + \Delta T \, \text{erf} \, \xi,$$

where

$$\Delta T = T_1 - T_0, \qquad \xi = z/2(\kappa t)^{1/2}$$

Define the thickness of the lithosphere, δ, such that $(T_1 - T)/\Delta T =$ constant. For example, if we choose the constant as 0.1 then

$$\delta = 2.32(\kappa t)^{1/2}$$

is the thickness of the zone in which 90% of the cooling of the top of the lithosphere occurs. If this zone is in isostatic equilibrium there will need to be a corresponding subsidence of the ocean floor of amount h, say. Ignoring details of the crust, then

$$h(\rho_* - \rho_0) = \int_0^\infty \rho_* \alpha_* (T_1 - T)\,dz,$$

where $\rho_0 \approx 1 \, \text{ton m}^{-3}$ is the density of sea water, $\rho_* \approx 3.3 \, \text{ton m}^{-3}$ is the density of mantle material and $\alpha_* \approx 3 \times 10^{-5} \, \text{K}^{-1}$ is the coefficient of cubical expansion of mantle material. Hence

$$h = \frac{2\alpha_* \rho_* \Delta T}{(\rho_* - \rho_0)} (\kappa t/\pi)^{1/2}.$$

This simple model, with for example $\kappa = 10^{-6} \, \text{m}^2 \, \text{s}^{-1}$, $\Delta T = 1200 \, \text{K}$, at first sight fits the observations surprisingly well —

especially when the rather extreme natural spatial variation is considered. Data for the Indian Ocean are shown in Fig. 4.1. The model fits the smoothed topographic data far better, however, than that of the heat flow.

Nevertheless, as has been pointed out by, for example, Crane and

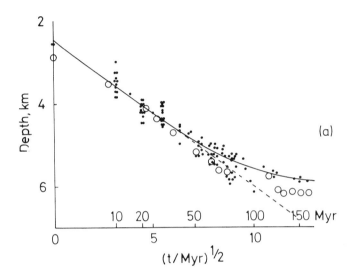

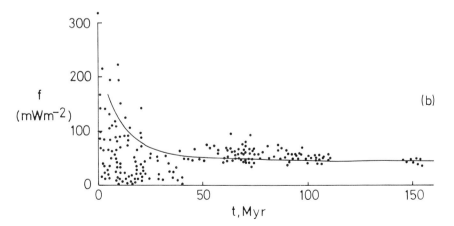

Fig. 4.1. Oceanic data as a function of age: (a) ocean depth (square root time-scale); (b) heat flux (linear time-scale) (Adapted from Anderson and Langseth, 1977).

Normark (1977), the actual topography in the innermost 10 km or so of the East Pacific Rise is considerably lower than predicted by this model, suggesting the presence of a cooled zone.

Note: The choice of κ, K, ΔT. The values used for κ and related quantities need to be consistent with $K = \rho c \kappa$. Thus, if we choose for upper mantle material $\kappa = 10^{-6}\,\text{m}^2\,\text{s}^{-1}$, $c = 10^3\,\text{J}\,\text{kg}^{-1}\,\text{K}^{-1}$, $\rho = 3.3\,\text{ton}\,\text{m}^{-3}$, then $K = 3.3\,\text{W}\,\text{m}^{-1}\,\text{K}^{-1}$. It should be noted however that in the model there is some arbitrariness in the choice. The model is linear in temperature, the development is controlled by κ, and K and ΔT solely in obtaining the heat flux. The fit in the $t^{1/2}$ region requires $K\Delta T = 6.10^3\,\text{W}\,\text{m}^{-1}$; for example $\Delta T = 1500°\text{C}$ and $K = 4\,\text{W}\,\text{m}^{-1}\,\text{K}^{-1}$.

Range of validity of the simple model

The simple model provides an acceptable fit to the observations over the middle range of the data. This is not the case otherwise.

(i) *Near field.* Near the source region, typically $t \lesssim 3$ Myr, although the subsidence fit appears good, most of the observed heat flows are much smaller than those of the model.

(ii) *Far field.* Distant from the source region, typically $t \gtrsim 50$ Myr, the observed heat flows tend to a non-zero constant value.

Thus we recognize four zones in which the influence of the upwelling becomes progressively muted:

(1) a central zone, out to about 0.5 Myr, above the source region, dominated by the upwelling, and in which (as we shall see) hydrothermal systems are common;

(2) a thermal aureole, out to about 3 Myr, in which the details of the conditions at the top of the slab are important;

(3) a "boundary layer" zone, out to about 50 Myr, dominated by the cooling slab;

(4) a "naked" zone, beyond about 50 Myr, in which the influence of the deeper mantle is dominant.

Modifications of the simple model

Far field

Measurements show that the heat flux $f \sim$ non-zero constant as $t \to \infty$, whereas the simple model gives $f \sim 0$. This suggests that the slab should be considered to have a finite thickness, h. A number of model curves for various h are shown in Fig. 4.2.(a). Near $t = 0$ all the curves are closely the same, the slab behaves as if it were semi-infinite, but as $t \to \infty$ and a steady state is approached the heat flux $f \sim K\Delta T/h$. For example, with $h = 100$ km we have an ultimate heat flux of 60 mW m^{-2}.

The conductive time-scale for transverse flow of heat across a slab of thickness h is $\tau = h^2/\kappa$. Depending on the boundary and initial conditions the thermal state of the slab will be a function of t/τ. For boundary conditions of fixed temperatures across the slab, about 90% of the temperature changes have occurred by $t/\tau \approx 1/4$. Thus a 100 km thick slab reaches a new equilibrium when $t \gtrsim \tau/4 \approx 80$ Myr. The observations then clearly show, solely on the basis of time to equilibrium, that a 50 km slab is too quick, a 200 km slab too slow and a 100 km about right. Further refinement of the estimate of the slab thickness is then best done by consideration of the equilibrium heat flux. If the equilibrium heat flux is $f(\infty)$ then

$$h = K\Delta T/f(\infty).$$

With the measured $f(\infty) = 50$ mW m^{-2}, we have $h = 120$ km.

A note of the model conditions on the slab base. The boundary condition $T = T_1$, a constant, on the base of the slab implies that we presume purely conductive heat flow through the slab and purely advective flow in the mantle beneath the slab. In other words the deep mantle can provide as much heat flux as the slab is capable of transmitting. With this in mind more elaborate conditions can be imposed on the slab base.

(i) A given heat flux, f_1 is of interest, and one such model result is shown in the figure for $f_1 = 100$ mW m^{-2} and an initially hot 100 km slab; and although the thermal history is similar to that of a slab with $T = T_1$ on the slab base for which $f(\infty) = f_1$, it is not a realistic possibility. It throws away the nature of the deep advective system.

(ii) A more realistic condition is to allow the slab thickness h to be a function of time, $h(t)$ determined by some condition on the depth to incipient partial melting. Thus we could take

$$dh/dt = (f_0 - f_1)/\rho L,$$

where $f_0 = K(\partial T/\partial z)_h$ and f_1 is given. In this case, with a small initial $h = h_0$ so that f_0 is large, h grows until $f_0 = f_1$ and a steady state is achieved. A representative model run is shown in Fig. 4.2(b).

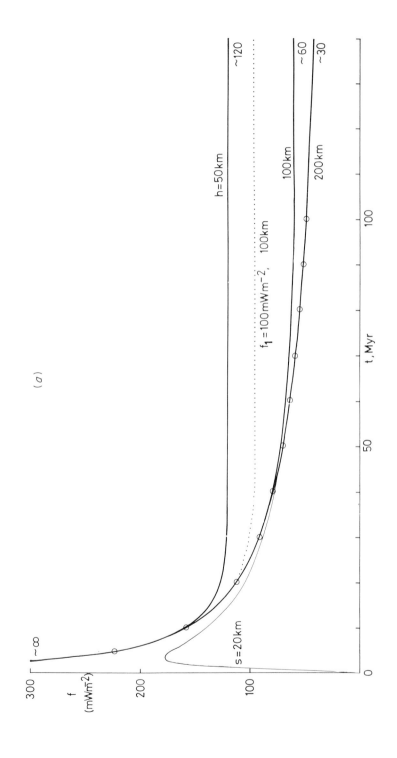

(a)

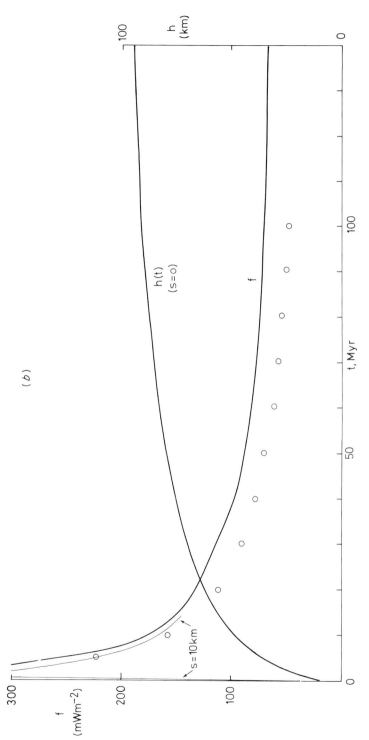

Fig. 4.2. Cooling slab model. (*a*) Slab of given thickness. Heat flux, f, as a function of time t, in Myr, for thickness h, in km. ○: 500 $(t/\text{Myr})^{-1/2}$. Also shown, for comparison, data from Fig. 4.3 with $s = 20$ km. (*b*) Slab of variable thickness. Heat flux, f, in mW m^{-2} and thickness h, in km as a function of time, t in Myr. Data used: $\rho = 3$ ton m^{-3}, $L = 400$ kJ kg^{-1}, $f_1 = 60$ mW m^{-2}, $h_0 = 10$ km.

Although these matters are of interest they are peripheral to those of this book.

Near field

Measurements show that the heat flux $\sim$ constant (very crudely, it is actually extremely variable) near $t = 0$, whereas the simple model has $f \sim \infty$. This suggests that an upper layer, or skin, of the slab is not hot in the source region. Consider therefore a skin of thickness s in which the initial temperature excess is zero. Thus the initial conditions at $t = 0$ are

$$0 \leqslant z \leqslant s, \quad T = T_0; \qquad s < z \leqslant h, \quad T = T_1;$$

and the boundary conditions are as before:

$$z = 0, \quad T = T_0; \qquad z = h, \quad T = T_1.$$

For a skin of thickness s, with boundary conditions of given temperature on one surface and one involving heat flux on the other, and time scale $\tau_s = s^2/\kappa$, equilibrium is reached in $t/\tau_s \gtrsim 1$. Thus for a skin of thickness $10\,km$ this is after about $3\,Myr$. Notice also that since $\tau_s \ll \tau$, the skin time-scale is much less than that of the slab, that during the time of order τ_s in which the skin effect lasts, the system behaves as if it were infinitely deep. In other words, the bottom of the slab is not felt in this time. The two effects, that of the skin and that of the finite slab thickness, are virtually independent.

A number of model curves for various s, all with $h = 100\,km$, are shown in Fig. 4.3. Note that all these curves tend to the same values. Notice the interval during which the surface heat flux is nearly zero. These do not give a critical determination of s but suggest that values near $10\,km$ are appropriate: the peaks values for $h = 5\,km$ are rather high, the model behaviour for $h = 20\,km$ is excessively muted. Incidentally, this suggests that the cold zone is somewhat thicker than that of oceanic crust.

4.2. ROLE OF OCEANIC HYDROTHERMAL SYSTEMS

The source region and hydrothermal systems

A skin of thickness $5-10\,km$ of relatively cool material is produced in the source region in spite of the manifest penetration of this zone

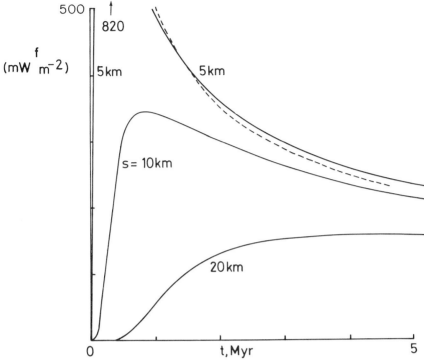

Fig. 4.3. Cooling slab model with initially cold skin. Heat flux, f, in m$\,$Wm^{-2} as a function of time t, in Myr, for various values of skin thickness, s, in km. Dotted curve: $f = 500(t/\mathrm{Myr})^{-1/2}\,\mathrm{m\,Wm^{-2}}$.

by basaltic magma. Clearly some vigorous process of apparently removing the thermal energy is at work. This indicates hydrothermal activity. Furthermore there is the evidence of the extreme variability of local heat fluxes. Recently, detailed measurements have produced direct evidence for oceanic hydrothermal systems. Before we look at this evidence it is of interest to continue the conductive models into the source region by allowing convective circulation of water in the surface zone. Such a system can readily transfer heat at rates of 1–$10^2\,\mathrm{W\,m^{-2}}$ so that an efficient mechanism for cooling the surface skin is available.

Consider therefore the following model. A slab of thickness, h, has a skin of thickness, s, with initial conditions the same as in the near field model except that on the base of the skin:

$$z = s, \qquad f = NK\theta/s,$$

where

$$N = (\theta/\theta_c)^3 \quad \text{for } \theta > \theta_c$$

$$= 1 \qquad \text{for } \theta \leqslant \theta_c.$$

This is a representation of the effect of the convective circulation of water enhancing the apparent thermal conductivity by a factor N, the Nusselt number, here related empirically to the interface temperature, θ. The form of $N(\theta)$ crudely takes into account not only the advective transfer of heat but also variation of viscosity and coefficient of cubical expansion of water-substance. (This representation is described in detail elsewhere). For interface temperatures less than θ_c convection is considered to be negligible and in this model, therefore, once θ has fallen below θ_c the system behaves just like the skin—slab conductive model. This is a steady state representation of conditions in the skin on the basis that the time-scale of the hydrothermal systems is very small compared to s^2/κ.

Note (as shown elsewhere) that provided $\theta > \theta_c$, f is independent of skin thickness. This arises since $\theta_c^3 \propto 1/s$.

Steady state model of the source region

Let us now treat the source region as a system in a steady state. It is as shown schematically in Fig. 4.4. A magmatic input drives a high level hydrothermal system which extracts the energy and leaves a cooled outward moving skin. Conservation of mass requires:

$$\rho w l = \rho_* u h,$$

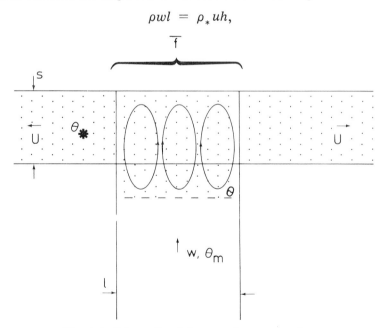

Fig. 4.4. Schematic of the oceanic source region.

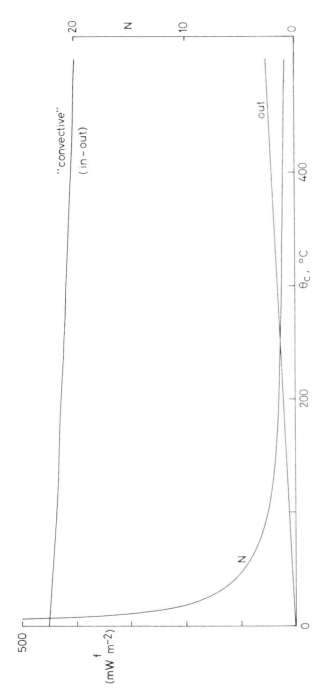

Fig. 4.5. Source region steady state model. Equivalent heat flux, f, in mWm^{-2} as a function of the critical temperature θ_c. Also shown values of the Nusselt number, N. Data used: $L_m = 400\,kJ\,kg^{-1}$, $c_m = c_* = 1\,kJ\,kg^{-1}\,K^{-1}$, $\theta_m = 1500°C$, $\rho = \rho_* = 3\,ton\,m^{-3}$, $u = 0.01\,m\,yr^{-1}$, $s = 5\,km$, $l = 20\,km$.

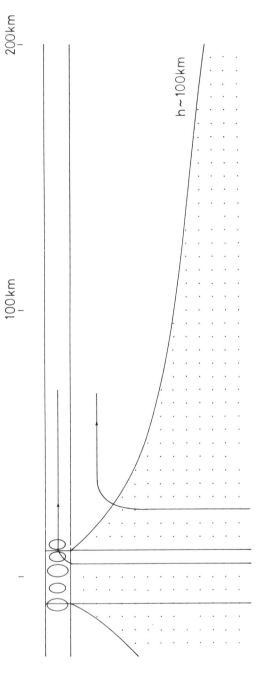

Fig. 4.6. Schematic diagram of the thermal representation of the oceanic lithosphere. Approximately to scale.

where $\rho w l$ is the rate of mass input as magma percolating upwards with volumetric flux w and $\rho_* u h$ is the rate of mass output of solid "crust". Conservation of energy requires:

$$l\bar{f} = \rho w l (L_m + C_m \theta) - \rho_* u h c_* \theta_*$$

$$= \rho_* u h (L_m + C_m \theta - c_* \theta_*)$$

where L_m, C_m are the latent and specific heats of the magma; θ_* is the mean skin temperature; $\bar{f}$ is the mean surface heat flux in the source region and arises solely from free convection of water in the crust.

Note that the actual mean magma velocity q relative to its ambient solid matrix of "porposity" ϵ is $q = w/\epsilon$. Hence

$$q/u = \rho_* h / \rho l \epsilon,$$

which, for example, with $\epsilon = 0.01$, say, and $h/l = 1/4$, gives $q/u \approx 30$.

A particular case is shown in Fig. 4.5. Over a wide range of θ_c the bulk of the input energy is transmitted by the hydrothermal systems. Furthermore, although the Nusselt number, N, varies over a wide range, the convective heat flow varies only slightly with θ_c. Thus if convection of water occurs, the skin will be substantially cooled.

Summary of the discussion of conductive lithosphere models

The situation revealed by the oceanic heat flow data, supported by topographic data, is sketched, in Fig. 4.6. The essential feature is the cooling by conduction of an initially hot slab being displaced sideways at a more or less uniform rate. The effective thermal thickness of the slab increases from an initial value of about 10 km to a final steady state of about 100 km. A cold surface skin of thickness 5—10 km is generated in the source region above the upwelling by the action of vigorous hydrothermal activity, the bulk of the heat supplied to the surface zone being discharged in heated sea water.

Direct observation of oceanic hydrothermal systems

In recent years some notable studies have provided data on the existence and nature of oceanic hydrothermal systems. A selection of these results is shown in Figs. 4.7 to 4.9. The hydrothermal systems predominate within times of order 0.1 Myr (that is for example, 10 km at 0.01 m yr^{-1} either side of the axis).

East Pacific Rise. On land a thermal area may be strikingly apparent

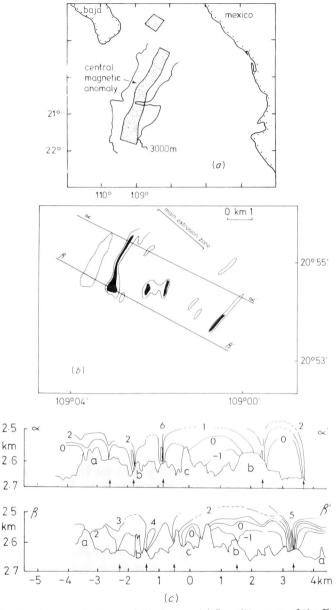

Fig. 4.7. Sea floor hydrothermal discharge. (a) Locality map of the East Pacific
Rise. (b) Distribution of thermal anomalies and section lines. Shaded areas: light,
1.62—1.63°C; heavy, >1.63°C. Local ambient (potential) temperature 1.60°C at
nominal depth 2570 m. (c) Temperature cross-sections. Position of discharge,
revealed by thermal plumes, indicated by arrows. Contour values $= 100(T°C -
1.60°C)$ (Adapted from Crane and Normark, 1977).

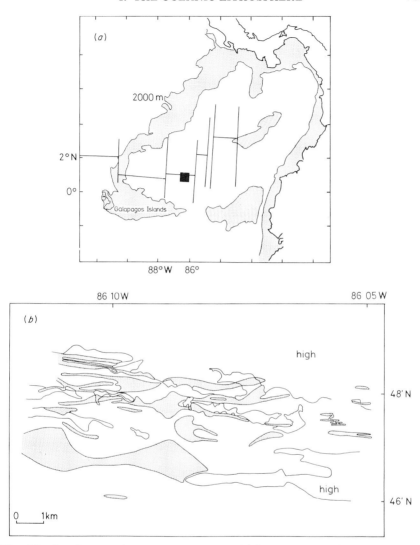

Fig. 4.8. Sea floor hydrothermal discharge. (*a*) Locality map of the Galapagos region. (*b*) Distribution of thermal anomalies. Shaded region is for potential temperature $>1.86°C$ at 2400 m depth (ambient temperature about $1.80°C$), roughly averaging $40\,MW\,km^{-2}$. Dotted line, contour 2480 m deep, the tick marks point to higher ground (shallower ocean). (Adapted from Crane, 1977).

from the plumes of steam rising from small vents and areas of steaming ground. On the ocean floor somewhat similar observations are possible by measuring the distibution of temperature in a zone near the bottom. Thus, as shown in Fig. 4.7 in patches of ground there are

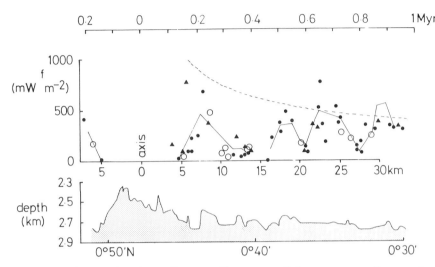

Fig. 4.9. Conductive heat flux profile across Galapagos spreading centre (Williams, 1974; Williams *et al.*, 1974).

distinct areas of discharge of hot water — this is especially clear in the sections (*c*). The temperature rises encountered are of course greatly muted as sea water is rapidly entrained into the discharge.

Galapagos Region. Data from a particularly detailed study are shown in Fig. 4.8. Here the thermal areas occur in the inner rift zone of an "oceanic spreading centre". They appear to be confined to the source area.

On land intense thermal areas are recognized often by the absence of certain plants. Oceanic areas, by contrast, provide partially warmed bottom water and stimulate abundant life. Typically, within a patch of diameter 100 m, there are numerous small vents, 0.1—0.3 m across, near which cluster life forms.

Discharge from individual vents is typically of order 1 kg s^{-1}, the total from a single patch being of order 100 kg s^{-1}. Gross average discharge is about $50 \text{ kg s}^{-1} \text{ km}^{-2}$.

These observations are nicely highlighted by the conductive heat flux profile shown in Fig. 4.9. Convective units of extent about 5 km are revealed out to 30 km or so from the ridge axis.

One cannot help forming the impression that oceanic thermal areas will be found to be more or less ubiquitous features of oceanic source areas, and indeed be more numerous than land thermal areas.

Part II

LITHOTHERMAL SYSTEMS

Embedded within the open reservoir of the upper mantle which is refreshed every 100 Myr or so are lithothermal systems — localized mobile regions of partial melting. Some of these lithothermal systems extend from depth to the surface and dissipate their matter and energy directly. Others remain trapped below the surface but are able to transfer part of their energy to higher levels interacting with hydrothermal systems maintained by the circulation of ground water.

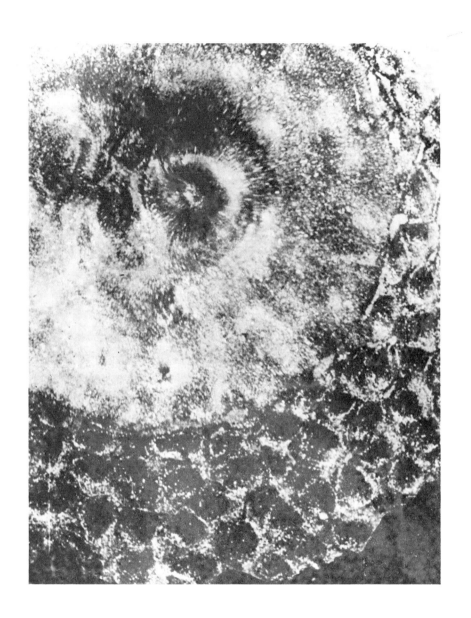

PLATE. Laboratory model of a hypothetical lithosphere: plan view, looking downward into the thin surface layer of a deep container of viscous fluid. A localized intermittent heat source exists on the base of the container. At the moment of the photograph, the source has been on for sufficient time to develop a strong upwelling of hot fluid which spreads out laterally over the surface. Outside the region of fresh hot fluid is a shallow zone of cellular convection in the cooling surface layer of fluid produced by earlier upwelling pulses. This photograph shows just one of the many possible situations which arise in laboratory model studies of the generation of the lithosphere.

5. Magma Traps

For the matter and energy flowing from the interior of the earth to be most readily available for use by high-level geological systems or human exploitation, it needs to be retained temporarily in a high-level reservoir. Perhaps the simplest systems of this kind are deep sedimentary basins which are heated by normal, near surface, conductive heat flow. A good example is the Pannonian basin of eastern Europe. The behaviour of these rather passive systems is well understood. Such is not the case for active systems related to contemporary volcanism. Strong geothermal activity, as found in modern hydrothermal systems, is undoubtedly related to ambient volcanism. Yet it is curious that often only minor parasitic hydrothermal systems are directly associated with active volcanic systems. This leads us to consider the very simple idea that in such systems the bulk of the energy has been dissipated at the surface and to raise the question of the circumstances under which a magmatic system can retain part or all of its energy at depth. We are lead to the notion of geothermal or magma traps.

A variety of trapping mechanisms have already been identified. Trapping can occur at quite deep levels simply because the rising magma volume freezes. Certain granitic systems, because of the chemical buffering of water, are especially prone to this (Harris *et al.*, 1970). Trapping can occur at high levels in many ways, a notable example being in recent subglacial volcanism of Iceland because of the enhance porosity and permeability of the deposits (see for example, Fridleifsson, 1976).

In this study, however, we consider the simplest trapping mechanism of all — namely the static effects of the ambient stratified crust and mantle on the magma column. The key idea is not new, and it has been used by several authors (for example, Holmes, 1965; Ramberg, 1963, especially pp. 53ff; Modriniak and Studt, 1959). Its testing against data from the central West Greenland Upper Cretaceous—Lower Tertiary sedimentary and volcanic area, described below, is the most detailed study at the time of writing. The study is incomplete in so far as purely static constraints are imposed on the model. The next step would be a study of reservoir dynamics.

The global distribution of known active volcanoes shows a number of striking features.

(1) The outstanding feature is how few there are within the interior of continents.

(2) Volcanism within the interior of continents, as in East Africa, is closely associated with strong crustal rifting.

(3) There is a pronounced grouping in great arcs marginal to continental areas.

(4) In oceanic areas distant from continental margins, volcanoes are apparently fewer, with a distinct grouping along the vicinity of some mid-oceanic ridges. Studies of the topography of the sea floor, however, suggest that sub-oceanic volcanism is actually the most common.

Within the Pacific basin there are of order 10^5 abyssal hills less than 1 km high, and 10^4 seamounts greater than 1 km high but confined below sea level (Menard, 1964). This identification is based on the form of these mounds which resemble subaerial shield volcanoes. These structures are found throughout the Pacific with pronounced groupings; they are, for example, abundant between the Murray and Molakai fracture zones, but much less abundant in the blocks either side.

It is possible crudely to quantify the discharge rates. The various estimates are shown in Table 5.1. They are derived from the following data: 5×10^7 km^3 supraoceanic volcanics of age less than 100 Myr and 10^7 km^3 of continental interior volcanics (Menard, 1964); erosional time-scale, 600 Myr (Elder, 1976); nominal areas of continental interiors, margins and oceanic basins as 10^8, 10^8, 3×10^8 km^2. The appropriate erosional time-scale is very uncertain especially since inactive cones are largely eroded in 10 Myr. In spite of this and other uncertainties the volumetric flux of mantle does appear to be much smaller within continental interiors than elsewhere.

There is a further outstanding aspect of the global distribution of volcanism. Oceanic volcanism is generally basaltic while marginal or continental interior volcanism is generally andesitic and in a sufficiently developed system, rhyolitic.

These differences of rate and type are broadly correlated with setting. This suggests that the rock ambient has a powerful effect on the development of a magmatic system. This idea is now explored in some detail.

We are particularly interested in those circumstances in which the

Table 5.1. *Average continuous discharge rates of magma.*

	Total ($km^3 yr^{-1}$)	Rate/area ($10^{-9} km yr^{-1}$)
Oceanic	0.5	2
Margins	0.1	1
Continental interior	0.02	0.2

Note: Compare estimate of recharge rate of new oceanic crust at $12 km^3 yr^{-1}$ in Table 2.2. The most voluminous historical single discharge was of about $12 km^3$ of basalt from the Laki fissure, 1783 (Thorarinson, 1970).

matter and energy of a lithothermal system is trapped below the surface — rather than being discharged directly at the surface — and thereby able to provide the heat source for hydrothermal systems.

The role of the buoyancy barrier of the stratified crust in controlling the ascent of magmas is represented by a model which operates close to lithostatic equilibrium and in which fractional crystallization of the magma or partial melting of the ambient rock can occur. The density structure of the crust has a power effect in trapping magma and thereby in controlling the occurrence of high-level geothermal systems.

In this chapter only theoretical matters are considered. The testing and application of the ideas is treated in subsequent chapters.

5.1. LITHOSTATIC CONTROL OF THE CRUST

Consider those aspects of lithothermal systems which are mechanically controlled by variations of the ambient crust. The emphasis is not so much on the magma but on the gross features of its container. For the want of a word we refer to lithostatics.

If the solid earth were a homogeneous body, production of a sufficiently voluminous partially molten region at depth would always lead to a surface discharge since, the magma density being less than that of its solid parent, the magma column, even if it reached the surface, would have less weight than a similar body of solid rock. But all magmas do not reach the surface. This is simply because the crust is stratified, with material near the surface not only less dense than that at depth but also less dense than some magmas. For these magmas, the uppermost layers of the earth present a buoyancy barrier which can be penetrated or punctured only under special circumstances.

Given a magmatic source, a number of simple constraints are immediately apparent.

(i) An extrusive system, namely one with sufficient initial head, grows until it chocks itself off even if the source still exists.

(ii) An intrusive system, namely one with insufficient initial head to puncture the crust, rises until the column is in lithostatic equilibrium with the ambient rock. A sill may form at the head of the column.

(iii) A dyke may partially penetrate the crust and remain mobile for a time determined by the lateral heat loss.

A simple way of looking at systems of the type considered here is to regard them as depending on mass rearrangement processes. This occurs in two fairly distinct forms.

(i) *One-shot rearrangement*. Matter is simply moved to a different place. An example is extrusive basaltic volcanism for which it is presumed that a liquid derived from partially melted mantle is moved upward through the crust possibly forming a deposit at some intermediate level owing to partial fractional crystallization.

(ii) *Multi-shot rearrangement*. Matter may be reprocessed several times. As an extreme we can envisage an extended period of andesitic/rhyolitic volcanism together with local erosion and sedimentation. The net effect on the mass of the local crust may be quite small. It is important to notice in lithostatic models of this type that the end result will be independent of the details of the local density distribution provided the total mass is unchanged. The only way of identifying the detailed crustal structure is through the particular sequence of events.

One of the paradoxes of the known kinematics of the global surface is the occurrence of vigorous volcanism above the cold recharge regions of the mantle (so-called subduction zones). This apparent paradox arises because of a preoccupation with the temperature needed to produce partial melting. There is a further vital factor in producing surface volcanism, namely the necessity of the magmatic system, even if it exists, to penetrate the buoyancy barrier of the crust. Thus in a cold recharge region the local isotherms will be depressed so that the melting point will be met rather deeper. But this is just what we want. Greater source depth leads to greater excess pressure and the possibility of surface volcanism.

All the versions of the model presented here presume a given source region and that the production is not limited by the ability of the source to produce magma. Discharge is limited by the buoyancy barrier of the crust. The model assumes that geothermal traps occur at the level of interfaces within the magma column, whether produced by fractional crystallization or partial melting of the ambient rock. Finally it is necessary to remind ourselves that having a trap does not necessarily mean that anything will be trapped, this will occur only if the trappable matter comes that way.

Manometric models

The key idea used in the quantitative discussion is the notion of static equilibrium. Isostatic readjustments are geologically very rapid, with a time-scale of order 10^4 yr (Haskell, 1937). Hydrostatic models of crustal and upper mantle systems with time-scales greater than 10^4 yr provide therefore good approximations to the overall distribution of mass. All the crustal models treat the system under discussion as if it were a simple manometer constructed from two vertical limbs: for example, one limb may be the unchanged crust the other limb the thinned crust. The essence of the calculations is that since the manometer is close to static equilibrium, equating the two pressures at the base of each limb is sufficient to determine the situation.

The structure of a volcanic system will also be treated as a manometer operating close to hydrostatic equilibrium. Clearly the lava pile will continue to grow locally until it reaches lithostatic equilibrium. Only this terminal or equilibrium state is used here for the identification of the volcanic manometer.

Rearrangement of the crust and upper mantle occurs on a time-scale of order 100 Myr. Hence, volcanic systems can be regarded in the gross aspects of their behaviour as quasi-steady systems embedded in an ambient system which is slowly changing. Although the volcanic system remains close to equilibrium, this does not imply that it also necessarily changes slowly, as we shall see below.

Crustal layers

Consider a layered crust, as sketched in Fig. 5.1, made up of a stack of layers of thickness h_i and density ρ_i, where, counting the layers downwards 0 to $n + 1$: layer 0 represents the ocean; layer $n + 1$ represents the deepest layer of interest, part of the mantle.

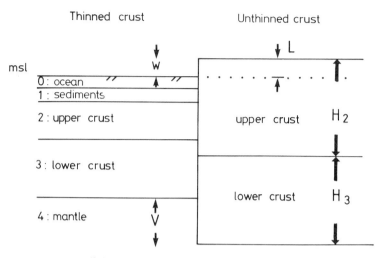

Fig. 5.1. Schematic representation of the crust.

If a layer is absent we simply set its corresponding thickness to zero. Here only the following are considered, with $n = 3$:

0. the ocean of density $1.03\,\mathrm{ton\,m^{-3}}$, taken as $1\,\mathrm{ton\,m^{-3}}$ in all calculations;
1. unconsolidated sediments of density $2.5\,\mathrm{ton\,m^{-3}}$;
2. upper crust of consolidated sediments and gneissic basement of density $2.7\,\mathrm{ton\,m^{-3}}$;
3. lower crust of density $3.0\,\mathrm{ton\,m^{-3}}$, gravitationally indistinguishable from basalt;
4. mantle of density $3.3\,\mathrm{ton\,m^{-3}}$.

Reference crust

It is particularly useful to have a standard or reference crust. This can be used as the basis of a variety of models. Although the choice is somewhat arbitrary, at least it is only arbitrary once. Here we make a choice by requiring that the ocean and continental parts of the reference crust are: (i) mutually in static equilibrium; and (ii) separately close to what is known of the structure of stable cratonic areas and stable abyssal oceanic areas. Thus, if we choose the layer thicknesses, H_j for $j = 0, \ldots, 3$, in km, as:

oceanic crust: 5, 2, 0, 5

cratonic crust: 0, 0, 25, 10

then the land elevation of the cratonic crust is 1.03 km. This figure is a little higher than would be suggested by the global average land elevation of 0.8 km (see, for example, Woollard, 1969 and Elder, 1976). A cratonic lower crust of 8 km gives a land elevation of 0.8 km. Changing the cratonic upper crust to 20 km, gives a land elevation of 0.12 km. This latter case is for example the sort of situation suitable for representation of young or small continental type crustal areas as in New Zealand.

Reference sea-level crust

It is also useful to have a standard continental crust with its upper surface at sea-level. Assuming this is obtained by eroding upper crust from the standard continental crust we find the layer thicknesses to be, in kilometres: 0, 0, 19.3, 10.

Relative to the reference sea-level crust the land elevation, L is given by $H_2 = 19.3 + 5.5L$, all quantities in kilometres.

Local crust

The crust in which the system of interest is embedded will be represented by a set of layers of thickness h_j, $j = 0, \ldots, n$. The depth below the local crust of an object of interest will be written as h_{n+1}. These quantities can be stated separately.

In many cases it is convenient and possible to relate the local crust to a reference crust. The local crust may be produced by a modification of a reference crust. For example, in the discussion below of tertiary volcanism in the "rift" zone of central West Greenland the local crust is considered as being produced by thinning the old continental crust and is represented by:

$$h_1 = \xi h_{11}, \qquad h_2 = (1 - \xi)H_2, \qquad h_3 = (1 - \xi)H_3 + \xi H_{31},$$

with $0 \leqslant \xi \leqslant 1$. Here a single parameter, ξ, allows representation of a range of crusts from $\xi = 0$, unmodified continental crust, to $\xi = 1$, oceanic crust. Note that if the depth of an object of interest is given in terms of H_{n+1}, the quantity h_{n+1} will also be a function of ξ.

Isostasy

Apart from a few isolated zones such as oceanic trenches the upper layers of the earth are close to lithostatic equilibrium. Where departures from equilibrium occur the local crust returns to equilibrium after the disturbing stresses are removed with a time-scale of order 10^4 yr. Systems of much greater time-scale embedded in the crust can be treated as if the ambient crust was permanently in lithostatic equilibrium. Hence we have:

$$\sum_{j=0}^{n} (H_j - h_j) = v + w, \qquad \sum_{j=0}^{n} \rho_j(H_j - h_j) = \rho_{n+1} v,$$

where w is the difference of levels of the tops of the two crusts and v is the difference of levels of their bases. In the simplest case of given H_j and h_j, these relations determine v and w. In the case of a continental reference crust the relation of the interfaces to sea level is given by reference to the land elevation, L.

It is more usual for one or more of the h_j to be the unknown. Consider the commonest situation where this is h_0, the water depth. For convenience write:

$$h_a = \sum_{j=0}^{n} H_j, \qquad p_a = \sum_{j=0}^{n} \rho_j H_j,$$

$$h_t = \sum_{j=0}^{n} h_j \equiv h_0 + h'_t, \qquad p_t = \sum_{j=0}^{n} \rho_j h_j \equiv \rho_0 h_0 + p'_t,$$

$$\tilde{v} = (p_a - p'_t - \rho_0 h_0)/f_{n+1} \qquad \tilde{u} = h_a - h'_t - w - \tilde{v}$$

$$\tilde{w} = h_a - h'_t - \tilde{v}.$$

(a) *Continental reference crust*

Two cases arise. Notice that $\tilde{v} > 0$.

(i) $h_0 \equiv 0$. Isostasy requires:

$$w + v + h'_t = h_a, \qquad \rho_{n+1} v + p'_t = p_a,$$

whence

$$v = \tilde{v}, \qquad w = \tilde{w}.$$

The local surface is $(L - w)$ above sea-level. This is the solution provided $\tilde{w} \leq L$. Otherwise $h_0 > 0$.

(ii) $h_0 > 0; w \equiv L$, given. Isostasy requires:

$$w + v + h_0 + h'_t = h_a, \qquad \rho_{n+1} v + \rho_0 h_0 + p'_t = p_a,$$

whence, by first eliminating v, we have

$$v = \tilde{v}, \qquad h_0 = \rho_{n+1} \tilde{u}/(\rho_{n+1} - \rho_0).$$

In either case the elevation of all the interfaces is then determined.

(b) *Oceanic reference* $(L \equiv 0)$

Two cases arise here also and can be treated in a similar fashion. Notice that now $\tilde{v} < 0$.

(i) $h_0 \equiv 0$. We find:

$$v = -\tilde{v}, \qquad w = -\tilde{w}.$$

The local land elevation is w. This is the solution provided $\tilde{w} \leqslant 0$.

(ii) $h_0 > 0$, $w \equiv 0$. We find:

$$v = -\tilde{v}, \qquad h_0 = \rho_{n+1}\tilde{u}/(\rho_{n+1} - \rho_0).$$

5.2. STATICS OF THE MAGMA COLUMN

As the magma rises from depth its character changes. In models of the type discussed here the information of interest is $\rho(z)$, the density of the magmatic fluid as a function of depth z below the top of the lava pile. Clearly $\rho(z)$ is a piecewise continuous function but its detailed form is virtually unknown. In the simplest case I represents $\rho(z)$ as a single step function:

$$z \leqslant z_*, \qquad \rho = \rho_b,$$
$$z > z_*, \qquad \rho = \rho_a,$$

with $\rho_b < \rho_a$ and z_* defined, for a given magmatic source, by the "characteristic pressure" p_* at $z = z_*$, as shown in Fig. 5.2 (a). It is important to appreciate that ρ_a, ρ_b and p_* are not independent parameters of the models described here. They only occur in combination to evaluate the pressures within the magma column. Thus the pressure, $p(z)$ as a function of depth, z in the magma column, where g is the acceleration of gravity is given by:

$$z < z_*, \qquad p = g\rho_b z,$$
$$z = z_*, \qquad p = p_* = g\rho_b z_*,$$
$$z > z_*, \qquad p = p_* + g\rho_a(z - z_*).$$

The choice of a characteristic pressure is both convenient and reasonable. For a given magmatic source, a statement of ρ_a and ρ_b is essentially a statement of composition.

It is important to realize, however, that p_* is a phenomenological

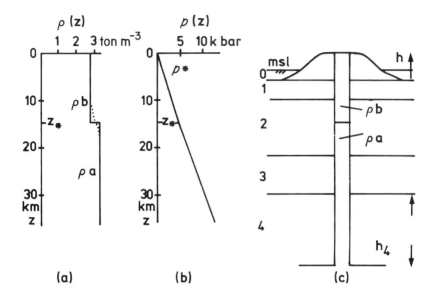

Fig. 5.2. Schematic illustration of the conditions in an open magma column. (a) Density $\rho(z)$; and (b) pressure $p(z)$ as a function of depth below the upper surface of the column. The dotted line shows a possible actual density profile, together with the step function approximation drawn as a solid line. (c) Schema for a hydrostatic model for determining the thickness of a lava pile.

variable solely related to a step function approximation to $\rho(z)$. Nevertheless it can be thought of, rather crudely, as representing the pressure level at which the greatest rate of change of composition with depth occurs. Hence, p_* is regarded solely as a property of the source magma, and not of its surroundings.

An independent determination of ρ_a, ρ_b and p_* is difficult. Various petrogenetic models of the derivation of magmas have been proposed (Wyllie, 1971). Most of these schemes have sequential differentiation. For example in basaltic systems there is proposed middle level differentiation at 5—10 kbar and high-level differentiation at typically 0—3 kbar. Possible values for the magma densities are $\rho_a = 3.1\,\mathrm{g\,cm^{-3}}$ for the hyperbasic magma, and $\rho_b = 2.8\,\mathrm{g\,cm^{-3}}$ for the high-level magma. Evidence from oceanic island volcanism suggests that these upper levels are well within the oceanic crust, so that, for example, in the case of Hawaii with a 9 km pile on a 5 km oceanic crust the pressure p_* is rather less than 5 kbar.

Consider the system of a magma column penetrating a layered crust and producing a volcanic pile of ultimate elevation h. The lithostatic pressure, P, at the source level is:

$$P = \sum_{i=0}^{n+1} \rho_i h_i,$$

where h_{n+1} is the mantle depth of the source in the thinned region. The pressure, p, in the magma column at the source level is:

$$p = \rho_b (h + z_*) + \rho_a \left(\sum_{i=0}^{n+1} h_i - z_* \right).$$

Here both h and z_* are measured relative to the palaeosurface, or the base of the water layer. When the system is in equilibrium $P = p$, which is a relation for h.

There is a further aspect of this model. Studies of oceanic islands built up by volcanism show large positive gravity anomalies, of order 100 milligal, centred on the islands (Malahoff, 1969). The anomalies can be interpreted as follows. The deep magma differentiates by partial crystallization in such a manner that the crystals form and are deposited in the volume of rock perfused by magma but they do not contribute to the pressure in the magma column. One needs to envisage the volume of rock perfused by the magma as a permeable medium, a densely interconnected dendritic structure possibly of quite small gross porosity. On the floors and walls of some of these channels the crystalline material is deposited.

Onset of volcanism

The very existence of extrusive volcanics allows us to make some useful deductions about the condition of the crust immediately prior to the onset of the extrusive volcanism. Consider a crust below which, within the mantle, a zone of partial melting is created and maintained. Owing to its buoyancy some of the magmatic fluid will migrate vertically upwards. When the system has again come into equilibrium, the height of the column perfused with magmatic fluid will be determined by two dominant conditions:

(i) Within the column the fluid pressure must at every level exceed or equal the lithostatic pressure of its surroundings.

(ii) At the top of the column the fluid pressure must exceed the lithostatic pressure by an amount equal to or greater than the finite strength, σ, of the material above, otherwise its advance will cease.

In addition, the crust and mantle are stratified in a manner such

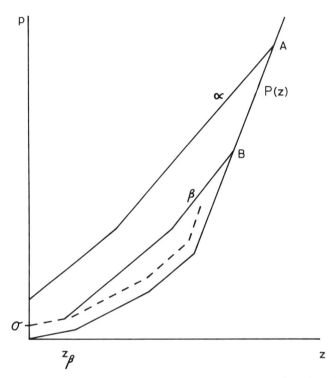

Fig. 5.3. Diagrammatic representation of the pressure as a function of depth, z, in a stratified crust: lithostatic pressure, $P(z)$; fluid pressure for two different magma columns α, β, for sources at A, B. Finite strength, σ.

that if $P(z)$ is the lithostatic pressure as a function of depth, the gradient dP/dz generally increases with depth. Thus, for example, as shown in Fig. 5.3 from a source represented by point A, the pressure distribution in the magma column is given by line α, so that a discharge is possible. But for a source represented by point B, and the pressure distribution in the magma column represented by line β, a discharge is not possible and the column is terminated at depth z_β.

The magnitude of the finite strength, σ, is not well known. A nominal value for crustal rocks of 1 kbar is frequently used and there is an estimate for basaltic lavas of about 0.1 kbar (Elder, 1976). The role of σ is largely in controlling conditions at and before the onset of eruptions. For values of order 0.1 kbar the effects are minor; for values of order 1 kbar most of the models do not even allow eruptions. For the want of anything better I used $\sigma = 0.3$ kbar in the numerical models.

It is worth noting that where wet sediments are encountered by an intruding magma there may be a type of phreatic eruption generated by the high gas pressure of the vaporized water. This effect is equivalent to having a lower finite strength — indeed, often a negative finite strength.

Consider, as sketched in Fig. 5.4, that the stratified crust and mantle is given as a sequence of layers of thickness h_i and density ρ_i.

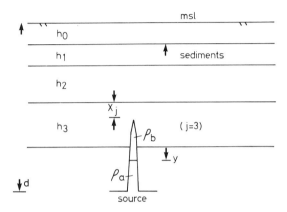

Fig. 5.4. Schematic arrangement for consideration of the conditions in the magma column which allow it to rise. A layered crust with a magma column reaching from a source in the mantle into layer j. The magma column has an upper portion of density ρ_b and a lower portion of density ρ_a.

To be determined are both the disposition of the magma column in vertical extent and the level z_* between the deep denser magma of density ρ_a and the shallow lighter magma of density ρ_b, if any.

(i) The source depth below sea level:

$$d = \sum_{i=0}^{n+1} h_i.$$

The lithostatic pressure at the level of the source is:

$$P = \sum_{i=0}^{n+1} \rho_i h_i,$$

which includes the contributions from the sea layer if present. Note that in both relations we use h_4 for the depth of the source below the base of the thinned crust. The quantity d, the source depth, is taken as fixed throughout the development.

(ii) the depth y below the top of the sediments at which the characteristic pressure P_* is reached is:

$$y = d - h_0 - (P - p_*)/\rho_a.$$

(iii) When $y < 0$, the dense hyperbasic magma is above the top of the sediments.

(iv) When $y > 0$, the dense magma level is below the palaeosurface. If there is a zone of light magma the excess pressure p' at the top of the sediments is given by:

$$p' = p_* - \rho_b y.$$

(v) When first $p' > \sigma_1$, that is the excess presssure in the magma column at the top of the sediments exceeds the finite strength of the sediments, eruption of the light magma is possible. This eruption could be either subaerial or subaqueous. The consequent height of the erupted lavas is p'/ρ_b. Note especially that we do not subsequently require the magma to satisfy the finite strength condition in its own pile since the ultimate development of the pile is believed to be by central eruptions and associated flank eruptions.

(vi) If $p' < \sigma_1$, so that the excess pressure in the magma column at the top of sediments, assuming it reached that level, is less than the finite strength of the sediments, then clearly the magma column is entirely within the crust, and in particular no eruption is possible.

(vii) The level of the top of the magma column is then obtained by finding the level at which the lithostatic pressure equals magma pressure plus finite strength of the rock at the top of the column. It is generally necessary to evaluate the two possibilities: (a) both magmas in the column; and then if a level cannot be found, (b) only the denser magma in the column. Thus the pressure excess, p_c over the finite strength at depth x_j in layer j in the magma column is:

$$p_c = p' + \rho_b x_j.$$

The lithostatic load is:

$$P_1 = \sum_{i=0}^{j-1} \rho_i h_i + \rho_j \left(x_j - \sum_{i=0}^{j-1} h_i \right).$$

We search successively with $j = 1, 2, 3, \ldots$, till we find $p_c = P_1$ and a corresponding value of x_j. If this fails, the column contains only the denser magma and we replace p_c above and repeat the search with:

$$p_c = p_* - \rho_a y + \rho_a x_j.$$

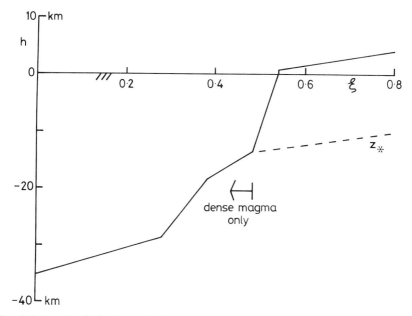

Fig. 5.5 Level of the top of the magma column, h, as a function of crustal thinning parameter ξ. Negative values of h refer to depth below the original surface. Values in kilometres. The level z_* is also shown.

The behaviour of the magma column is illustrated in Fig. 5.5 using the data of Table 2 with $p_* = 5$ kbar and $\sigma = 0.3$ kbar.

(1) For a given source, there will be a thickness of crust for which no discharge is possible. The magma column will rise above the source as the crust is thinned, that is as ξ increases from zero, but there is a range of ξ for which the magma column remains below the land surface.

(2) The striking aspect of this model is the sharpness with which, at a critical crustal thinning, eruptive volcanism commences. The sharp start is controlled by the non-uniform density stratification of the crust and its finite strength.

Thus we have a very diagnostic observation. There is a sharply defined crustal thickness. In the case of a crust gradually being thinned there would be negligible surface manifestation until the critical condition was reached, whereupon a convulsive start would be made of an eruptive phase which would continue till the lava pile or plateau was built or the source turned off.

In most of the models presented here the entire system is assumed

Table 5.2. *Model data.*

ρ_0	$1.00\,\mathrm{g\,cm^{-3}}$	Ocean water density, nominal value.
ρ_1	$2.5\,\mathrm{g\,cm^{-3}}$	Unconsolidated sediment density.
ρ_2	$2.7\,\mathrm{g\,cm^{-3}}$	Upper crust, or consolidated sediment density.
ρ_3	$3.0\,\mathrm{g\,cm^{-3}}$	Lower crust density.
ρ_4	$3.3\,\mathrm{g\,cm^{-3}}$	Mantle density.
L	$0.6\,\mathrm{km}$, $(0.4\,\mathrm{km})$	Land elevation of ice free continental interior. Initial estimate and revised estimate.
H_2	$25\,\mathrm{km}$	Upper crust thickness, unthinned.
H_3	$10\,\mathrm{km}$	Lower crust thickness, unthinned.
H_{31}	$5\,\mathrm{km}$	Oceanic crust thickness (completely thinned lower crust).
ρ_b	$2.8\,\mathrm{g\,cm^{-3}}$	High-level magma density.
ρ_a	$3.1\,\mathrm{g\,cm^{-3}}$	Deep-level magma density.
σ	$0.3\,\mathrm{kbar}$	Fracturing strength of rock.
P_*	$5\,\mathrm{kbar}$, $(3\,\mathrm{kbar})$	Characteristic pressure, initial estimate and revised estimate.
$\bar{\xi}$	0.55	Crustal thinning parameter, model design value.

Notes:
(a) $H_0 = 0$, dry continental interior; $H_1 = 0$, no unconsolidated sediments on continental interior remote from the embayment.
(b) The actual density of ocean water is $1.03\,\mathrm{g\,cm^{-3}}$ but the nominal value of $1.00\,\mathrm{g\,cm^{-3}}$ has been used in all calculations here.

to be in static equilibrium. In other words the new crust, including any surface deposits and internal mass rearrangement, is in isostatic equilibrium with the reference crust. It is important to appreciate that this requires the areal extent of the altered parts of the new crust to be sufficiently large to affect the gross pressure distribution in the system. This will not always be the case. A notable exception will be where surface volcanism is confined to localized volcanic mounds rather than volcanic plateaux.

6. Sub-Continental Lithostatics

The role of the buoyancy barrier of the stratified crust in controlling the ascent of magmas has been represented by a model which presumes that the system operates close to lithostatic equilibrium. The model is now tested against data from the central West Greenland Upper Cretaceous--Lower Tertiary sedimentary and volcanic area. There the idea works well and encourages its application to other fields. A sketch of an application to trapping within a region of acidic volcanism is given in the following chapter.

6.1. GEOLOGICAL SETTING

Direct evidence for the hypothesized control of the rise of the magma column by the buoyancy barrier of the crust is shown in a study of the development of the crustal structure in central West Greenland during Cretaceous—Lower Tertiary time.

As seen today, the volcanic region of central West Greenland, apart from the relatively small oceanic basin of Baffin Bay, lies on and within a large block of ancient crust of continental scale: see Fig. 6.1. The basement of the embayment is composed of granodioritic gneisses of generally amphibolite facies of the Rinkian mobile belt which was active 1680—1870 Myr ago and lies just to the north of the slightly older Nagssugtoqidian mobile belt (Escher and Pulvertaft, 1976). As this block migrated northward since before 200 Myr ago (Roy, 1973) there has been an intra-continental zone of extensive rifting (Henderson, 1974) and the development of small temporary ocean-covered basins but in which the crust remained largely continental except for a progressive thinning which reached a localized peak of activity in the central West Greenland area about 60 Myr ago. Subsequent tectonic activity has been minor so that much of the crustal structure developed at the time has been preserved both on the land and off-shore.

The broad features of the development of the region of interest

Table 6.1. *Time Chart A. Development of the America-Europe block, t, Myr ago; ϕ, north latitude of North Disko; in brackets the total northward displacement relative to Europe of North America. Scenario from Roy (1973), unless stated otherwise.*

t (Myr)	ϕ	
0	70	
	—	
	—	
	—	
	—	Labrador sea floor spreading ceases (about 40–47 Myr ago.)[†]
	—	
	67.5	North Atlantic volcanism and sea floor spreading strongly
	—	established.
	—	Labrador sea floor spreading starts (about 80 Myr ago.)[‡]
	—	Crustal thinning in central West Greenland.
100	61	
	—	
	—	
	—	Distinct Greenland element, moving faster than.N. America.
	52	Dyke swarm in south West Greenland. North America
	—	slowing. (16)
	—	Atlantic opening.
	—	Intra-continental basins developing.
	—	
200	42	Sinistral strike slipping continues, especially along East
	—	Greenland solely latitudinal. (20)
	—	Pacific coast orogeny, plutonic phase.
	—	
	—	
	—	Block rupturing north–south, sinistral strike slipping
	—	commences. (3?)
	—	
	—	
300	15	Block intact, with northward polar wander.

[†]Keen et al. (1974).
[‡]Le Pichon et al. (1971).

are displayed in "time-chart" A of Table 6.1 This temporal scenario is taken in the first instance from Roy (1973), in particular the palaeolatitudes are derived from his reconstructions. Although the evidence is not very detailed, it is consistent with the reconstruction of Le Pichon *et al.* (1971) which proposed a two-phase opening of the Labrador Sea and as a consequence the creation of Baffin Bay. The oceanic magnetic anomalies in the Labrador Sea suggest that the bulk of the opening of the Labrador Sea occurred between 80 and 60 Myr ago, followed by smaller displacements which terminated about 50 Myr ago. The net effect was a movement of Greenland, relative to North America of about 5°.

6.2. THE CRUSTAL TROUGH

Sediments have been deposited extensively over the intra-continental zone during at least the past 120 Myr (oldest sediments are Barremian age, 118 Myr ago) to thicknesses of several kilometres. Off-shore, both in the bathyl region of Baffin Bay and on the intra-continental slopes there are thick deposits over broad areas and especially within minor marginal basins (see for example Keen et al., 1974). The off-shore sediments are of uncertain age and have been delineated only rather crudely. On land, however, in the central West Greenland area there is an extensive and well developed succession of terrestrial and marine sediments.

The form of the crustal trough, otherwise referred to as the embayment, is shown in Fig. 6.1.(b). The stratigraphy on land is known in great detail (Henderson, Rosenkrantz and Schiener, 1976). A transverse schematic section as seen today is shown in Fig. 6.1(c). This section exposes all the major formations of the area. An axial schematic, as reconstructed for about 60 Myr ago and much simplified, is shown in Fig. 6.1(d). This shows most clearly the sequence of subsidence which produced the basin. In more recent time much of this material has been uplifted by up to at least 2 km. A thick sequence of terrestrial sediments deposited in a subsiding basin was followed by marine transgression and subsequent deposition of shallow marine sandstones and shales. The initial volcanism was at least in part subaqueous, pillow breccias of thickness 0.5 km being typical. The rate of development of the lava pile exceeded that of the subsidence since the bulk of the laval pile is subaerial. A summary of the events is displayed in "time-chart" B of Table 6.2.

The sequence of events, revealed by stratigraphic and structural studies on land (Henderson, 1973), is: onset of rifting; continuation of rifting with progressive overall deepening of the sedimentary trough; the onset of surface volcanism; and finally an overall uplift. This sequence of events and related details can be described in terms

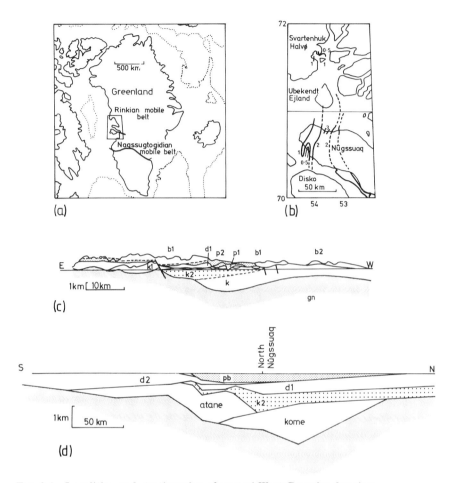

Fig. 6.1. Localities and stratigraphy of central West Greenland region.

(*a*) Greenland and its surroundings. The dotted contour shows ocean floor 2000 m below sea level. The box encloses the central West Greenland region.

(*b*) The locality of the Cretaceous–Tertiary sediments and Tertiary volcanism. The sediment trough, shown as contours of depth, in kilometres, below mean sea level of the base of the 3.2 km s^{-1} sedimentary layer, determined by refraction and reflection seismic measurements.

(*c*) Schematic representation of the stratigraphy in an 80 km section along northern Nûgssuaq, shown as profiles projected on the east-west vertical surface $70°45'\text{N}$. Viewed looking south, vertical exaggeration 2:1.

Formations:
  b2:   upper basalts, feldspar-phyric;
  b1:   lower basalts, olivine-rich;
  p2:   pillow breccia, olivine basalt;
  p1:   highly altered pillow breccia, olivine basalt;

of one major process, progressive thinning of the continental crust within the intra-continental zone.

6.3. THE LAVA PLATEAU

The most prominent feature of the area and that which is of interest here is the lava plateau. This structure and its rocks as seen on land have been studied by many workers (Clarke and Pedersen, 1976). The areal extent on land is well exposed and recent magnetic surveys, together with dredging, have been made to delineate the outline off-shore. Two such compilations are included in Fig. 6.2. The area shown in outline is about 5×10^4 km.

These flood basalts have been named as the West Greenland Basalt Group (Hald and Pedersen, 1975) with formations:

Lower basalt: Vaigat Formation, tholeiitic olivine-rich and picrite basalts;

Upper basalt: Maligât Formation, plagioclase—porphyritic tholeiitic basalts together with a thin Hareøen Formation of transitional olivine—porphyritic basalts.

Figure 6.1. (*continued*)

d2: Upper Danian, marine sandstone (small isolated 100 m thick patches in this section — not shown);

d1: Lower Danian, marine shale and sandstone;

k2: Upper Cretaceous (Coniacian to Maastrichtian), shale and sandstone of mixed origin;

k1. Middle Cretaceous (Barremian to Turonian), non-marine shale and sandstone;

k: Middle to Lower Cretaceous (presumed) and older (?);

gn: gneiss basement, identified seismically, or as seen in coastal outcrop.

Two topgraphic profiles also shown: (i) on $70°45'$N; (ii) along the high ridge behind the coast. (Drawn from GGU geological maps, 1:100 000, 70 V.1 and 70 V.2)

(*d*) Gross sedimentary stratigraphy, shown as a schematic reconstructed axial section, at the time of onset of subaerial basalts. Drawn by projecting known sections on the line NE Disko to East Svartenhuk Halvø with the known or supposed subaerial basalt level taken as a level surface — the top surface of the diagram. Vertical exaggeration 20:1. The formations are: the gneissic basement; Kome, Lower Cretaceous (Barremian-Aptian, 118—106 Myr ago); Atane, Cretaceous (Upper Turonian-Coniacian, 94—82 Myr ago); k2, marine and terrestrial formations, Upper Cretaceous; d1, Lower Danian; d2, Upper Danian; pb, basaltic pillow breccia. (Adapted from a diagram prepared by E. J. Schiener for the book Geology of Greenland GGU 1976, personal communication. See Henderson, Rosenkrantz and Schiener (1976).)

Table **6.2.** *Time Chart B. Development of the embayment. Time indicated in Myr ago.*

t (Myr)		
50	—	Uplift of land.
	—	
	—	
	—	
60	—	
	—	Volcanism. Marine connection to Europe.
	—	
	—	
70	—	Extensive block faulting, abundant conglomerates.
	—	
	—	
	—	Terrestrial and marine sedimentation.
80	—	
	—	
	—	Marine deposition. Marine connection northward into
	—	Sverdrup basin.
	—	
90	—	
	—	Intermittent marine connections, black shales.
	—	
	—	
100	—	Rifts widening.
	—	
	—	
	—	
110	—	
	—	
	—	Non-marine sediments.
	—	
120	—	
	—	
	—	
	—	
130	—	Land locked rifting.

Duration estimates from dating measurements

Volcanism seems to have started in the Danian. For example, Jürgensen and Mikkelsen (1974) have identified coccoliths in Danian andestitic tuffs at Marrait kitdlit (south Nûgssuaq) and at Kangilia (north Nûgssuaq), an age range of 59.5—63 Myr. These tuffs can also be correlated with early native iron-bearing lavas suggesting that they

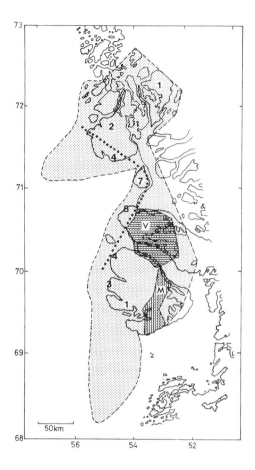

Fig. 6.2. Gross outline of the central West Greenland volcanic pile now. Eastward land exposure indicated by fine lines, strippled area drawn "headland to
headland". Off-shore outline given by Denham (1974) and in the extreme south
by Park *et al.* (1971) Some of the material in the north off-shore may be ice-
rafted deposits, identified north and west of Svartenhuk Halvø by Baker and
Friedman (1973). The line of crustal thinning is shown dotted.
   Also shown, basaltic breccia occurrence, in the southern part of the volcanic
area, from 69° to 71°N. The outlines have been drawn "headland to headland"
assuming continuity across the Vaigat, using the 1:500 000 Sondre Strømfjord—
Nûgssuaq, Geological Map, GGU 1973. V, Vaigat formation breccia (sub-aqueous
lower basalt); M, Maligât formation breccia (sub-aqueous upper basalt).
   The numerical labels indicate lava pile thickness, in km, based on a compilation by Elder (1978) from data given by: Münther (1973), Drever (1958),
Hald (1973), Hald and Pedersen (1975), various others and the author.

were deposited contemporaneously with the lower picrite lavas and breccias of the Vaigat Formation (see Clarke and Pederesen, 1976, p. 373).

An estimate of the duration of the episode of lava production can be made from the observation that of the lavas so far measured for direction of magnetization all but a few are the same — reversed. The notable exception is the well established zone of normal magnetization low in the Vaigat Formation (Athavale and Sharma, 1975). In view of the rather scattered and unsystematic sampling of the lava pile, particularly poor in its upper levels, this observation is by no means certain, but let us accept it for the moment. Then, provided some other evidence gives us an approximate age of the lavas, we can look for a suitable known interval of reversal. Whereas the earth's magnetic field has had normal and reversed intervals averaging about 0.3 Myr in the recent past, before 45 Myr ago the intervals were generally longer (Heirtzler *et al.*, 1968). For example, during the period 55—65 Myr ago the normal interval averaged 0.6 Myr, the reversed interval 1.2 Myr. In Danian time there was an exceptionally long period of reversal from 60.5 to 62.7 Myr ago. This event fits with the "Danian" reversed volcanics. Thus, even allowing for the undoubted errors in the dating of the magnetic time-scale, this suggests that these lavas were extruded in this period of reversal — a total time of about 2.2 Myr. The interval of 2.2 Myr is a lower bound for this interval of lava production.

The bulk of the lavas could have been produced in a few million years, perhaps about 3 Myr. The time of development of the lava pile was short compared to that of the embayment as a whole so that the production of the laval pile can be treated as a system isolated in time although spatially embedded in the embayment. A summary of the lava pile development is displayed in "time-chart" C of Table 6.3.

The pillow breccia thickness

The deposits of pillow breccias, indicated in Fig. 6.2, are within the area of the sediment trough. This sedimentary and breccia trough can be regarded as an elongated depressed block, probably composed of numerous smaller blocks, which was subsiding during the sedimentation. It is important to realize however, that the breccias are not a regional chronostratigraphical unit. They are locally early lithostratigraphic elements. In a wider context they are aqueous facies:

Table 6.3. *Time Chart C. Development of the volcanics. Time indicated in Myr ago. Very schematic, see text.*

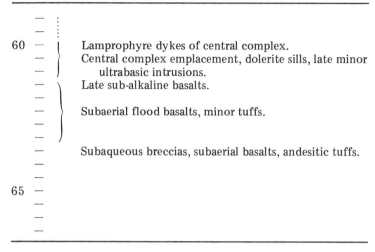

60 — Lamprophyre dykes of central complex.
— Central complex emplacement, dolerite sills, late minor ultrabasic intrusions.
— Late sub-alkaline basalts.

— Subaerial flood basalts, minor tuffs.

— Subaqueous breccias, subaerial basalts, andesitic tuffs.

65 —

of the Vaigat Formation in Nûgssuaq, and of the Maligât Formation in south Disko. Thus the presumably deeper parts of the trough, in Nûgssuaq were filled with sub-aqueous basalts and then covered with sub-aerial basalts before the onset of sub-aqueous volcanism in south Disko. In the model presented below this is interpreted as due to a continuing subsidence within the trough controlled by the amount of local crustal thinning. The thickness of these breccias reaches 0.7 km with typical thicknesses of 0.5 km. This thickness is an important quantity in calibrating the models below.

The lava pile thickness

The thickness of the lava plateau provides one of the dominant quantities in the identification of the manometer system. The lava pile is extensively exposed on land so that reasonably reliable estimates of the thickness can be made; these are indicated on Fig. 6.2. The pile is roughly lens-shaped, more or less flat-bottomed with a domed upper surface reaching a maximum thickness in excess of 7 km. The average thickness is about 5 km.

We have a pile of original average thickness about 5 km spread over

Table **6.4.** *Lava pile thickness in* km.

1.	Svartenhuk Halvø, north	2.0	Münther (1973)
2.	Svartenhuk Halvø, south	3.5	Pulvertaft and Clarke (1969)
			Münther (1973)
3.	Ubekendt Ejland	7	Drever (1958)
4.	Nûgssuaq	4.5	Hald (1973)
5.	Disko, north	4	Hald and Pedersen (1975)

Notes:
1. Lower basalt including breccia, west of Simiutapkua, 1.0 km. Upper basalt, see (2).
2. Lower basalt, vicinity of Tartûssaq, 2.5 km. Upper basalt, estimated thickness from GGU map 72V Pangnertoq, range 0.5—1.0 km, taken as 1.0 km throughout.
3. Lower basalt, 5 km. Upper lavas 1.5—1.8 km which include 0.75 km tuff, 0.3 km acid lavas, 0.5—0.8 km basalts.
4. Lower basalt, 1.5 km, Upper basalt, 3.0 km.
5. Lower basalts, 2 km, Upper basalts, 2 km.

an area of about 5×10^4 km. Thus the original pile had a volume of 2.5×10^5 km^3, somewhat bigger than Hawaii with 1.4×10^5 km^3 and comparable in volume with that of other lava plateaux. A summary of the data used is shown in Table 6.4.

Amount of subsidence

It is possible to make an estimate of the amount of lowering of the palaeosurface, at the base of the pillow breccias. Sub-aqueous breccias form part of the Maligât Formation, but were proceeded, in Nûgssuaq and north Disko by 1.5 km and 2 km, respectively of the Vaigat Formation. The Mâligat Formation breccias are typically up to 0.5 km thick. Clearly the situation will be complicated by any differential vertical displacements. If these are ignored, the above figures suggest a subsidence of the palaeosurface of at least 2—2.5 km, with 3 km quite possible. A typical figure of 2.5 km is used in the models below.

Post-volcanic uplift

The most casual inspection on land of the sedimentary basin shows that there was a substantial overall subsidence before and during the volcanism. And yet it is clear that subsequently the land had risen up. For example, in north Nûgssuaq we see pillow breccias at 1500 m above sea level. This, together with the measurement of breccia uplift, together with the estimate of the elevation of the original palaeosurface, $L = 0.6$ km, suggests a net vertical displacement of about 2 km.

The elevation of the land in and adjacent to the embayment after the volcanism is clearly related to regional processes. An uplift of 2 km could arise from the combined effect of a continued crustal thinning of about 0.5 in the intra-continental zone leading to the formation of Baffin Bay together with a mean erosion of about 1.5 km

6.4 THE CRUST

Gravity data, converted to a relative Bouguer anomaly map, is shown in Fig. 6.3. In the eastern part of the area shown there is an extensive region of roughly uniform gravity, this region being bordered by the -60 milligal contour, further east of which gravity is somewhat lower with a typical level of -70 milligal. In the western part of the area shown, there is another extensive region of gentle variation, this being bordered by the $+60$ milligal contour, with a typical level immediately outside the volcanic region of $+70$ milligal. The total range of the change in gravity across the structure reaches extreme values of about 170 milligal, but is more typically about 120–140 milligal.

Between these two regions of relatively uniform gravity there is a pronounced westerly increase across an area which occurs in a strip 40 km wide at its narrowest, trending very roughly north–south through the area. This inidcates a dramatic crustal change in which the western crust has been substantially modified.

The gravitational effects of lateral variation of the uppermost clastics and volcanics are disregarded in the gravity interpretation. The gravity data is already corrected to sea-level as if upper crust extended to sea-level, so that in the gravity model $h_0 = h_1 = H_0 = H_1 = 0$. We have the situation sketched in Fig. 6.4(a). Define $\xi = \xi(s)$ as the proportion of crustal thinning, then $0 < \xi < 1$ where s is the distance, measured from a reference point (here the Tuperssuartâ base, north Nûgssuaq) along the trend of the gravity data, taken to be the line zero milligal shown in Fig. 6.3. Hence for a chosen value of the thicknesses of the pre-existing crust the gravity data can be used to obtain the proportion of crustal thinning, $\xi(s)$, as a function of position along the trend.

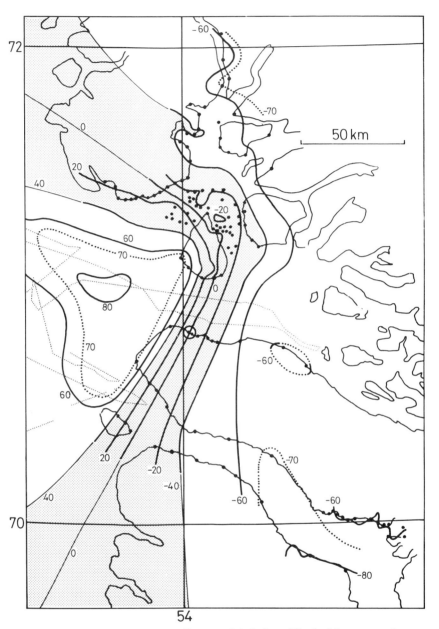

Fig. 6.3. Gravity anomaly map. Contour labels in milligals. Measurements converted to Bouguer values *relative* to the value at the Tuperssuartâ base, north Nûgssuaq, shown as a heavy circle, based on land and sea-ice stations. *Note:* The contours do *not* show absolute Bouguer values. To obtain the absolute Bouguer values *add* 20 milligal to the values shown.

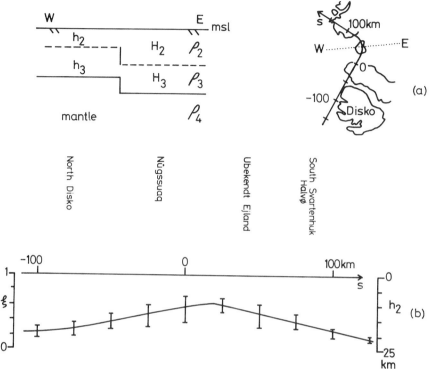

Fig. 6.4. Gravity interpretation. (*a*) Simple gravity model, vertical section drawn normal to the line of crustal thinning. Map showing line of crustal thinning, distance *s* and a possible "profile" line. (*b*) Proportion of crustal thinning, as a function of position, *s*, along the line of crustal thinning, measured relative to The Tuperssuartâ base, north Nûgssuaq. Distances in kilometres. Also shown on the right-hand side is the thickness of upper crust to the west, h_2. The vertical bars indicate the range of ξ for the fit to the profile at the plotted *s*, drawn as ± twice the root mean square difference between the model and the actual profiles.

The variation of the proportion of crustal thinning, $\xi(s)$, is shown in Fig. 6.4(*b*). There is strong thinning in the region from Disko to Svartenhuk Halvø and extreme thinning is found between north Nûgssuaq and south Ubekendt Ejland where the thinning exceeds 0.7 and the upper crust is a mere 8 km thick. The present data give the location of greatest thinning at $s = 20$ km, a few kilometres south of Ubekendt Ejland. Also shown are the ranges of values of ξ. The range crudely indicates the local variability of the crustal thinning, whereas the solid line gives a smoothed value. The amount of crustal thinning is rather variable on the local scale.

In later discussion it is useful to have a typical value of ξ for use in the initial design of various models. The value chosen is $\bar{\xi} = 0.55$. This is an average value over the central part of the region of thinning where the full stratigraphic sequence is measurable.

6.5. LITHOSTATIC MODEL

The crustal restructuring is considered to be determined by a single process operating on two scales:

(i)   an intra-continental zone of progressive crustal thinning, determined by processes on a continental scale which, in particular, is responsible for the post-volcanic uplift;

(ii) a local region of rapid short-lived crustal thinning, super-imposed on the slower but long-duration continental-scale thinning, and determined principally by local conditions beneath the embayment.

We ignore the details at the margins of the two parts and treat them as two distinct layered blocks. The requirement of static equilibrium then imposes constraints on the thicknesses and densities of the various layers.

We have considered the development of a basin as a consequence of thinning the crust below the basin and the development of a magma column and the lava pile within a given basin. In this combined model, as sketched in Fig. 6.5, both processess are at work. As

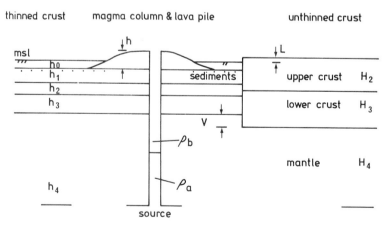

Fig. 6.5. Schema and nomenclature of the combined model of crustal thinning and evolution of the lava pile.

the crust thins and sinks down relative to the remaining undisturbed crust, the magma column experiences a change in ambient pressure distribution which modifies its behaviour. It should be noted, in this model, that the depth of water is controlled by competition between the crustal thinning, which leads to a lowering of the crust, and the sedimentation, which leads to an effective raising of the palaeosurface. Thus the water depth is an important factor in the identification of the model.

Since the initial development of the lava pile, immediately after onset of the extrusive volcanism, is rapid compared to all the other processes, in particular the sedimentation, once onset has been reached in the model, further sedimentation is cut off. In reality some sedimentation will still occur but the total extra thickness will be a very small proportion of the lava pile thickness. It is worth remarking that the very small amount of intercalated sediment in the subaqueous breccias and lavas is clear evidence of the rapidity of the initial extrusion phase. The effect of turning off the sedimentation has the effect in the model of allowing the magmatic system to rise slightly higher than it would otherwise do since there is less lighter material to act as a buoyancy barrier, and to give a slightly greater water depth owing to the somewhat greater crustal load.

The model has been adjusted to provide a fit to the following data.

(i) The rate of production of sediment, a parameter of the model, must be such that at onset of volcanism the local terrestrial sediment thickness is that found by field measurement in the range 0—3 km and typically 2.0 km.

(ii) The depth of water at onset must be such that the initial volcanics are locally subaqueous, and remain subaqueous until a breccia pile of measured thickness develops, in the range 0—0.7 km and typically 0.5 km.

(iii) The total thickness of the lava pile must reach the estimated values in the range 0—8 km, and typically 5 km.

(iv) The depth to the palaeosurface at the end of the volcanic phase must be in the estimated range 0—3 km and typically 2.5 km.

Although the model contains several free parameters, they are severely constrained to rather narrow limits. Many possibilities can be immediately discarded as having, for example, no discharge, zero water depth or no basalt. The method of finding acceptable combinations has been iterative, by varying successively one parameter at a time to find the best range for that parameter, and then successively repeating the scan of the parameters to obtain the best fit. The criterion for the goodness of fit was the value of the root mean square of the proportional differences· (mode value − field value)/ (field value) of the above four quantities.

Take as given the data of Table 5.2 and the ultimate upper crust thinned to the design value $\xi = \bar{\xi} = 0.55$. The model is then defined by choosing: sediment thickness parameter h_{11}; mantle depth, H_4; pressure scale, p_*; and ambient land elevation, L. We start with extreme values in the ranges: h_{11}, 0—10 km; H_4, 0—100 km; p_*, 0—10 kbar; L, 0—1 km; and search for some intermediate values for which there is an absolute minimum in the root mean square of the fit. A minimum root mean square of the fit of 3.1% is found with h_{11} = 7.9 km. H_4 = 41 km; p_* = 3.0 kbar; L = 0.43 km. The sensitivity of the fit can be demonstrated by varying these three parameters separately near their optimum values. The root mean square of the fit is better than or equal to 10% for values varied singly in the range:

$$h_{11} \; = \; 6.8\text{—}8.5 \,\text{km}; \qquad H_4 \; = \; 37\text{—}42 \,\text{km};$$

$$p_* \; = \; 2.4\text{—}3.4 \,\text{kbar}; \qquad L \; = \; 0.38\text{—}0.50 \,\text{km}.$$

The minima for h_{11} and L are sharp, but those for H_4 and p_* are very flat-bottomed. The very deep, flat, root mean square minima for variation of H_4 and p_* indicate that values outside the stated ranges produce quite unacceptable models.

The effect of choosing one of these parameters different from its optimum value is as follows.

(i) H_4: Low values give a thin pile and thick sediments and initial eruptions in deep water. High values give a thick pile but eruptions throughout on dry land.

(ii) p_*: The effects are less pronounced than those of H_4. Low values give a thinner pile and initial eruptions in deeper water. High values produce the opposite effects.

(iii) h_{11}: Low values which correspond to thin sediments, produce a thicker pile and a deeper final water depth. High values which correspond to thick sediments produce a thinner pile.

(iv) L: Low values produce larger water depths, thin sediments and a thicker pile. High values produce the opposite effects.

The model of best fit requires a source of the hyperbasic magma at a depth of 75 km bmsl, which differentiates at a depth below the discharge surface of 11 km.

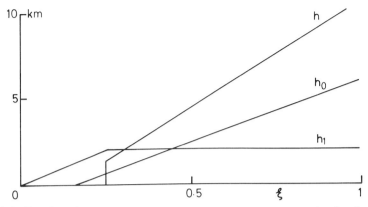

Fig. 6.6. Results of the best combined model. Water depth, or depth of paleo-surface, h_0; sediment thickness, h_1; lava pile thickness, h. Values in kilometres as a function of the proportion of crustal thinning ξ.

The behaviour of the model with the parameters of best fit is illustrated in Fig. 6.6. Let us consider what happens as the crust is progressively thinned.

(1) $\xi < 0.15$: During this interval terrestrial sediments accumulate on top of the subsiding old crust.

(2) $\xi \approx 0.15$: In spite of the accumulation of terrestrial sediments the crust has subsided sufficiently to allow the transgression of ocean water over the land surface of the developing embayment.

(3) $0.15 < \xi < 0.25$: Subaqueous marine sediments are now deposited. In reality the entire embayment was not inundated at a single instant, rather a progressive marine transgression occurred. The evidence suggests that this transgression was predominently from the north and west, but that the subsidence was jerky rather than uniform (see also Henderson, Rosenkrantz and Schiener, 1976). The model here gives only a gross average representation.

(4) $\xi \approx 0.25$: Throughout the preceding interval the magma column has been rising higher. It now erupts through the top of the sediments into shallow ocean water to produce the subaqueous pillow breccias. The use in the model of a finite strength of the pre-existing rock leads to the sharp occurrence of the initial eruptions.

(5) $\xi > 0.25$: The lava pile now builds steadily in response to the continuing crustal thinning. Early in the life of the pile its thickness exceeds the depth of oceanic water and a regression of oceanic water occurs. As with the earlier transgression, the regression in reality is not sudden. Indeed there is good evidence for a minor succession

of local regressions and transgressions, particularly noted on north Disko (Pedersen, 1975).

Once the regression has occurred it is no longer appropriate to refer to the water level, rather the quantity h_0 is now the depth of the palaeosurface on which the lavas were originally deposited.

The lava pile continues to grow until the local crustal thinning ceases.

In essence, these models treat the system as a kind of manometer imbedded in the crust and upper mantle. A number of simple but distinctive and diagnostic features allow us in effect to read the manometer at various instants during the development of the system. These readings are more than sufficient to calibrate the fossil manometer and thereby obtain an understanding of the essential simplicity of the overall behaviour of the system.

The correlation of crustal thickness and lava thickness

We have assumed that the source region was distributed beneath the volcanic zone at a uniform depth and deduced that there will be

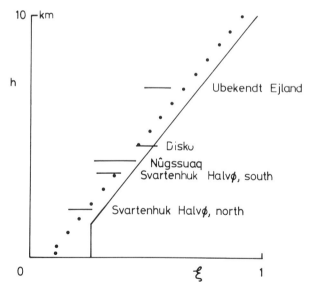

Fig. 6.7. Correlation of lava pile thickness, h, with proportion of crustal thinning, ξ. The short bars indicate the range of ξ corresponding to the rather broad regions over which the thicknesses have been measured. The solid line is obtained from the model.

a relation between crustal thickness and lava thickness. If this is correct, it implies that the source region was created and maintained solely by large-scale processes arising in a region much greater in area than the volcanic area itself. There is indeed a strong relation between the local crustal thickness deduced from the gravity data and estimates of local laval pile thickness as shown in Fig. 6.7.

If the field data used to calibrate the model were precise it would be possible to obtain an estimate of the finite strength σ. The dotted line shown is for $\sigma = 0.1$ kbar and the heavy line for $\sigma = 0.3$ kbar. In general there is a tendency to overestimate the local pile height as measured in the field by an uncertain amount, so that $\sigma = 0.3$ kbar is probably about right.

This correlation suggests that the development of the embayment, and especially its associated volcanism, arose from a single localized event at a uniform depth in the mantle.

As a final remark, it is worth noting that the different ages of the pillow breccias, mentioned above, suggests that the crustal thinning proceeded more or less uniformly from zero to its local final value. The crust of south Disko, potentially less thinable, needed to continue its thinning for longer before the onset of volcanism as compared to that of Nûgssuaq. Thus the younger breccias of the Maligât Formation occur in a region less thinned than those of the older breccias of the Vaigat Formation.

7. Orogenic Lithostatics

The lithostatic control of volcanism is readily described for a simple crustal structure, but in orogenic regions with extreme lateral variations in crustal structure together with the temporal variability near changing crustal boundaries, the situation is complex. We now concentrate attention on a single region, that of the south-west Pacific near New Zealand, not only because a crustal boundary passes from oceanic into continental crust, thereby providing an extreme crustal variation, but particularly because this is the region in which occurs the Taupo volcanic zone, the hydrothermal systems of which are a major feature of interest in this study.

7.1. GEOLOGICAL DEVELOPMENT

Let us now consider the case of acidic volcanism as found today, in, for example, the North Island of New Zealand. A scenario of the sequence of events has been given by Kear (1959) and more recently by Ballance (1976). Briefly the sequence of events has been as follows.

(i) In Northland: 20—15 Myr ago, subaqueous basalts; 18—6 Myr ago, andesites followed by dacites; 6—3 Myr ago, rhyolites; 3—0 Myr ago, basalts.

(ii) In Rotorua-Taupo area: 3—0.75 Myr ago. rhyolites and ignimbrites; 0.75—0 Myr ago, continued rhyolites plus andesites.

(iii) In Mt. Egmont area: 3—0 Myr ago, high potash andesite.

The details are quite complex especially as it would appear as if at least two systems have been in operation. Here I merely wish to use this type of situation as a background for a first step in considering the overall creation of geothermal and magma traps in a high-level continental or orogenic system. The kernel of the idea has already been sketched by Modriniak and Studt (1959) in providing an explanation for the sequence of volcanics in the Rotorua-Taupo area.

116

If the column is terminated within the crust below the solid surface, even though the matter is trapped, the heat may not be. This heat is potentially available to partially melt the ambient rocks. I believe that the evidence is overwhelming that high-level partial melting of this type occurs (see, for example, Steiner, 1958). For the want of a better hypothesis I will assume that basaltic magma is the ubiquitous product of the partial melting of the upper mantle and probably subsequent high-level fractionation, and further that it supplies the heat to produce high-level acidic magmas by partial melting of the crust. In the model sketched below it is also assumed that the melting of the crust is confined to the level at the top of the basaltic magma column. This assumption is one of many that are possible — melting of crust may well occur throughout the vertical extent of the basaltic magma column. Most crustal melting will occur where the bulk of the basaltic magma is trapped and the numerous pod-like gabbro bodies suggest that this is high in the basaltic system. Nevertheless, these questions cannot be resolved until a study of reservoir dynamics has been made and will be put aside here. Thus what follows is a theoretical sketch which merely suggests the sort of possibilities that arise from the control of such a system by a stratified crust.

Kinematics of the south-west Pacific region

A scenario of the tectonic evolution of the south-west Pacific region can be constructed from a study of bathymetry and marine magnetic anomalies. The summary presented here is from Molnar et al. (1975) with modifications based on studies of magmatic evolution by Ballance (1976). The gross features are sketched in Fig. 7.1. The presumed sequence of events in the vicinity of the North Island of New Zealand is strongly influenced by the Alpine fault, with its horizontal displacement of order 500 km and the rather sudden onset of volcanism from a late basaltic phase in Northland to the concurrent activity in the Taupo district. A series of five marginal basins with intervening ridges occurs to the north and northwest of New Zealand. The basins are newer from west to east, thus: Tasman Basin, 80—60 Myr; New Caledonia Basin, about 60 Myr; Norfolk Basin, age unknown; South Fiji Basin, 20 Myr or older; Lau-Havre Trough, about 10 Myr. The basins are of more than one type. The Tasman opened by symmetrical spreading about a mid-ocean rise and has crust of oceanic character and little volcanic detritus in its sediment fill. The New Caledonia and Norfolk Basins

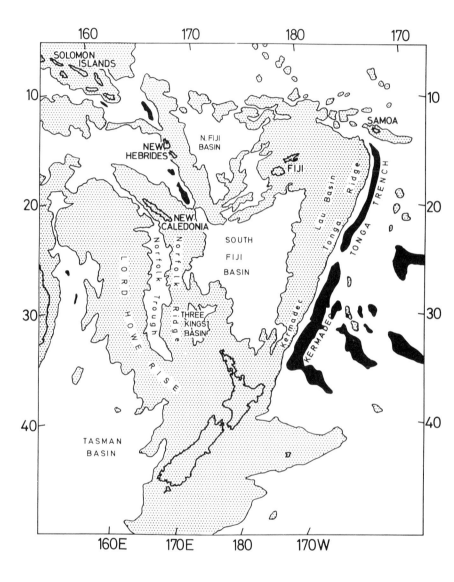

Fig. 7.1. Relationship of the New Zealand region to the current global systems of the south-west Pacific. Depth: light shading, 3 km +; black, 6 km +.

have smooth magnetic profiles with low amplitude. The South Fiji and Lau-Havre Basins have magnetic profiles with considerable relief and moderate to high amplitude, are adjacent to active or remanent arcs, and have abundant volcanic detritus in their sediment fill.

The intervening ridges are also of more than one type. The Lord Howe Rise is considered to be the Palaeozoic foreland of the Mesozoic New Zealand—New Caledonia Geosyncline; it foundered in the late Cretaceous. The Norfolk Ridge is thought of as the orogen of the New Zealand—New Caledonia Geosyncline, and it has an extensive cover of Pliocene basalts. The Three Kings Rise is unknown apart from its magnetic anomalies, which are up to 500 gammas, the maxima corresponding to topographic highs. The Lau—Colville Ridge is thought to be a remanent arc older than Miocene, while the Tonga—Kermadec Ridge is the active arc.

The following sequence of events occurred.

(a) Prior to about 35 Myr ago, the area is reconstructed by straightening the North Island. The western part of the North Island is placed 230 km south and the Norfolk Basin is closed.

(b) Beginning 35 Myr ago, the Norfolk Basin opened wide and the Cook and Vening Meinesz Fracture Zones developed as sinistral transcurrent faults, isolating the Loyalty Islands Ridge and the Three Kings Rise.

(c) At about 20 Myr ago, a big change occurred and the magmatic arc jumped southwards into Northland and a trench developed. The South Fiji Basin formed with generation of new crust in an inter-arc or behind-arc setting. This was probably accompanied by eastwards migration of the magmatic arc and by sinistral movement on the Hunter Fracture Zone between the New Hebrides and Fiji.

(d) From 20 to 3 Myr ago, minor changes occurred in the arc within New Zealand, and the Lau—Havre inter-arc basin to the north began to open after about 10 Myr ago.

(e) At about 3 Myr ago, displacement between eastern and western North Island was accompanied by anticlockwise bending of western North Island, reversal of movement on the Vening Meinesz Fracture Zone, partial closure of the southern Norfolk Basin and slight distortion of the Norfolk Ridge.

Development in New Zealand

(1) At about 38 Myr ago, an active tectonic zone developed through New Zealand.

(2) This zone traversed the western side of South Island as a linear zone of subsiding basins, which may have continued northwards to the west of North Island to join the Norfolk Basin. Activity in this zone in the South Island was extensional 38—10 Myr ago.

(3) At about 21 Myr ago, transcurrent movement began on the Alpine Fault in South Island, while arc volcanicity and associated tectonism spread south into the North Island. The magmatic arc may have spread southwards from Tonga to New Zealand suddenly, rather than in stages through the lower Cenozoic.

(4) At about 10 Myr ago, with another change of pole of rotation, the calculated motion of the Pacific and Indian regions became obliquely compressional at the Alpine Fault. The magmatic arc in the North Island appears to have been very stable at this time.

(5) Transcurrent motion between western and eastern North Island accelerated about 3 Myr ago, and anticlockwise lateral bending of western North Island began about the time of the Kaikoura orogenic climax in the western South Island and eastern North Island.

Stratigraphy of Cenozoic volcanism in New Zealand

Two distinct current volcanic phases can be recognized. They are distinct in time, space and petrology. The first phase, started about 20 Myr ago in Northland, is now predominantly basaltic; the second phase, started about 3 Myr ago in the Taupo—Rotorua region, is predominantly rhyolitic. The development of these two phases strongly suggests that two distinct portions of crust of somewhat different thicknesses are involved. The following summary is taken from Kear (1959) and Ballance (1976), whose scenario infers the location of structures from the presumed relationship of magmatic type to local crustal structure: namely, arc front low-K rhyolite/ andesite to arc rear high-K andesite/basalt.

The onset of volcanism was preceded by strong fault movements dated at Waitakian (late Oligocene or early Miocene). They took place throughout the North Island, the faults being normal in the western half, and largely transcurrent in the eastern half; the ensuring volcanism was confined to the western half. This event may mark the propagation through New Zealand of the Indian—Pacific boundary in its present form. The boundary to the north of New Zealand is recorded by early Tertiary arc volcanism in Tonga, Lau, Fiji, New Caledonia and the New Hebrides, but there is no record of arc volcanism in or south of New Zealand prior to Waitakian.

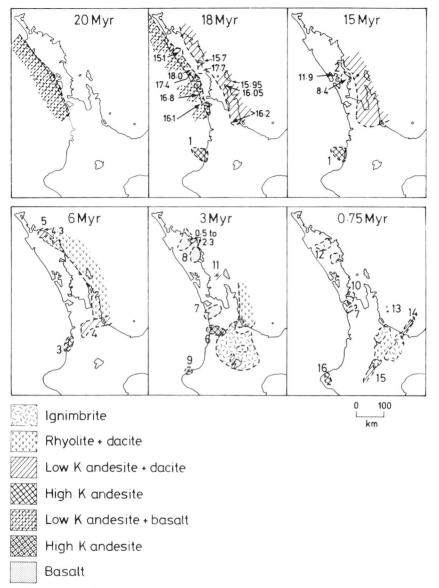

Fig. 7.2. Evolutionary reconstruction of Upper Cenozoic volcanism in NW New Zealand since 20 Myr ago, proposed by Ballance (1976).
1 = Mohakatino volcanics; 2 = Ti Point basalts and Tokatoka olivine nephelinite; 3 = Orangiwhao and Whareorino; 4 = younger Kiwitahi and Beesons Island volcanics; 5 = Whangaroa 'Wairakau' andesites; 6 = Alexandra volcanics; 7 = Franklin basalts; 8 = Horeke basalts; 9 = Paritutu; 10 = Auckland basalts; 11 = Little Barrier andesites; 12 = Taheke basalts; 13 = Mayor Island pantellerites; 14 = White Island; 15 = Ruapehu-Ngaruhoe-Tongariro; 16 = Mt. Egmont. Ages in Myr arrowed.

Six successive geographically separated zones can be distinguished. These are indicated in Fig. 7.2.

Waitakere zone, 20—15 Myr ago

The Waitakere zone extended along the present west coast and continental shelf of northern North Island. Its extent is determined partly from off-shore magnetic anomalies. Early activity is recorded in volcaniclastic sediments of Waitakian/Otaian age (ca. 20 Myr), and subsequent activity migrated gradually eastwards. The Waitakere zone was sited above a substantial structural depression. Lava types were mixed alkaline, tholeiitic and calc-alkaline — mainly andesite, andesitic basalt and basalt, with rare dacite. Activity was dominantly submarine and the rocks mainly fragmental. No plutons are exposed, mineralization was very slight, and no behind-arc activity is known.

Northland zone, 18—15 Myr ago

The Northland zone was initiated apparently during the Otaian Stage. It was parallel to the Waitakere zone and between 10 and 75 km east of it. The dominant lava types were calc-alkaline andesite and dacite with minor rhyodacite and basic forms. The andesites of this and the succeeding Coromandel zone consist predominantly of epiclastites, including lake deposits, with subordinate pyroclastic breccias and flows. Eruption was sub-aerial, through a horst of basement rocks, which had been subsiding rapidly during deposition of the adjacent Waitemata flysch body. Small plutonic bodies of gabbro, quartz diorite, and granodiorite are exposed.

Andesitic volcanism began to the south of the Waitakere zone, from an isolated and now offshore centre at about 16 Myr ago.

Coromandel zone, 15—6 Myr ago

Block faulting in northern New Zealand, at approximately 15 Myr, coincided with the extinction of the Waitakere zone. Andesitic and dacitic volcanism continued in the south of the Northland arc in the Coromandel. Dacite tended to follow andesite at any one locality.

The first clearly behind-arc activity occurred during this stage. In the north, undersaturated basic volcanics are dated at 11.9 Myr, and the Ti Point basalt field at 8.4 Myr; in the south the continuation of Mohakatino andesitic volcanism was now clearly behind-arc.

Whitianga zone 6—3 Myr ago

Although parallel to, and overlapping, the preceding Coromandel zone, the Whitianga zone shifted some kilometres eastwards. A local erosional hiatus of uncertain duration followed. The Northland arc continued, shifted some kilometres eastward with: (1) the first voluminous production of rhyolites and ignimbrites; (b) the longitudinal differentiation of the arc into a central zone of predominantly acid volcanism, flanked to north and south by andesitic volcanism; and (c) extensive hydrothermal mineralization (the Hauraki Goldfields) and propylitization of the underlying andesites and dacites.

No behind-arc basalts are known from this period, but high-K behind-arc andesites occur.

Tauranga zone, 3—0.75 Myr ago

At about 3 Myr, a pronounced spatial separation of the volcanism into the two zones occurred. In the south, voluminous rhyolitic volcanism was dominated by ignimbrites. There were minor areas of andesite. In the west, behind-arc high-alumina to alkaline basalts occurred in three separate fields, and high-K andesites were again erupted. Basalt and high-K andesite appear to have overlapped in the Karioi-Pirongia field.

Taupo zone, 0.75—0 Myr ago

At approximately 0.75 Myr, further southwards extension occurred. The Taupo zone contains a central zone of rhyolite and dacite, with voluminous ignimbrites, flanked to north and south by andesite. Minor high-alumina basalts occur. Behind-arc activity has continued as before.

The Rotorua-Taupo depression

The Rotorua-Taupo graben, indicated in Fig. 7.3, is one of the major Tertiary features of the North Island. In upper Waitotaran time (2—3 Myr ago), the first rhyolitic products became incorporated in the marine sediments of the Taruarau Saddle, indicating that the explosive phases of the Rotorua-Taupo volcanism had started. The full impact of this volcanism was not evident until late Nukumaruan-Castlecliffian time about 1 Myr ago, when rhyolitic tuffs were spread in quantity; the volume of volcanic dust in single pumice bands is often several cubic kilometres. The present shape of the graben

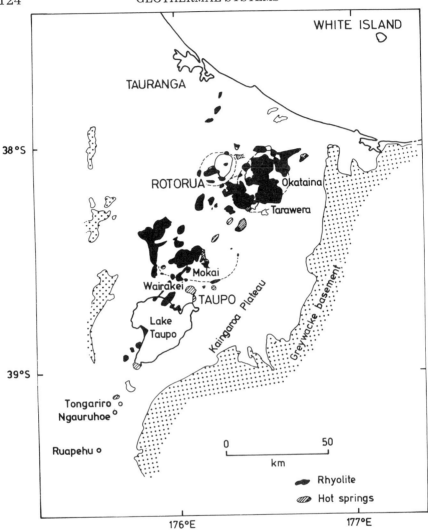

Fig. 7.3. Taupo volcanic zone of New Zealand, distribution of greywacke basement and acid volcanics.

apparently started to develop about 3 Myr ago, and was fully developed in late Castlecliffian time about 0.6 Myr ago.

Steiner (1958) has investigated the 1954 Ngauruhoe lava and its xenoliths. Volcanicity in the National Park area started with basic lava which became mildly contaminated with basement material, forming a basalitic-dacitic sequence. During the development of this sequence, a separate, acid magma was being formed from transfusion of acid gneiss in prolonged contact with olivine basaltic magma at great depth. This acid magma produced the acid tuffs and ignimbrites of the Rotorua-Taupo graben. The graben developed contemporaneously.

Geophysical surveys along the fringes of the Rotorua-Taupo area have revealed that greywacke rock is above sea-level in large areas, but overlain by a thin cover of ignimbrite and pumice (see below). The thickness of the volcanic strata in the graben is about 5 km. Gneisses, granites, etc., have been brought up by lavas. There are: biotite quartz diorites in Coromandel; biotite granodiorites at Atiamuri; granite boulders at Huka Falls; gneissic xenoliths in the Ngauruhoe lava.

In order to put all this information in perspective, it is useful to have as a summary some hypothetical vertical sections, as shown in Fig. 7.4. These emphasize the importance of the crustal barrier to volcanism, a barrier which here is lens-shaped and rides on a downwelling part of the mantle. This view in section, rather than in plan, is what will now preoccupy us.

7.2. CRUSTAL ACCUMULATION MODEL

Suppose an active portion of oceanic crust is transgressed by a thickening wedge of cratonic material (or what is the same in effect, a portion of active oceanic crust passing under a wedge of cratonic material). The wedge can be represented, for example, by:

$$h_1 = H_1 + \xi h_{11},$$
$$h_2 = \xi h_{21},$$
$$h_3 = H_3 + \xi(h_{31} - H_3).$$

Here ξ is a descriptive parameter, a measure of distance into the wedge, such that $\xi = 0$ corresponds to original oceanic crust and $\xi = 1$ to a cratonic crust upon which is a sedimentary deposit of thickness $(H_1 + h_{11})$. These sediments may be of terriginous or volcanic origin. The quantity ξ is simply a measure of the thickness of the crust. In a static model it is irrelevant how this arises. It could come from a direct local crustal thickening, or from a wedge of crust passing over the source region.

What would be appropriate values of these parameters? With New

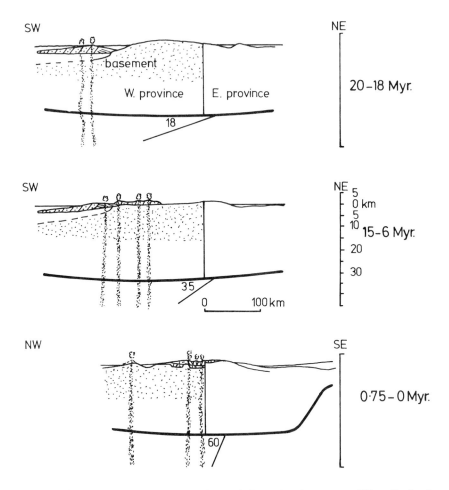

Fig. 7.4. Cartoon showing cross-sections of the volcanic region of New Zealand since 20 Myr ago. Adapted from Ballance (1976).

Zealand and the Taupo depression in mind, for example, and $h_{21} = 20$ km, $h_{31} = 10$ km, then a land elevation of about 0.5 km and sedimentary layer of 5 km or so requires $h_{11} \approx 5$ km with $\xi = 0.75$ ($h_{11} = 5$ km, $\xi = 0.75$ gives $L = 0.49$ km, $h_1 = 5.57$ km and a total crustal thickness of 29.5 km).

A more flexible approach to the accumulation of the crust is to maintain the same total mass for the fully built crust but otherwise to allow the proportions to be set explicitly.

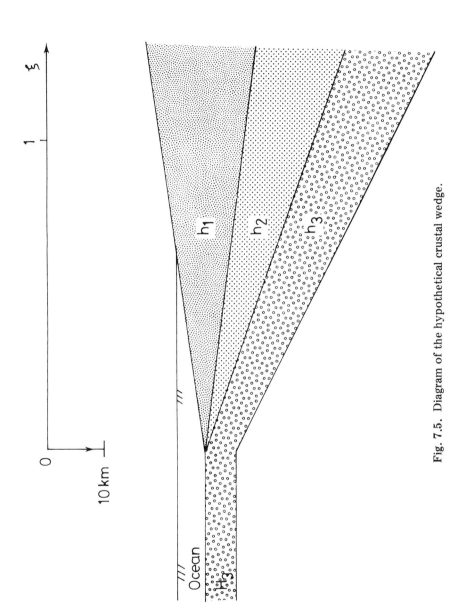

Fig. 7.5. Diagram of the hypothetical crustal wedge.

128   GEOTHERMAL SYSTEMS

Thus, let us take,

$$h_1 = \xi\lambda h_{11}$$
$$h_2 = \xi(1-\lambda)h_{21}$$
$$h_3 = H_3 + \xi(h_{31} - H_3)$$

as sketched in Fig. 7.5. Here λ represents the relative proportions of the thicknesses of the accumulating sediments and upper crust. If we choose h_{11}, h_{21} so that the crust has the same mass/unit area as before for $\xi = 1$ and all λ, then with $h_{21} = 20$ km, as before, we need $h_{11} = (\rho_2/\rho_1)h_{21}$, namely 21.6 km, with standard values.

Purely for simplicity of presentation the magma density of the crustal melt will be written $(\rho - \Delta\rho)$ where ρ is the parent crustal rock density and given by stating $\Delta\rho/\rho$.

The behaviour of the whole system will now be evaluated. Some typical results are shown in Fig. 7.6 for $\lambda = 0.1$, $\Delta\rho/\rho = 0.2$. First consider the case with $H_4 = 40$ km.

(i) For $\xi < 0.93$ the paleosurface is below sea level.

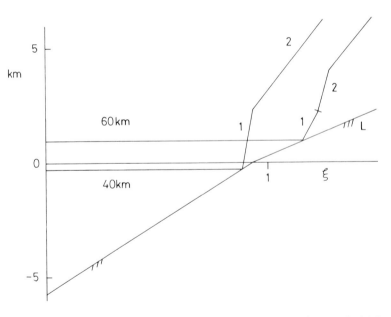

Fig. 7.6. Crustal melting model behaviour as function of crustal thickness parameter ξ. $\lambda = 0.1$, $\Delta\rho/\rho = 0.2$. L = Level of paleosurface. Lines for two source depths: 40 km, 60 km. 1 = upper crust melt (rhyolite); 2 = lower crust melt (andesite).

(ii) For $\xi < 0.88$ basaltic magma penetrates the crust to produce suboceanic volcanism reaching about 0.3 km below sea level. The basaltic pile will be thickest for $\xi = 0$ and of zero thickness for $\xi = 0.83$.

(iii) For $0.88 < \xi < 0.90$, and thereafter, there is a rapid build up of acidic lavas, the top of the basalt column is in the upper crust and volcanics from upper crustal melts are extruded, here presumed to be rhyolitic.

(iv) For $\xi > 0.9$ the top of the basalt column is in the lower crust and lower crustal melts are extruded, here presumed to be andesitic.

A much deeper source, here with $H_4 = 60$ km, gives entirely subaerial lavas and a somewhat extended interval of rhyolites. With $\lambda = 0.1$, i.e., rather small, the crust has only a thin cover of sediments and near $\xi = 1$, it is rather like that in the western part of the North Island, New Zealand. If we took the volcanoes Egmont, 2.5 km, and Ruapehu, 2.8 km, as representative this suggests $\xi \approx 0.95$, with the paleosurface barely above sea-level and a total crustal thickness of 29–30 km.

Notice that for smaller values of $\Delta\rho/\rho$ the thickness and elevation of the acid volcanics is smaller — they have less buoyancy. On the other hand, larger values of source depth H_4 produce a thicker basaltic pile, but a thinner acidic pile — simply because the basaltic volcanism continues to larger values of ξ.

Similar data, for the case $\lambda = 0.5$, $\Delta\rho/\rho = 0.2$ is shown in Fig. 7.7. The overall behaviour is similar to that with $\lambda = 0.1$, but here, with a rather large value of λ, the sedimentary cover is rather thick and rhyolitic volcanism occurs over an extended range of ξ. Further, there is a pronounced discontinuity in the pile elevation when switching from rhyolitic to andesitic volcanism. This discontinuity arises because the andesitic magma is more dense than the rhyolitic magma and with equal geometry the andesite column must be less tall. Thus if we take the model as it stands, if ξ is increasing, there is an interval of ξ at the end of rhyolitic volcanism before andesites can be extruded: in the data here with $H_4 = 40$ km from $\xi = 0.94$ to 0.99; with $H_4 = 60$ km from $\xi = 1.35$ to 1.48.

These systems are relatively insensitive, as far as elevation is concerned, to density changes produced in the crust since the effect is proportional to $\Delta\rho/\rho_4$ where $\Delta\rho$ is the density change and ρ_4 is the mantle density. As illustration, consider as standard a crust with a 25 km layer of density 2.7 g cm^3 from the lower 5 km of which acid

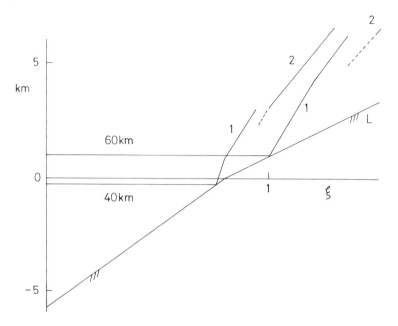

Fig. 7.7. Crustal melting model as for Fig. 7.6 but with $\lambda = 0.5$.

volcanics of thickness 5 km and density $2.5\,\mathrm{g\,cm^{-3}}$ are produced and deposited on the surface. Consider two cases: (i) the crustal layer is unaltered — the new elevation is 1.21 km above that of the standard crust; (ii) the bottom 5 km is altered to density $3.0\,\mathrm{g\,cm^{-3}}$ — the new elevation is 0.76 km above that of the standard crust. The difference between the two cases of 0.45 km is $5\Delta\rho/\rho_4$ km where $\Delta\rho = 0.3\,\mathrm{g\,cm^{-3}}$ and $\rho_4 = 3.3\,\mathrm{g\,cm^{-3}}$.

Acid magma densities. We can turn the argument on its head by posing the question as follows. Assuming that the acid volcanics are produced, at least in part, from partially melted crust what are the limits on the density of the magma to allow extrusive volcanism?

The extreme case would be material derived from the base of the lower crust. Using our standard cratonic crust, if ρ is the magma density then the acid lava pile height is:

$$h = (\rho_2/\rho - 1)h_2 + (\rho_3/\rho - 1)h_3.$$

For $\rho = 2.6$, 2.7, $2.8\,\mathrm{g\,cm^{-3}}$, then $h = 2.5$, 1.1, -0.2 km. In the case of the andesitic volcanoes of New Zealand reaching 2—3 km above the paleosurface, clearly we need $\rho \lesssim 2.6\,\mathrm{g\,cm^{-3}}$.

A similar argument applied to rhyolitic discharges originating at the extreme of the base of the upper crust gives a pile height:

$$h = (\rho_1/\rho - 1)h_1 .$$

In the case of the rhyolite domes of New Zealand rarely reaching 1 km above the paleosurface we need

$$(\rho_1 - \rho)/\rho_1 \lesssim 0.04.$$

This value is far too small! Clearly some other factor must be responsible for inhibiting the development of the extruded rhyolites. Plainly these systems do not reach hydrostatic equilibrium. This could arise owing to: (i) extreme slowness of development of a very viscous material; (ii) the consequent dominance of erosion; (iii) more frequently, because of the wide dispersal of the material as ignimbrite eruptions.

We can retain the use of a static model by the device of imposing a critical condition on the most viscous magmas. Various possibilities suggest themselves. For example, rather than restrict, say, the andesitic source to the head of the basalt column if and only if it is located in the lower crust, we could allow in addition partial melting of lower crust whenever rhyolitics holdup occurs. The rhyolites may leave their source region merely as buoyant blobs and not under the control of a plumbing system. This possibility is consistent with the occurrence of rhyolites in ring structures, typically 40 km in diameter within which there are numerous domes and a coherent pattern of volcanism.

The behaviour of the system is sensitive to λ for values of ξ near unity. Thus, as the system develops, relatively small changes in crustal thicknesses will lead to pronounced changes in behaviour. Furthermore, very good possibilities of magma trapping will occur. One is led to the picture of a rather sporadic system quite different from that of a simple one-shot oceanic lithothermal system.

The aim here has been to give some idea of the circumstances under which a magmatic source can be implanted in the crust. Given such a source, that is to say a localized lump of hot matter, we now leave behind all questions of origin and structural control to consider how the heat is extracted.

8. Lithothermal-Hydrothermal Interaction

The behaviour of hydrothermal systems, the focus of this book, cannot be described in isolation from the active environment in which they are embedded. The central problem is to be able to relate the inputs to the hydrothermal systems to their sources at depth. Modern studies of ore deposits, many of which are seen to be arte-facts of fossilized hydrothermal—lithothermal systems, have opened the door to understanding this problem (see, for example, Gass, 1977; Barnes, 1979). Our purpose here is to sketch some of these ideas in order to show how hydrothermal systems arise and are supplied from depth. A great variety of hydrothermal ore deposits are now recognised. Here attention is centred on the porphyry ore bodies found in orogenic—volcanic regions, particularly the copper porphyry bodies. The form of these bodies is sketched in Fig. 8.1.

There is a sub-vertical core of potassium-rich highly metamorphosed rock with minor ore, sheathed by lower grade metamorphic rock within the inner parts of which ore mineral concentrations — largely pyrite, of order 10%; with the copper ores, 1—3% concentration (of economic interest) lining the sub-vertical portions of the sheath.

The ore bodies are usually associated with calc-alkaline intrusive and volcanic bodies, some in batholithic bodies but commonly in small epizonal intrusives. The economically valuable copper ore is a disseminated deposit of typically $10—10^3$ Mton of concentration 0.04 or more of copper ore (Gustafson and Hunt, 1975). The source whole rock copper concentration was originally possibly of order 10^{-5}. Deposits of this form are numerous.

A great variety of models have been proposed for the production of an ore body from a magmatic intrusive body.

(i) Simple exhalation of juvenile magmatic water vapour (for example, Rose, 1970).

(ii) Convection of liquid ground water through a subsolidus intrusion (see, for example, Norton and Cathles, 1976).

132

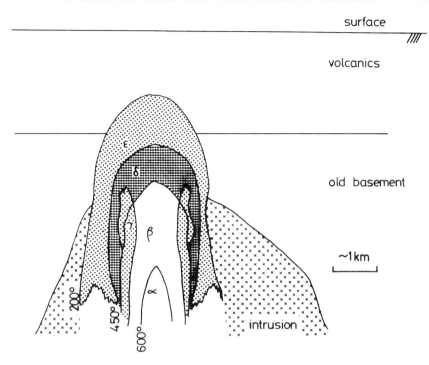

Fig. 8.1. Schematic drawing of the distribution of alteration and deposits in a porphyry ore deposit. (Adapted from Guilbert and Lowell (1974), based on the San Manuel — Kalamazoo body). The various zones are not always as distinct as indicated here. Zones, with typical maximum concentrations:

α = Sericite innermost core: quartz, sericite, chlorite, K-feldspar, minor ore.
β = Potassic core: quartz, K-feldspar, biotite, minor ore.
γ = Ore shell: pyrite, 1%, copper ores, 1—3%; minor molybdenum.
δ = High pyrite cap (phyllic zone): pyrite, 10%; copper ores, 0.1—3%.
ε = Low pyrite cap (propylitic zone): pyrite, 2%.

(iii) Convection of ground water, involving multi-component multi-phase conditions (for example Henley and McNabb (1978)). A small portion of the water could be juvenile.

The evidence from isotope studies is unequivocal, the great bulk of the water is of meteoric origin drawn through the system at the time of emplacement of the ore body. The evidence from fluid inclusion studies, too, is unequivocal that multi-phase, multi-component conditions are an essential feature of the system, especially during its vigorous phase and in particular that two quite distinct phases are involved.

The outer envelope of phyllic-prophyllic alteration develops at 350°C or so and involves exchange with meteoric ground water. The potassic core contained dense, high salinity fluids, of densities up to 1.3 ton m^{-3} and temperatures of 350–700°C, together with a low salinity vapour — as found in coexisting fluid inclusions.

8.1. HYDROTHERMAL ORE GENESIS

Ores are rocks which within localized volumes contain minerals that are quite different in kind or amount from those in most rocks.

From a global point of view the crust itself is a great "ore" deposit, being exceptionally rich in minerals such as quartz, and feldspars, which are believed to be largely absent from the mantle. Although the idea is a useful one, it is an overextension of the notion, since clearly an orebody is recognized by its existence within a volume not only small compared with the volume of the crust but all of whose dimensions are small compared with the thickness of the crust. On the other hand, neither do a few localized grains of, say, rutile in a high-grade metamorphic rock represent an orebody.

Whereas the great bulk of rocks, more than 99% by volume, are composed of the elements O, Si, Al, Fe, Ca, Na, K and Mg, and the relative abundance of all other elements is generally much less than 1%, there are some rocks for which they are rich. In comparison with the crustal average, relative enrichment by a factor fo 10^2 to 10^3 is typical.

The central question is how these localized concentrations were formed. Clearly, there are two extreme types of such bodies.

(1) The segregation of material may occur essentially locally within the ambient rock. We include here most of the deposits closely associated with igneous, metamorphic and sedimentary rocks, including sinter, pegmatites, small-scale metasomatism deposits, placers, laterites and so on. These processes are in some form operating more or less continually throughout the crust and are most readily described as variants of the petrogenesis of ordinary rocks. Except where the two types of segregation processes work together, systems of this type are not our present concern.

(2) The segregation of material may occur by its being moved to a quite different part of the ambient rock. The notable example of this type are the so-called hydrothermal deposits. The physical processes within such systems are our present concern.

The discussion proceeds with some elementary estimates of the order of magnitude of the water flow mass rate and the scale of a deep hydrothermal system. A reservoir model is outlined. Next, the features that control its initiation and maintenance are discussed. The main technical aspect is the analysis of the conditions in the gas column; this is followed by a consideration of the extraction of matter and energy from the source. The main result is obtained. The surface zone, driven by the power transported in the gas column, is identified through its characteristic temperature. The transport of soluble matter, in particular the effects of the variation of solubility with temperature, is briefly treated.

Formulation of the problem

Whereas the mantle is rather dry, with water content possibly less than 0.1%, water-substance is a ubiquitous constituent of the crust. This is most dramatically and directly seen in the regions of surface mass discharge in geothermal and volcanic areas. Among the studies of these phenomena, those which describe the role of water-substance in modern hydrothermal systems are the most detailed. Freely circulating bodies of hot water-substance have been identified within the upper 10 km of the crust. Mass transport at the rate of 1 ton s^{-1} is typical. These systems are the uppermost parts of deep systems, which ultimately draw their energy from the mantle. The mass and assumed chemistry of the so-called meso and hypothermal ore deposits strongly suggests that they are formed by transportation in a deep-seated system of water-substance circulation. In effect, an ore deposit is a fossil hydrothermal system in which we obtain a glimpse of a deeper part of the system. Whatever the details, water is vital. Its existence in the crust is taken for granted here. All I am concerned to demonstrate is that the system can move enough of it.

The essential questions

There are two extreme possibilities for the mode of transportation of the ore material in the hydrothermal fluid.

(1) The ore material concentration is high. This is the least demanding on the required transport of water. Undoubtedly, this is the case for some pegmatitic deposits; for hydrothermal deposits, fluid inclusions, although they often show high concentrations of, for example, sodium chloride, do not give any indication of high ore material concentration.

(2) The ore material concentration is low, being of the same order as that in the source rocks. This situation is the most demanding on

the required transport of water. Nevertheless, even in this extreme case, ample water can be available; indeed, it is the main point of this chapter to show in some detail how this enormous flow of water is generated and maintained.

A simple calculation, which I am sure has been done by everyone who has ever thought about this problem, exposes the three key questions of solubility, flow rate and time.

Consider that an orebody of volume, V, is produced from a source volume, W, as follows. Let the concentration ratio of ore material in the deposit to that in the source be n. If ore material is taken at the concentration in the source and transferred to the deposit volume and all dumped there, then $V = W/n$. This will require a total equivalent liquid water transport of order volume W. The time, t, required at mass rate m, is $t \sim$ (total mass of the source volume)$/m$. By way of illustration consider: concentration $\sigma = 10\,\text{PPM}$; $m = 1\,\text{ton}\,\text{s}^{-1}$, $t = 10^4\,\text{yr}$, $n = 10^3$. Then a source volume of $10^3\,\text{km}^3$, from which all the ore material is extracted and deposited within a volume of $1\,\text{km}^3$, produces a deposit of 1% concentration and of ore material mass $3 \times 10^7\,\text{ton}$. The total mass transport of water is equivalent to a flow of $3 \times 10^{12}\,\text{ton}$ through the source. Are these quantities at all reasonable and could such a system work? As we shall see, these are all acceptable values and a variety of suitable mechanisms can be quantitatively identified.

8.2. SCHEMATIC MODEL

A hydrothermal ore extraction system is envisaged as a cascade of reservoirs interconnected by permeable conduits (pipes) as illustrated in Fig. 8.2 in a manner analogous to that in an elaborate distillation column. In the analysis presented here only three reservoirs are considered.

(1) The source is the largest reservoir and here, for the sake of concreteness, is considered as a granitic mass of typical volume $10^3\,\text{km}^3$, partially molten at the time of its emplacement.

(2) The primary deposit — a much smaller reservoir — is the region in which most of the ore and associated gangue material is deposited. Here it is considered to exist near the main interface in the system, the boundary between the gaseous deep system and the liquid surface zone.

(3) The hot water mushroom: the hydrothermal system in the

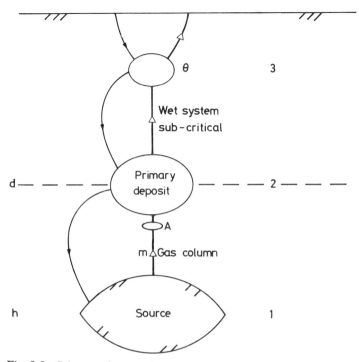

Fig. 8.2. Schema of reservoir-pipe model of hydrothermal ore system.

surface zone has a temperature distribution shaped like a mushroom, within the head of which is a rather discrete body of hot water and rock of characteristic temperature typically about 250°C for a vigorous system.

The localization of the primary deposit reservoir at the phase boundary may at first sight seem rather arbitrary. It is very important to appreciate that it is not the change of state of water-substance and the possible consequent effect on its solvent capability that is dominant. By far the most powerful effect arises through the permeability.

In the vicinity of the interfacial zone two phases of water-substance can be present. This has the effect of drastically reducing the effective permeability of the medium. For a single phase, the permeability is solely determined by the void geometry and not at all by the fluid. But when two phases are present they restrict each other's passage.

The localization of the head of the mushroom is not so sharply defined. It is limited strongly only in its relation to the "boiling point with depth" (BPD) relation (see Appendix). If the heat flow is sufficiently high for temperatures locally to exceed the BPD, phreatic eruptions are likely. The surface zone hydrothermal system becomes (certainly for a time) a one-pass eruptive system without the ability to create a deposit.

The models discussed here do not impose any strong constraints on the source of the water-substance, since the bulk of it may be recirculated many times or it may be derived directly from the surface zone.

In a rather schematic way the approach here recognizes three main types of hydrothermal ore deposits, which are called in the classification of Lindgren (1933): hypothermal (our primary deposit); mesothermal, here considered to be deposited in the surface zone "mushroom"; epithermal and telethermal, virtually ignored here, but thought of as arising from the recharge—discharge part of the surface zone system.

Source constraints

The initiation and maintenance of a hydrothermal extraction system imposes three conditions on the source.

(1) In the mode of emplacement, the main concern here is the time-scale of the emplacement. If the emplacement is rapid, the analysis of the problem will be much more complex. On the other hand, if the emplacement is sufficiently slow, the analysis can be treated as a quasi-steady model — indeed, this is the only case considered here.

(2) Two distinct regions are envisaged for access of the hydrothermal fluid to the ore material within the body of the source. (a) While the source is sufficiently hot for it to remain partially molten, the hydrothermal fluids presumably have direct access only to the outermost portions of the source region, otherwise being transported through the source dissolved in the magma. Provided that there is sufficiently rapid mixing within the source, this will provide ready access to the interior. The time-scale of this process will be relatively short. (b) When the source is sufficiently cooled for it to be substantially solid it behaves as a porous medium; to the flow, apart from its permeability, it is indistinguishable from the ambient rock. The time-scale of this process will be relatively long.

(3) As we shall see, the most effective mode of transportation of material from the source to the deposit volume is in a column of gas (of water-substance). Only under certain conditions can a gas column be initiated and maintained, and it is during the life of the column that the bulk of the material transport occurs.

Emplacement time-scale

Consider the rise of a molten body. We can estimate the rate of rise for two extreme sets of circumstances.

If the rate of rise is controlled by a balance between the buoyancy of the body and the viscous drag of its surroundings, the vertical velocity is:

$$w \sim \tfrac{2}{9} a^2 g'/\nu, \qquad g' = (\Delta\rho/\rho)g,$$

where $\Delta\rho/\rho$ is the proportional density contrast. This is the well known Stokesian expression. For the typical values $a = 10\,\mathrm{km}$, $\Delta\rho/\rho = 0.1$ and $\nu = 10^{16}\,\mathrm{m^2\,sec^{-1}}$ we have:

$$w = 0.07\,\mathrm{m\,yr^{-1}}.$$

This body would rise a distance equal to its diameter in $3 \times 10^5\,\mathrm{yr}$.

If the rate of rise is controlled by the rate at which the advancing face can melt the ambient rock and refreeze material at its base, it is readily shown that:

$$w \sim \kappa\theta N/aE,$$

with

$$N = (A/A_c)^{1/3} \quad \text{and} \quad E = L + c\theta,$$

where L is the latent heat of freezing, c is the specific heat and θ is the temperature contrast. This expression is derived by equating the rate of transfer of heat to the interface by the internal convection to the rate of supply of enthalpy to melt the interface. For the typical values used above and $a = 10\,\mathrm{km}$, $c = 10^3\,\mathrm{J\,kg^{-1}}$ and $\theta = 100\,\mathrm{K}$ we have

$$w = 30\,\mathrm{m\,yr^{-1}}.$$

This body would rise a distance equal to its diameter in $700\,\mathrm{yr}$.

For this later case, the time-scale of emplacement is comparable with that of the development of the hydrothermal system. This suggests, even though we ignore this possibility in the subsequent discussion, that some hydrothermal extraction systems will be dominated by the emplacement of the source region. For somewhat smaller bodies, however, and if the interaction of the hydrothermal system on the source region during its emplacement is also taken into account, conditions will not usually be so extreme.

Freezing of the source

A source entering a water-saturated zone may have its rise terminated prematurely. The interaction with the hydrothermal system will lower its temperature. There is, however, an even more powerful constraint (for details see Harris *et al.*, 1970). A dry melt onset of freezing occurs on a liquidus with $dP/dT > 0$ so that if it were hot enough it could rise to the surface and remain a magma. A melt with excess free water has a portion of its liquidus with $dP/dT < 0$ because of the buffering of water-substance by hydrous minerals. For wet granitic melts, the range of $P—T$ conditions for which partial melts can occur is narrow. Thus a granitic melt can freeze well below the surface: for example, at about $700°C$ and 2 kbar, that is at a depth of about 7 km.

The time-scale of this process, $\tau \sim \rho VL/F$, where F is the rate of loss of energy. For a granitic body with initial $F \sim 10^4$ MW, see below, radius $a = 10$ km and $L \approx 400$ kJ kg^{-1} then $\tau \sim 3000$ yr. For the most intense bodies with $F \sim 10^5$ MW, then $\tau \sim 10^2$ yr.

Circulation within the source

If the source region, at the time of its emplacement, is largely molten, the time-scale τ of the internal mixing process is given by:

$$\tau \sim (A_c /A)^{1/3} a^2 /\kappa,$$

with

$$A = \gamma g \theta a^3 /\kappa \nu,$$

where a is a length-scale, the mean radius for a blob-like source, γ, κ, ν are, respectively, the coefficient of cubical expanison, thermal diffusivity and kinematic viscosity of the magma, θ is the temperature difference between the source interior and the immediate surroundings, A is the Rayleigh number and $A_c \approx 10^3$, an empirically determined constant (see, for example, Elder, 1976, pp. 60—77). The derivation of this expression assumes that the interior motion creates a boundary region of extent

$$\delta \sim a(A_c /A)^{1/3}$$

of time-scale

$$\tau \sim \delta^2 /\kappa.$$

As an example consider a large granitic pluton with $a = 10$ km, $\theta = 100$ K, $\kappa = 10^{-6}$ m^2 s^{-1}, $\nu = 10^2$ m^2 s^{-1} and $\gamma = 10^{-5}$ K^{-1} for which $A = 10^{14}$ and $\tau = 10^2$ yr. This source is being rapidly mixed — indeed, motion will persist until $A \sim 10^3$, when $\nu \sim 10^{14}$ m^2 s^{-1}, which is well into the freezing range for granite.

Subsequently, therefore, we shall assume, that the time-scale of the mixing within the source region is small in comparison with the time-scale of the hydrothermal system, typically 10^4 yr (see below). In other words, mixing within the source is sufficiently rapid to allow almost complete access of the hydrothermal fluid to the material of the source throughout the life of the hydrothermal system.

Existence of the gas column

The problem of the existence of the gas column can be treated by means of a pipe model (detailed in a later chapter). Thus, if water leaves the surface zone, of thickness d, with density ρ_0, passes through the source volume at depth h, rises as a gas, of density ρ_1, to heat the surface zone to a local density ρ_2, it can be shown that the mass rate of flow m is given by

$$m/kSg = [(\rho_0 - \rho_1) - (\rho_2 - \rho_1)d/h]/[\nu_1 + (\nu_2 - \nu_1)d/h],$$

where k is the permeability, S is the pipe cross-sectional area and, although they are readily accounted for, the variation of permeability along the pipe is ignored and the recharge resistance is taken to be zero.

We are here principally interested in the range of mass transport. Consider the two extremes: (1) an erupting system has $d/h = 0$ and $m(\text{erupt})/kSg = (\rho_0 - \rho_1)/\nu_1$; (2) a collapsed system has $d/h = 1$, namely no gas column, when $m(\text{collapse})/kSg = (\rho_0 - \rho_2)/\nu_2$. Thus, the mass rate ratio is:

$$\xi = m(\text{collapse})/m(\text{erupt}) = (\rho_0 - \rho_1)\nu_2/(\rho_0 - \rho_2)\nu_1.$$

Taking representative values, $\rho_0 = 1 \text{ ton m}^{-3}$, cold surface zone water, $\rho_2 = 0.96 \text{ ton m}^{-3}$, surface zone water at $100°C$, $\rho_1 = 0.3 \text{ ton m}^{-3}$, near-critical gas, and $\nu_2/\nu_1 \approx 1$, and we have $\xi \approx 1/20$. In this case we can expect that the bulk of the possible mass transport will have occurred before collapse sets in.

Peculiarity of a super-critical hydrothermal system

There is an important distinction between a wholly sub-critical (wet) hydrothermal system and that which is in part of wholly super-critical. In a wet system, even though liquid water-substance is compressible, the dominant effect of the density arises from variations of temperature; moreover, these density variations are principally effective through their role in generating buoyancy forces. Furthermore, this is the basis of the analysis of such systems by means of the Boussinesq approximation in which density variations are ignored, except insofar as they lead to the generation of buoyancy forces. In a super-critical hydrothermal system the working fluid is a very compressible gas, so density variations arise from variations of both temperature and pressure. We cannot therefore directly use data from a wet system to aid the analysis of a dry system; nor can we use the Boussinesq approximation, but this is a very minor point in a pipe-like model.

8.3. ROLE OF SOLUBLE SALTS

A ubiquitous feature of these bodies is the highly saline fluid found in inclusions. The main constituents of this fluid are water and NaCl. There is now clear evidence of the ability of a H_2O + NaCl system to transport heavy metals as a variety of complexes. This together with the isotope evidence suggests that a convecting system of predominantly H_2O + NaCl, in particular its thermodynamic structure, is the key element in the mechanism of formation of these bodies (see especially Henley and McNabb, 1978).

Thermodynamics of the system: H_2O + NaCl

The pressure and temperature relationships as a function of the mass fraction of salt are shown in Figs. 8.3—8.5.

The presence of the salt makes a qualitive change to the thermodynamics of that of pure water substance alone. Whereas pure water has two distinct phases only below its critical point of $374°C$, 220 bar, salt opens up a two-phase region at considerably higher temperatures and pressures. The occurrence of this region is a function of the proportion, n, of soluble salt, the critical point for example reaching $700°C$, 1240 bar for $n \approx 0.3$. Above the critical point, where for pure water only a gas is present, there are two very distinct phases. The two phases are: (1) a highly saline, dense liquid; and (2) a vapour of moderate or low salinity. At lower temperatures (in the example, below $500°C$), liquids of a range of composition from pure water to NaCl-saturated water are possible, with densities, as shown in Fig. 8.5 for a pressure of 500 bar, which cover a wide possible range, both of higher and lower densities than those of typical near pure ground water.

Thermodynamic structure

As sketched in Fig. 8.6 consider the convective system in water-saturated ground above a magmatic source. The source contains of the order of 1% soluble salt. Let us follow the path of a typical fluid particle. The corresponding thermodynamic path is indicated in Fig. 8.6(b).

(a) Warm ground water, at perhaps $250°C$, is drawn into the system or is already part of the recirculating flow, as a result of the horizontal pressure gradient arising from the presence of the low-density convective column.

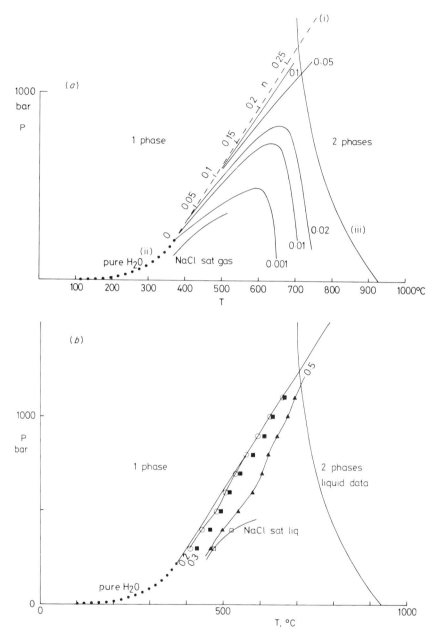

Fig. 8.3. Pressure and temperature data for an equilibrium mixture of $H_2O +$ NaCl (data originally from Sourirajan and Kennedy, 1962), drawn as curves of constant NaCl mass fraction, n; pressure, P, bar; temperature, $T,°C$ (a) Gas phase; (b) liquid phase. Also shown are: (i) critical curve, above which only a single phase is present — values of n are shown; (ii) saturated vapour pressure curve for pure water; (iii) the cotectic minimum curve for the system ab-or-qtz, representive of the freezing conditions of granitic magma.

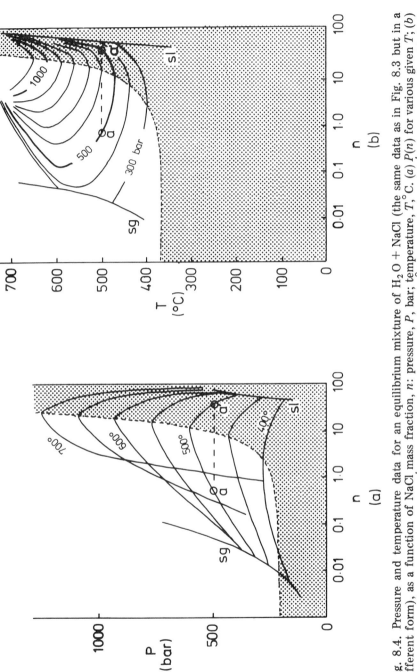

Fig. 8.4. Pressure and temperature data for an equilibrium mixture of $H_2O + NaCl$ (the same data as in Fig. 8.3 but in a different form), as a function of NaCl mass fraction, n: pressure, P, bar; temperature, T, °C. (a) $P(n)$ for various given T; (b) $T(n)$ for various given P. The points aa′ show the fluid state at 500 bar, 500°C; a, vapour phase; a′, liquid phase. sg = NaCl − saturated gas; sl = NaCl − saturated liquid. Shaded area = "liquid"; Unshaded area = "vapour"; boundary = critical curve. Note that for $n \lesssim 0.01$, critical $(P, T) \approx (P_c, T_c)$ of pure water.

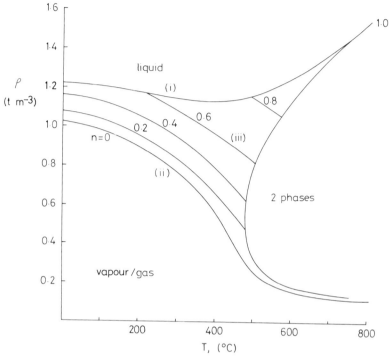

Fig. 8.5. Representative data for density, ρ, in ton m^{-3} as a function of temperature, T, in $^\circ$C at a pressure of 500 bar, of the two phases of an equilibrium mixture of H_2O + NaCl. Also shown are curves for: (i) NaCl-saturated solution; (ii) pure water, both below and above its critical point; and (iii) single-phase solutions of NaCl mass fraction n. Data from Ellis and Golding (1963) and others.

(b) The source is entered and the fluid is heated and dissolves salt. Provided the source is not exhausted and the concentration of salt in the source is sufficient, the net effect on the density may be to increase it. Already between b and c boiling commences. At c and d, the maximum temperature of the fluid particle is reached but it is now in two phases at c and at d. The saline liquid at c is trapped in the source since it is too dense to rise. The slightly saline low density gas d leaves the source.

(e) The gas column expands upwards into regions of lower pressure. Its temperature falls. There is progressive condensation of relatively small amounts of saline liquid.

(f) The fluid becomes sub-critical and is a liquid of moderate salinity.

(g) The saline liquid cools further losing its heat to circulating

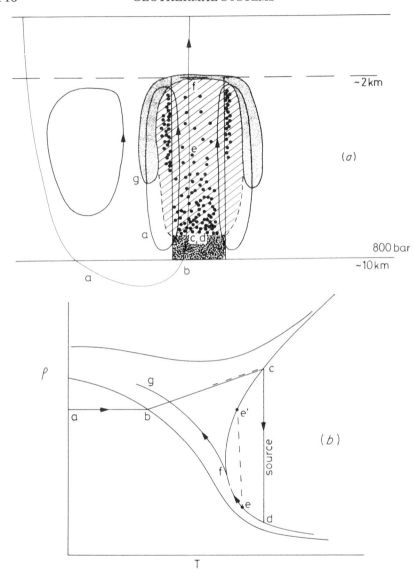

Fig. 8.6. Diagramatic representation of a convective $H_2O + NaCl$ system; (a) vertical section; (b) density—temperature section. A particular flow path is indicated.

ambient ground water and its density increases, possibly to values exceeding those of the ambient ground water.

The distribution of soluble salts within this structure is controlled by both the thermodynamic properties of the phases and their relative permeabilities. The source region will be filled with near-saturated salt solution. In the two-phase region the condensed droplets of saline fluid will be impeded in their upward movement both by the low relative permeability and by their excess density. Superimposed on this effect will be that of a radial pressure gradient tending to move droplets away from the central region of the column. Where droplet concentration is sufficiently high, this dense fluid will drain downwards and some may return to the source. In the upper parts of the gas column, where it begins to condense through mixing with local cool ground water, the salinity will be small.

The strong role of relative permeability will be pronounced: at the outlet of the source, as the liquid proportion falls; at the top of the column where advent of ground water increases the liquid proportion.

The bulk of the ore, in this view, is deposited in the region g in which the column fluid mixes with the ambient ground water. The copper ores are deposited in the dense droplet zone on the outer portion of the two-phase column.

8.4. THE GAS COLUMN

The key element in the extraction process is the gas column, of effectively super-critical water-substance, which connects the source region — of the magma body and its immediate surroundings — to the surface zone of (generally) sub-critical water-substance. As has already been discussed, for an interval the gas column is close to equilibrium. It is therefore straightforward to describe its properties in some detail. We shall first treat the gas column in isolation and subsequently consider its effect on the source region.

Field equations

The mecahnics of the gas column can be described by the following relations:

$$\rho = P/\alpha T \tag{a}$$

$$\mu = \beta T^{1/2} \tag{b}$$

$$w = m/\rho S \tag{c}$$

$$dP/dz = \rho g + \mu w/k \tag{d}$$

$$dT/dP = 1/\rho (C_p + \alpha) \tag{e}$$

The properties of the fluid are described by (a) and (b) and are discussed elsewhere. The conservation of fluid mass is expressed by (c), where m is the rate of mass flow through every section of the column of local cross-sectional area S. The pressure gradient in the column, represented by (d), arises from two effects: the weight of the fluid itself, together with the resistance experienced by the fluid in permeating the ambient ground, represented by the relation of Darcy. Finally, (e) represents the energy relation, in which it is assumed that there is a balance between the rate of working of the decompressing gas and the upward advection of heat. For the present discussion, unless stated to the contrary, the following quantities are taken as constants for a given gas column: α, β, m, S, k and C_p. There is, however, no difficulty in treating the more general cases; for example, in which $S = S(z)$, $k = k(z)$.

Boundary conditions

The gas column is represented by two first-order differential equations. Provided, therefore, that two conditions are imposed, P,T are defined. For the moment it is convenient (and realistic) to consider the pressure and temperature given at the top of the column or, rather, the base of the sub-critical surface zone. Thus, at some level $z = d$ we take $P = P_d$, $T = T_d$. The equations can then be integrated (numerically) downwards to obtain $P(z), T(z)$. Notice that there is some redundancy in the use of parameters, since the effect of m, k, S is determined in this model solely through the combination m/kS. It is convenient to use and state these values separately, but it should not be forgotten that they occur in combination. Further, the relative importance of the flow resistance is measured by $(m\mu/gk\,S\rho^2)$.

Temperature distribution

A typical set of profiles of $T(z)$ in a gas column is shown in Fig. 8.7 for various values of m. For small values of m, here somewhat less than $0.1\,\mathrm{ton\,s^{-1}}$, the gas column is nearly isothermal at the temperature of the base of the sub-critical zone. For large values of m, here those greater than about $1\,\mathrm{ton\,s^{-1}}$, the gas column is strongly non-isothermal, especially in the region of expansion near the top of the column. High temperatures are required at depth to maintain the column, and these temperatures are strongly related to m. For this column the operating range is about $0-30\,\mathrm{ton\,s^{-1}}$, the upper limit being set by the temperatures of basaltic magmas.

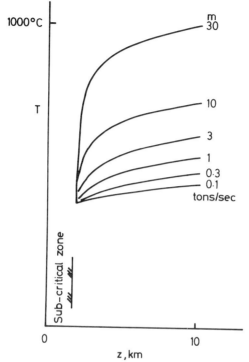

Fig. 8.7. Temperature $T,°C$, as a function of depth, z, km, in gas column with $d = 2$ km, $P_d = 0.2$ kbar, $T_d = 400°C$, $S = 0.01$ km^2 and $k = 1$ darcy, for various values of rate of mass flow, m, ton s^{-1}.

Given temperature at depth

So far the rate of mass flow, m, has been treated as a free parameter. It is, of course, determined by the system. Because m is strongly related to the temperature in the gas column at depth, which is itself controlled by conditions in the source region, it is appropriate to specify a temperature T_h at the base of the gas column $z = h$. We then have boundary conditions imposed at two levels, which are sufficient to determine $T(z)$, $P(z)$ and m. This is performed iteratively by integrating the equations downwards from d to h with an assumed value of m, obtaining the base temperature, adjusting m accordingly, and repeating until the base temperature is equal to T_h with the required precision ($\pm 0.1\%$ used here).

Thus, the state of the column is determined by its geometry, the quantity m/kS, the properties of water-substance and the two temperatures T_d, T_h.

Flow velocity

It is important to realize that the volumetric flow velocities are very small. For the archetypal model described here, w lies generally in the range 10^{-4}–10^{-3} m s^{-1}. The system operates entirely within the range of validity of the Darcy relation.

8.5. EXTRACTION FROM THE SOURCE

Given the mechanics of the gas column, we have a method of transporting matter to another place. But there remains the essential problem of first extracting that matter from the source and, even more to the point, whether enough matter can be so extracted. Provided that the time-scale of adjustments of the gas column is rapid in comparison with all the other rate-dependent processes in the system, we can make a quasi-steady model of the extraction process.

Consider a source volume V_1 of rock substance of density ρ_* and specific heat c_* in contact with fluid circulating at the mass rate m at recharge temperature T' and discharge temperature T_h. Then we have the following relations:

$$F = mc(T_h - T') \qquad (f)$$

$$dT_h/dt = F/\rho_* c_* V_1 \qquad (g)$$

$$dM/dt = m/\rho_* V_1 \qquad (h)$$

$$dQ/dt = -\sigma m \qquad (i)$$

Relations (f) and (g) express the conservation of energy as a balance between the rate of loss of the total source energy and the net rate F transported by the flow. For simplicity, the recharge temperature T' will be taken as T_d, the temperature at the base of the surface zone, unless stated to the contrary. Relation (h) merely gives the total mass of fluid, M, as a proportion of the source mass which has been in contact with the source. The total amount of "salt", Q, in the source is depleted at the rate σm, as expressed in (i), where σ is the concentration of "salt" in the discharge. It is known that $\sigma = \sigma(P, T)$ and, in fact, varies strongly with P, T; but for the moment we will shut our eyes to this and treat σ as a constant, so the total amount of "salt" extracts is $\sigma \rho_* V_1 M$.

Collapse of the gas column

The situation is actually somewhat more restrictive than has so far been described. The gas column cannot persist indefinitely. In order to maintain itself its internal pressure must at every depth exceed the water pressure in the adjacent rock, otherwise it will collapse and the gas phase is over. The inclusion of this feature does not greatly alter the total amounts transported for the more intense systems, so a very simplified approach is adequate.

Let the adjacent ground have an effective mean fluid density ρ_a. Then the adjacent ground pressure is

$$P_a = P_d + \rho_a g(z - d).$$

We then require simply that for the pressure at the base of the gas column $P_h > P_a$. If this condition is satisfied, in most cases the gas column pressure will at every level exceed that in the adjacent ground. Depending on the circumstances envisaged, we may choose ρ_a in the range 0.3–3 ton m^{-3}, namely from that of water-substance at the critical point to a lithostatic load.

Total mass flow

Data from a particular model shown in Fig. 8.8 give values of the total mass flow ratio $M(t)$. The initial mass extraction is rapid. In this

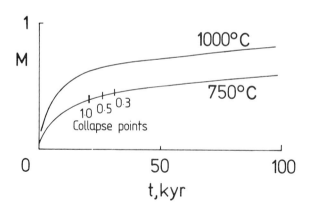

Fig. 8.8. Total mass flow ratio of water-substance, M, passed through gas column as a function of time, t, for case described in text.

example it takes about 1 Myr to transfer an amount $M = 1$, but already half of this is transferred in 5×10^4 yr. The ultimate rate for a long-life system without collapse is $M \sim \log t$, a continually decelerating rate.

Collapse points are indicated for various values of ρ_a. Even for the extreme case of $\rho_a = 1\,\text{ton}\,\text{m}^{-3}$, so that collapse is relatively early, the bulk of the possible mass flow has been transferred.

The mass of fluid available to flush out the source region is at most about equal to the mass of the source region. This result applies to a very wide range of systems — basaltic and granitic volcanism, extreme variation of permeability. Furthermore, this mass of fluid is largely produced in a relatively early part of the life of the system in an interval typically of the order of 10^4 yr. In other words, the extraction process is geologically rapid. We thereby deduce that the essential feature of the production of an orebody is the extraction of the material from a large volume and its deposition in a small volume.

Power extracted

Although our immediate concern is the total mass transport, it is of interest and provides a link with the surface zone (see later) to consider the power extracted from the source region (see Fig. 8.9). The power level during the first few thousand years is very high, being of the order of 10^4 MW, falling to the power level found in, for example, the Taupo systems after 5000–10 000 yr, thereafter remaining rather vigorous for an interval of the order 10^5 yr, through-

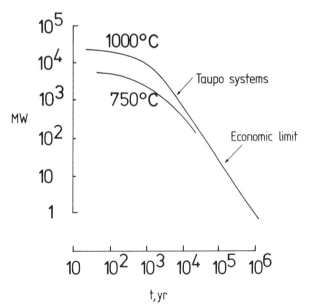

Fig. 8.9. Power, MW, transmitted in gas column as a function of time for initial source temperature of 750 and 1000°C.

out which time geothermal power exploitation would be economical with current technology, and after which the power is rather small. Note, however, that the initial high power levels are not available at the surface, since the bulk of this power is absorbed in heating up the ambient hot water and rock. A power of 10^4 MW is sufficient to heat a cylinder of rock of area 30 km^2 and depth 10 km in 10^3 yr.

Comment on the time-scale

The time-scale, of the order of 10^4 yr, suggested in this work is for a single hydrothermal event. This must not be confused with the time-scale of a geothermal or a volcanic area. Within an orogenic belt, for example, volcanism and plutonism may persist for times of the order of 100 Myr, but individual elements of such a system have much shorter lives. In the Taupo district, as observed at the surface and at depth from aeromagnetic surveys, it is possible to identify of the order of 10^2 rhyolitic intrusions produced over the last million years of activity, which suggests a rate of production of about 10^{-4} yr^{-1}. An extensive area of mineralization will have been produced by a multitude of individual short-lived events.

Comment on the gas column cross-section

It is important to point out that the model described here is a phenomenological model. The discharge is, for a given system and boundary conditions, determined by m/kS. In principle, k could be estimated for a given system by a detailed inspection of the deposit and its surroundings to determine the void geometry. But until such studies are made, k is virtually unknown, and the choice of 1 darcy is simply that found as an upper limit for Taupo systems as determined by pumping tests on bores about 1 km deep. This value is of the same order as that estimated from fracture abundance in the Mayflower mine Utah, 0.1—10 darcy (Villas, 1975). Even so, there remains the cross-sectional area, S. I have been unable to devise a satisfactory method of independently estimating S from the flow mechanics (for a wet system $S \sim$ (heated area)/40). At least S is more or less known from direct observation for some orebodies. Fortunately, however, the gross thermodynamics of the gas column is such that the total mass transport ratio, M, is determined largely from the thermodynamic boundary conditions. This cannot be said for the time-scale.

8.6. FLOW IN THE SURFACE ZONE

Above the gas column, within the wet surface zone, there will be a freely circulating body of liquid water. The mechanics of these wet hydrothermal systems are reasonably well known. Here we are concerned with only the grossest features in order to show how the deep and the surface systems are linked together and thereby provide further methods for identification of the whole system.

Characteristic temperature of surface zone

We have, of course, no direct evidence for any of the above description of the deep system. It is, however, possible to make a partial identification as follows. Let us regard the surface zone as a kind of calorimeter into which the heat transported in the gas column is deposited and from which heat is lost by the convective process in the surface zone. Then, by noting the amplitude and timescale of the surface zone temperatures and comparing these with field information, we can check our description.

As was shown by Elder (1966), the power F' transferred by a wet hydrothermal system, allowing for the variation of density and viscosity of liquid water with temperature, is related to the characteristic temperature θ of the liquid water (and rock) in the "mushroom" of the hydrothermal system by

$$F' \approx \xi \theta^4 ,$$

where ξ is a fixed quantity defined by the system geometry. For data from the Taupo hydrothermal systems we find $\xi \approx 0.2\,\mathrm{W\,K^{-4}}$. Equating F' to the power F transferred in the gas column allows us to calculate θ. Typical results are given in Fig. 8.10. The data suggest that individual Taupo systems with $F' \sim 10^3$ MW have been active for of the order of 10^4 yr, the whole area having been active very much longer, as is indicated by the most deeply exposed ignimbrites of age at least 10^5 yr. Older systems reach temperatures less than $100°$C in times of the order of 10^5 yr, comparable with the system of hot springs in the Hauraki area.

The evidence is flimsy, but is reasonably convincing.

Also shown are data for the heating up of the surface zone for various values of V_3/V_1, the ratio of the volume of the sub-critical convection region to that of the source. The heavy line represents the

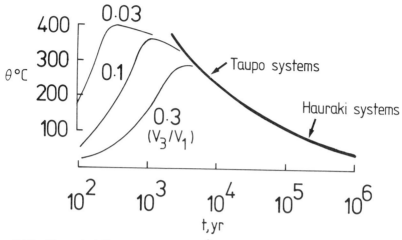

Fig. 8.10. Characteristic temperature, θ, °C, as a function of time, t, yr. Initial heating-up phase also represented for various values of V_3/V_1.

characteristic temperature for $V_3/V_1 \ll 1$, so that the surface zone is always in equilibrium with the gas column. Otherwise, the energy balance for the surface zone requires

$$\rho_* c_* V_3 \, \mathrm{d}\theta/\mathrm{d}t \; = \; F - \xi\theta^4 \qquad\qquad (j)$$

During the early life of the surface hydrothermal system the bulk of the energy supplied by the gas column is used to heat up the surface zone "calorimeter". After an interval, for example, of the order of 250 yr for the Taupo systems, the behaviour is no longer dominated by the thermal capacity of the surface zone and the characteristic temperature approaches that for an equilibrium system.

A remark on the early vigour of the surface zone

The time-scale of the early "calorimeter" phase of the surface zone system is roughly proportional to V_3/V_1. Thus, for the archetypal model, as is shown in Fig. 8.10, the time to reach maximum temperature is about $10^4 V_3/V_1$ yr. F falls little during this time. Smaller surface systems not only achieve their maximum temperature earlier but also reach a higher temperature. Not only do these temperatures approach those at the top of the gas column but they also approach BPD temperatures above which the surface zone is unstable to an adiabatic decompression, and phreatic eruptions arising from depths of as much as a few kilometres are possible. Such eruptions are a feature of the Taupo systems.

A note

In a schematic manner, I have shown how the operation of the upper mantle and trapping of magma drive a hydrothermal system. This point is a watershed in the book. Up to now the centre of attention has been the upper mantle and certain processes which occur in it. From now on attention concentrates on the operation of hydrothermal systems presented as distinct more or less isolated entities. To this point we have been, as it were, looking upwards from a viewpoint a few hundred kilometers inside the mantle. Now from the surface we look downwards a few kilometers into the crust.

Part III

HYDROTHERMAL SYSTEMS

In regions where magma is trapped below the surface and the ground is sufficiently permeable, convective circulation of ground water produces hydrothermal systems. These high-level crustal boilers may discharge hot water or steam at the surface to produce an intense geothermal area.

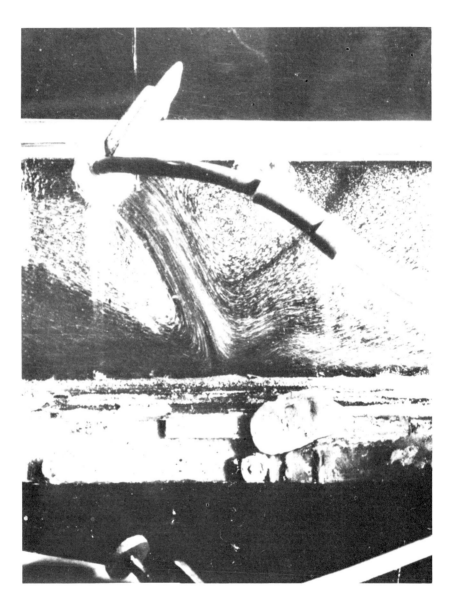

PLATE. Laboratory model of a hydrothermal system: view, looking horizontally, at a thin vertical layer of fluid contained between transparent vertical walls. The fluid is heated over part of its impermeable base, cooled over its entire upper impermeable surface, but is open to cold recharge fluid entering from reservoirs (to left and right) and is discharging fluid through a siphon attached near the surface. The equivalent porous medium Rayleigh number is about 10^3. Notice: (i) the broad flow of cold recharge fluid; (ii) the plume rising above the heated region of the base of the fluid; (iii) the lense shaped region attached to the head of the plume in which hot fluid is recirculated; and (iv) note the sharp boundary between the recirculated fluid and the recharge fluid.

9. Components

A hydrothermal system is a heat transfer mechanism in the earth's crust and upper mantle relying for its operation on the transport of water substance but not necessarily the discharge of water at the earth's surface, and producing at the surface an area (the so-called thermal area) in which the heat flow is *different* from normal. Within the thermal area there will be places where the heat flow is greater than normal but it is also important to note that in, and adjacent to, a thermal area there are often places where the heat flow is less than normal and perhaps zero.

The bulk of the phenomena exhibited by a hydrothermal system can be described merely in the terms of a hot water reservoir or deposit at depth. The total heat flow would be considered to arise solely from the energy stored in the reservoir, in which case the heat flow would be proportional to the discharge. The increase in heat flow produced by exploitation is simply due to increase in discharge. The natural heat flow would be determined simply by the rate at which fluid and heat could leave the reservoir, a rate dominated by the local topography and permeability.

9.1. THE ORIGIN OF THE WATER

In the natural state the water level is stationary and yet fluid is continuously removed from the reservoir. A recharge system must then exist. The recharge fluid may be meteoric, originate at depth as juvenile water, or be a mixture of both meteoric and juvenile water. Our argument closely follows White (1957).

Natural waters can be identified from their characteristic proportions of the isotopes of oxygen, ^{16}O and ^{18}O, and hydrogen, ^{1}H and ^{2}H ($= D$, deuterium). Let ξ be the isotopic ratio $^{18}O/^{16}O$ in a sample, and ξ_* the ratio in a standard. We write:

$$\xi/\xi_* = 1 + 10^{-3}\delta_O,$$

161

such that δ_O, in units of parts per thousand, represents the isotopic variation of oxygen in the sample relative to the standard. Similarly, we define δ_D for the isotopic ratio, D/H of deuterium to hydrogen. For natural waters a suitable standard, defined by (Craig, 1961) is "standard mean ocean water" = SMOW. The total natural variation of δ_O is about -50 to 10; the total natural variation of δ_D is about -300 to 50.

Meteoric water

Provided the water is freely able to circulate in the atmosphere—hydrosphere and does not become trapped in partially closed systems there is a distinctive relationship for all natural waters, namely

$$\delta_D \approx 8\delta_O + 10,$$

indicated in Fig. 9.1. The data shown has been obtained from water, ice and snow, in rivers, lakes, oceans and ice caps. The correlation reflects the variation of vapour pressure of the two molecular species — the vapour pressure of the molecules containing the heavier isotopes is lower and they tend to condense first. It provides a global isotopic label for meteoric water. In particular, any meteoric ground water supplied to a hydrothermal system will be marked by this characteristic relationship.

Surface waters depart from this relationship where they are strongly affected by evaporation, especially in closed basins, typical examples being waters of the Nile and East African lakes.

Ground water

There are broadly two types of water discharge at the surface; (i) near neutral or slightly alkaline, usually chloride-bearing, water, typically with pH of 6—8 or so; (ii) and so-called acidic waters, the acidity of which arises typically from superficial, near surface oxidation of, for example, H_2S.

The neutral discharge waters show a generally pronounced "oxygen shift" to higher values of δ_O compared to local meteoric water but with negligible "hydrogen shift". The shift is typically about 5 units. Some data are shown in Fig. 9.2. This behaviour arises because of oxygen exchange between the percolating water and the ambient rock. The water is enriched in ^{18}O. The rock is correspondingly

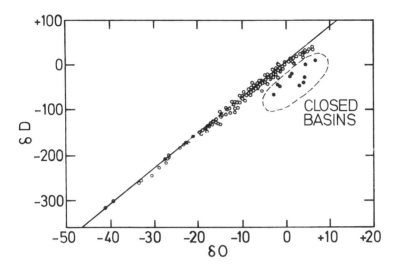

Fig. 9.1. Isotopic variation in meteoric waters, parts per thousand variation relative to SMOW, standard mean ocean water. (After Craig, 1963.)

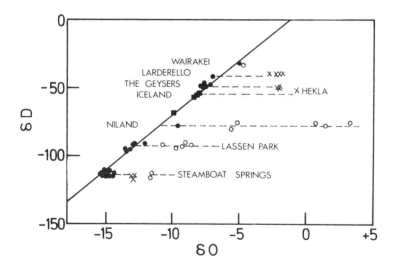

Fig. 9.2. Isotopic variation in near neutral thermal water: ●, local meteoric water; ○, hot spring water; ×, geothermal steam. (After Craig, 1963.)

depleted but in general this will be relatively small because of the large proportion of oxygen in the rock relative to the water — nevertheless reverse "oxygen shifts" in the rock have been measured (Clayton, 1963). Since, however, there is little hydrogen in the ambient rock compared to that in the percolating water there is little change in δ_D.

The evidence seems unequivocal that, within experimental error, the neutral geothermal water is meteoric.

The acidic discharge waters show both an "oxygen shift" and a "hydrogen shift" both to higher values than the local meteoric waters. Shifts reach 20, 60 units respectively. Some data are shown in Fig. 9.3. The data for a particular area lie on lines $\delta_D \approx 3\delta_0 +$ constant. The trend arises from the effects of evaporation below the

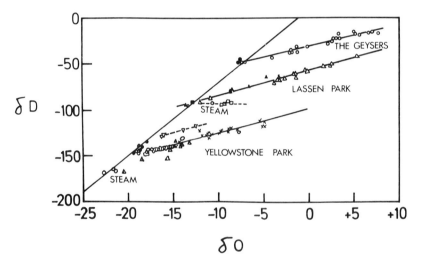

Fig. 9.3. Isotopic variation in acid thermal water: ●, local meteoric water; other circles, geothermal water.

boiling point. These lines have been duplicated in the laboratory simply by evaporating waters in pans (Craig, 1963). The effect involves purely meteoric water. Thermal springs and pools of this type must be heated by conduction or by minor amounts of deep hot chloride water or steam. (It is a pity that so much of the early work on hot springs, see for example Day (1939), was preoccupied with these superficial acidic springs, especially because of the mistaken idea that they provided the best samples of "magmatic water").

Possible juvenile water

Waters which have been in isotopic equilibrium with magmas would have values of δ_O of the corresponding rocks: granitic rocks 7 to 12; basic and ultrabasic rocks 6 to 7. Clearly these data for geothermal waters are incompatible with purely juvenile water. Suppose however, that the geothermal waters were a simple mixture of meteoric and juvenile water isotopic composition. This is not what we see; the lines are more or less parallel and certainly do not converge on a point somewhere along δ_O of 6 to 12. The accuracy of the method cannot rule out a small proportion of juvenile water, less than about 5%, but the position is most simply stated that while the quantitative evidence for geothermal water being solely meteoric is strong there is no quantitative evidence for the presence of any juvenile water.

9.2 EXISTENCE OF A HEAT (AND SOLUTE) SUPPLY

Since ground water has no excess energy or chemical content the recharge—discharge system will continuously flush energy and chemicals out of the reservoir till it is exhausted, unless it is replenished from depth. Consider a reservoir of volume V of water-saturated rock of porosity ϵ suddenly raised at $t = 0$ by temperature θ_i and given a salinity σ_i. Let a volume Q per second of cold nonsaline ground water pass through the reservoir and be discharged at density ρ to the thermal area, let the power supplied from depth by F, let mass rate of chloride supplied by Σ. Assume that the fluid in the reservoir is uniformly mixed and that the time-scales involved are small compared to the time-scale for transient conduction. Then at time t, the temperature, θ, and salinity, σ, of the reservoir are:

$$\theta = \theta_i \exp(-t/\tau) + F(1 - \exp(-t/\tau))/\rho_w c_w Q,$$

where
$$\sigma = \sigma_i \exp(-t/\tau') + \Sigma(1 - \exp(-t/\tau'))/(1 - \eta)\rho_w Q,$$

$$\tau = (1 - \epsilon)\rho_m c_m V/\rho_w c_w Q, \qquad \tau' = \epsilon V/(1 - \eta)Q,$$

and ρ_m, c_m are the density and specific heat of the saturated rock, ρ_w, c_w the density and specific heat of water and η is the proportion of the discharge which is steam.

Inserting field values for Wairakei: $\epsilon = 0.2$, $\rho_m c_m = 2\,\text{MJ m}^{-3}$, $\eta = 0.5$, and considering the reservoir to have a depth of 5 km and mean cross-sectional area A, we find fortuitously $\tau \doteq \tau'$, say, with $Q\tau/A = 2000\,\text{m}$.

(i) If Wairakei is considered as a separate system, estimating $A = 20\,\text{km}^2$, the area of the surface discharge, and $Q = 1\,\text{m}^3\,\text{s}^{-1}$ gives $\tau = 1250\,\text{yr}$.

(ii) If the entire Taupo zone is regarded as a single system, $A = 2500\,\text{km}^2$ and $Q = 5\,\text{m}^3\,\text{s}^{-1}$ gives $\tau = 3 \times 10^4\,\text{yr}$.

There can be little doubt that no gross changes in temperature have occurred at Wairakei in $10^3\,\text{yr}$ and substantial evidence exists that no gross change has occurred in $10^4\,\text{yr}$: Maori legends go back to 1200 AD; ash eruptions give dating layers calibrated by radio carbon from 800—400 years BP (Baumgart and Healy, 1956). Thus by (i), with $\tau = 1250\,\text{yr}$, the transient terms will be negligible for a life t probably in excess of $10^4\,\text{yr}$. Indeed as shown later the reservoir at Wairakei is probably $20\,\text{km}^3$ rather than the $100\,\text{km}^3$ assumed in (i), so that $\tau = 250\,\text{yr}$ and there can be no doubt that the transient terms are negligible.

For the Taupo zone as a whole there is geological evidence of activity existing for the last million years, although there is some evidence for a peak of volcanic activity 1.5×10^5 years ago (Grindley, 1965). Thus by (ii) again the transient terms are negligible. Further, if each of the Taupo systems are similar to Wairakei the total reservoir should be $500\,\text{km}^3$ rather than the assumed $1.25 \times 10^4\,\text{km}^3$, so that in (ii) again $\tau = 250\,\text{yr}$.

Thus at the present time, all the estimates of τ ($250\,\text{yr}$ is to be preferred) require the steady state.

If the system were not in the steady state, even though there was a large energy discharge this would not imply the existence of a heat supply since all the energy could come from that stored in the reservoir. But here with $t/\tau \gg 1$ the reservoir would be exhausted were it not provided with a heat supply.

9.3. THE SOURCE OF THE CHLORIDE

To put the problem of the source of the chloride in perspective, consider the volume of rock required to produce the chloride solely by leaching. Take a chloride discharge of $1\,\text{kg}\,\text{s}^{-1}$ for a mean life time of $10^4\,\text{yr}$. This is a total of 3×10^8 ton. The amount of chloride in rocks ranges from 10 PPM for some sandstones to 200 PPM for some shales and basalts, but is typically 100 PPM. With this value the volume required is $10\,\text{km}^3$; quite small.

Consider the schematic arrangement shown in Fig. 9.4. Two reservoirs of masses M_1 and M_2 are supplied with water. The total

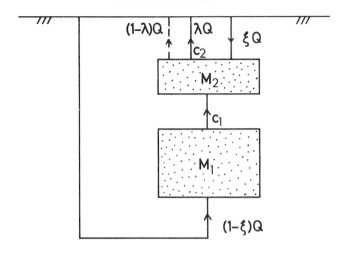

$$0 < \lambda < 1 \; ; \; 0 < \xi < 1$$

Fig. 9.4. Schematic diagram of chloride leaching model; two reservoirs 1 and 2 with total throughput Q.

discharge and recharge mass rate is Q, of which a proportion, ξ, passes only through the shallow reservoir and a proportion, λ, of the discharge is liquid water, the remainder being water vapour. Let the concentration of a potential solute in the reservoirs be C_1 and C_2. Take the initial concentrations in the reservoirs to be the same, C_0. For solutes of high solubility, assume that the concentration of the reservoir fluids is also C_1 and C_2. Then since solute material is conserved and taking the solute concentration in the recharge and the vapour discharge as zero we have:

$$dC_1/dt \;=\; -\alpha_1 C_1, \qquad dC_2/dt \;=\; -\lambda \alpha_2 C_2 + \beta \alpha_1 C_1.$$

so that

$$C_1/C_0 \;=\; \exp(-\alpha_1 t)$$

where

$$C_2/C_0 \;=\; (1-\gamma)\exp(-\lambda \alpha_2 t) + \gamma \exp(-\alpha_2 t;)$$

$$\alpha_1 \;=\; (1-\xi)Q/M_1, \qquad \alpha_2 \;=\; Q/M_2,$$

$\beta \;=\; M_1/M_2$, for the ratio of reservoir masses, and

$$\gamma \;=\; \beta(1-\xi)/[\beta \lambda - (1-\xi)].$$

Here, $0 < \lambda < 1$, $0 < \xi < 1$, and we restrict the discussion to $\beta \gg 1$, a larger deep reservoir.

A number of features are immediately apparent: see Fig. 9.5.

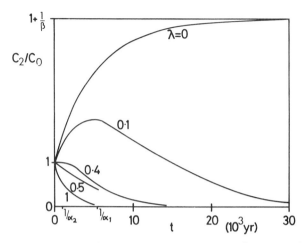

Fig. 9.5. Relative chloride concentration, C_2/C_0 as a function of time for a hypothetical case: $V_1 = 30\,\text{km}^3$; $V_2 = 10\,\text{km}^3$; $Q = 1\,\text{ton s}^{-1}$; $\xi = 0.5$; for various values of λ. Rock density $2.7\,\text{g cm}^{-3}$. The reservoir time scales, $1/\alpha_1 = 5100\,\text{yr}$, $1/\alpha_2 = 850\,\text{yr}$, are indicated. Note that with this ξ the curve for $\lambda = 0.5$ has a maximum (zero slope) at $t = 0$.

(i) $\lambda = 1$, no vapour discharge. Then, for $\beta > 1$ we have $\gamma > 0$ and both concentrations C_1 and C_2 diminish exponentially at rates determined by α_1 and α_2.

(ii) $\lambda = 0$, entirely vapour discharge. Then, $\gamma = -\beta$ and ultimately all the material is extracted from the deep reservoir and retained in the shallow reservoir with final concentration $(1 + \beta)C_0$, the time-scale being determined solely by α_1, that of the deep reservoir.

(iii) $\lambda = 1 - \xi$. There will be some intermediate value of λ for which the initial change of C_2 is zero. It is obvious by inspection that if $\lambda = 1 - \xi$ and initially $C_2 = C_2$, then initially the input of solute from reservoir 1 is balanced by the discharge of solute from reservoir 2.

The concentration in the deep reservoir monotonically diminishes, but for $\lambda > 1 - \xi$ that in the shallow reservoir has an initial rise to a peak value given by the time t' for which $dC_2/dt = 0$ namely:

$$(\lambda\alpha_2 - \alpha_1)t' = \log\left[-\lambda\beta(1 - \gamma)/\gamma(1 - \xi)\right].$$

If the potential solute in the upper reservoir is not shared by the rock and the water, but is preferentially held in the water (we assume material whose solubility exceeds any possible concentration in the fluid), then an upper limit to solute concentration is C_2/ϵ, where ϵ is the porosity of the reservoir.

Crude estimates of the above quantities can be made

(i) Wairakei, New Zealand. In the natural state we had, steam discharge $300 \, kg \, s^{-1}$, water discharge $600 \, kg \, s^{-1}$ (see elsewhere). Thus $\xi = 1/3$ and $\lambda = 2/3$. This is close to the critical condition $\lambda = 1 - \xi$, so that the concentration falls with ultimate time-scale $1/\alpha_1$, provided $\alpha_1 \ll \alpha_2$. Taking $V_1 = 20 \, km^3$ and $V_3 = 200 \, km^3$, the long-term time-scale is 2.8×10^4 yr. Assume that Wairakei is somewhat younger than this, then the concentration in the upper reservoir has not changed much. If the original concentration in the ambient rock was $100 \, PPM$ — possibly a low value in view of the measured $400 \, PPM$ in local rhyolites (Ellis and Wilson, 1960) — the spring discharge concentration of about $1000 \, PPM$ merely requires a porosity of 0.1.

(ii) Salton Sea, California. Deep bore water of this area carries about 25% of dissolved material, of constitution very different from the crustal average. Using the data given by Skinner et al. (1967) we find that many of the relatively insoluble elements are depleted compared to the crustal average but that several constituents are increased in concentration by factors: 1200, Cl; 30—50, Br, I, Ag; 6—9, Cs Pb, Zn, As; all others near unity or less. This suggests that the ambient rocks may have had a higher than average initial Cl, perhaps $1000 \, PPM$ or so and that a typical concentration ratio is 100. This requires $\lambda \approx 0$ and with a porosity of 0.1, say, a reservoir volume ratio of 10.

9.4. ORIGIN OF THE THERMAL ENERGY

Geophysical data on F are limited to: (1) the surface heat flow distribution, from which the value of F and a rough estimate of the horizontal length-scale of the system can be deduced; (2) a rough time-scale obtained by comparing different areas, such as the older, weaker Hauraki systems with the Taupo systems.

The total heat flow over the whole area affected by the hydrothermal system (and this area will be larger than the thermal area) may be normal or greater than normal. If the total heat flow is normal, the hydrothermal system is merely redistributing the normal heat flow which originates from depths down to the order of 100 km and is unaffected by a system penetrating only a few kilometres into the crust. But if the total heat flow is greater than normal an extra source of energy must be available from a thermal anomaly below the hydrothermal area. Such an anomaly can be produced by volcanism, the transport of rock or magma in the earth's crust or mantle. In principle all the possible types of hydrothermal system could occur in a normal area; the only justification for assuming that the heat flow of a particular thermal area arises from the thermal energy supplied by volcanism is that the total heat flow over the whole hydrothermal area is greater than normal. Nevertheless, it is rare to find the very intense thermal areas, where water is near 200—300°C and within 100 metres of the surface, except in regions of active surface volcanism.

We have now reached a different viewpoint from that of a reservoir, it is that of a system which is transferring heat from depth to the surface. At Wairakei this heat transfer occurs at the regional average of 40 times the rate at which the earth normally loses heat. This cannot be explained by a system which relies largely on conduction of heat: the temperature at the base of the Taupo depression, estimated from extrapolated bore data is near 400°C, compared with 200°C at the same depth of 5 km outside the depression; conduction alone will then give a heat flux twice normal, or if the base temperature were even 1000°C only five times normal. Clearly the heat is transferred by the flow of the fluid through the system, a hydrothermal system, and our search is for a flow mechanism of sufficient intensity.

10. Flow in an Extensive Medium

We can consider all hydrothermal systems as particular cases of the following simplified model which directs attention to three aspects of the system: recharge, heating, and discharge of the fluid.

(1)   Cold meteoric recharge water flows from the surface into the heating volume. This flow is maintained by a pressure gradient produced either by a physical difference in the levels of the recharge and the discharge areas (as in a cold spring) or by the pressure difference between the column of cold recharge water and the column of hot discharge water. In a steady system the mass flow rate of discharge water will equal that of the recharge. Notice, however, that in the case of a purely circulatory type of system a discharge or a recharge is not necessary, and for this case all the energy supplied to the system would be transported out of the system by conduction.

(2)   Within the heating volume the water is heated by energy supplied either by the normal flux of heat through the ground or from the energy of cooling magma injected by the local volcanism into the heating volume and its surroundings. The supply of energy may be sufficiently great to convert the recharge water to steam. Also, within the heating volume itself circulatory convective processes may occur.

(3)   Hot fluid (either liquid or gas) leaves the heating volume and passes to the surface. Near the surface, this fluid may mix with local, cold, near surface water or, in the case of steam leaving the heating volume, perhaps heating a (perched) body of surface water and then being discharged. This will be the most complex part of the hydrothermal system due to the variety of modes of discharge that are possible and their strong dependence on local conditions.

Two processes occur within a hydrothermal system: one is the transport of fluid through the system, the other is the heating of this fluid. Both of these processes are intimately connected, but when it

171

is desired to emphasize the transport process, or even more particularly the gross features of the entry of fluid into and the discharge of fluid out of the system, it is convenient to refer to the *discharge—recharge system*. On the other hand, when emphasis is placed on the heating process, reference is made to the *heating system*. That part of the discharge system which is directly affected by the presence of the ground surface is called the *surface zone*.

It is desirable to attempt to classify and discuss the great diversity of hydrothermal systems by means of a small number of features so that in a particular case the broad aspects of a system can be immediately discussed in terms of the selected features and, in addition, so that the remaining details may be clearly revealed as needing further investigation. Such an attempt has the extra merit of preventing the over-emphasis of features found in the study of a single system.

There is no reason why a classification of hydrothermal systems could not be based on considerations of the discharge—recharge system. However, the very great variety of possible discharge—recharge systems, especially in the surface zone, would lead to the need for a very complex and detailed classification. This complexity would lead to the need for a very complex and detailed classification. This complexity would tend to obscure the simple criteria to be derived merely from a consideration of the heating system.

The first and obvious feature is whether the fluid enters the discharge system wet or dry, that is, as water or steam. In permeable ground with a ready supply of water, the discharge will be wet, but in ground of poor permeability and high temperatures the discharge will be dry. More generally, this feature directs our attention to the thermodynamic state of the fluid at each point of its passage through the system.

The second criterion is not at all obvious; it is whether or not fluid is circulating by free convection in the porous medium and if such a free convective system of water in the porous rock of the earth's crust could transport sufficient energy to maintain an intense thermal area.

An important difference of emphasis will be found throughout the discussion. On the one hand, it is not only implied that a discharge actually occurs, but that the discharge is strong in the sense that we refer to *forced* convection. The fluid enters the system and passes directly through it once and the heat transfer mechanisms are dominated by this discharge-based flow. On the other hand, the emphasis is not at all on the discharge but rather on the continued circulation of the same volume of fluid; even if a discharge occurs, it is regarded merely as a disturbance to a system dominated by *free* convection.

10.1. PIPE FLOW AND DIFFUSE FLOW

Hydrothermal systems can be viewed in two extreme ways: the approximation of the system as a pipe or set of pipes, or as a diffuse flow in an extensive medium. At first these two view-points may seem to be in conflict, but in fact each merely emphasizes (and provides the simplest description of) certain important features.

Firstly, we look only at the gross features of a hydrothermal system, ignoring the details of the velocity or temperature distribution. If a local concept is introduced it is immediately replaced by a gross property, for example, the use of the total resistance rather than the permeability. The hydrothermal system is regarded merely as a pipe, rather like the pipe of a heat exchanger, through which fluid flows and is heated, but in which little consideration is given to details within the pipe; and even so interest is centred on variations along the pipe rather than over its cross-section.

Secondly, to proceed further it is necessary to consider that the hydrothermal system is produced not by the confined flow in a pipe but by a distribution of flow and temperature which occurs in an extensive medium (generally referred to as a porous medium). This medium is the partly porous rock. To describe it quantitatively requires two quantities: the thermal conductivity (conductivity to flow of heat), K; and the permeability (conductivitiy to flow of matter), k. Both of these quantities are assumed to vary in space almost everywhere and continuously throughout the medium. A specification of the nature of a hydrothermal system will then require, as well as a knowledge of the equation of state of water (and its viscosity), a statement of the distribution of K and k throughout the medium. It will then be assumed that, provided the temperature distribution and the mass flux are given on the boundary of the flow system, the behaviour of the flow system is determinate. The flow system can then be formulated mathematically by four partial differential equations: the continuity, momentum, energy, and diffusion equations. Except for the simplest non-isothermal conditions these equations are difficult to solve analytically but can be conveniently solved with scale models in the laboratory, or numerically.

It will be seen that the pipe-approximation is merely a special case of the diffuse-approximation in which either because of the extreme non-homogeneity of the permeability distribution or because of the extreme nature of the flow itself in the porous medium, the flow paths are largely localized to rather restricted or pipe-like zones.

Properties of the medium

Thermal conductivity

The thermal conductivity of most rocks is in the range 1—$2\,\mathrm{W\,K^{-1}\,m^{-1}}$, a value determined quite accurately by the mineral composition so that for related rocks the variation in thermal conductivity is small. Here we consider the net conductivity of the water-saturated porous rock K_m, a crude estimate of which can be obtained from a knowledge of the rock conductivity K_*, the water conductivity K_w, and the porosity ϵ (the void space/unit volume) by:

$$K_m = (1 - \epsilon)K_* + \epsilon K_w$$

Since K_w for water is about $0.6\,\mathrm{W\,K^{-1}\,m^{-1}}\,s$ and ϵ is order 0.1, $K_m \doteq K_*$. Representative values of K_m are difficult to determine to better than $\pm 20\%$ but any error here is insignificant compared with the large uncertainty in permeability.

Permeability

When a fluid permeates a porous rock the actual path of an individual fluid particle cannot be followed analytically, but the flow can be represented by a macroscopic relation discovered by Darcy (1856). If the mean volume of flow is q per second of a single-phase fluid of viscosity μ across unit area under a pressure gradient dP/ds where P is the pressure at the point s along a spatial mean streamline:

$$q = -\frac{k}{\mu}\frac{dP}{ds}.$$

This relation involves the permeability, k, a property of the rock and its voids (but not of the fluid in the voids). The original unit of permeability is the *darcy*, defined with all quantities measured in the c.g.s. system, except that viscosity is expressed in *centi*poise and the pressure in atmospheres. Thus 1 darcy $\doteq 10^{-12}\,\mathrm{m^2}$.

In so far as hydrothermal systems are dominated by the permeability, this book is a study merely of the thermal phenomena which can be understood in terms of the permeability concept.

The range of validity of the permeability approximation

The flux q is an average over an area sufficiently large so that enough void paths pass through it to give a meaningful average. Hence, a permeability is definable provided the masses of fluid involved in the flow system have a length-scale which is large compared with the scale of the porous structure, the joints or fissures of the rock. For the hydrothermal systems found, for example, in the 5 km deep Taupo depression it is obvious that the length of the porous structure of order 1 mm, or that of the joints of order 1 m are compatible with the proviso, but it may not appear so obvious that fissure structures with length-scales up to 1 km can also be dealt with by the permeability approximation.

In a porous medium which has a variety of different types of void paths, of widely varying length-scale, the permeability appropriate to the operation of a flow system or a part of a system will depend on the size of the system. Systems of extent rather less than the length-scale of the joint structure will rely on the permeability arising from the porous structure, rather larger systems will rely on the permeability of the joint structure, and if very large systems can exist they will be dominated by the permeability established by the fissure structure. Of course, a system may be pierced by a single element of the larger-scale porous structure. For the operation of that system the element cannot be treated by the permeability approximation, rather it must be considered as a pipe or channel embedded in the surrounding porous medium of the system. The approximation is also no longer valid nearer to a boundary surface than a distance about equal to a length-scale of the permeable structure.

Darcy's relation expresses the linear relation between the flux of fluid and the local pressure gradient. This linear relation becomes invalid when the flow velocities are high. More precisely, when the Reynold's number

$$R_m = q\Delta/\nu,$$

based on the effective grain size Δ of the porous medium, the flux velocity q and the kinematic viscosity of the fluid $\nu = \mu/\rho$, exceeds about 10 the linear relation is replaced by a quadratic law equivalent to

$$1/k \propto R_m$$

(Muskat, 1937). The origin of this effect was made clear by the interesting observation that the relation between the velocity and pressure gradient for a porous medium such as a bed of glass spheres, and for a tube in which there were spaced along it annuli normal to the flow, was the same. Then by simple observation of the flow in the tube (and with less clarity in the bed of spheres) one sees that the change in the form of Darcy's relation occurred at $R_m \doteqdot 10$ when the fluid in the voids of the porous medium became turbulent. For systems considered here, such as those in the Taupo depression, Darcy's relation will be valid except in the immediate vicinity of bores and the most active hot springs (in the hydrothermal system itself, e.g. Wairakei with $\Delta \doteqdot 1$ mm and a maximum $q = 10^{-4}$ cm s^{-1} (see below), R_m will be found to be everywhere less than 10^{-2} and only exceed 10 within a distance of 1 m from the most active bore). It should be noted, however, that the permeability approximation is still valid even if the void flow is turbulent.

Measurement of permeability

Measurements of permeability have been made with an apparatus in which rock samples were gripped in a polythene tube and the flow for a given pressure drop across the sample was directly measured. The working fluid was water. About thirty samples of rock typical of the Taupo area were measured, including rhyolites, ignimbrites, breccias, tuffs, and sandstones, and the rock permeability k_r found to be of order 1 millidarcy.

It has been possible to obtain some field measurements from bore holes at Wairakei. If the water level in a non-discharging bore is, for example, raised by pumping in cold water, measurements of the pumping rate against the steady level reached or by reading the level as a function of time after pumping has ceased allow the permeability of the rock surrounding the uncased portion of the bore to be calculated. Even though the data available were collected for other reasons and are often not very reliable, all the bores examined give a permeabilaity of $k_m = 0.05$—0.5 darcy. This is further confirmed by consideration of the output characteristics of discharging bores.

The conclusion cannot be escaped that, certainly for the upper 1 km of the Taupo area, the contribution of the porous structure of the rock elements themselves to the permeability is negligible; rather, the permeability arises from the flow in the joints and fissures of the rock. It is possible to show that this conclusion is not at all surprising. An order of magnitude calculation of the permeability of the joint structure $\tilde{k}$ can be made by considering a system in which the rock is a cubic array of blocks of side Δ_1 with a narrow space of width Δ_2 separating them. If the flow in these spaces is laminar it is easily shown that:

$$k_m = k_r + \tilde{k} = k_r + \Delta_2^3/6\Delta_1 .$$

We also write:

$$\epsilon_m = \epsilon_r + \tilde{\epsilon} = \epsilon_r + 2\Delta_2/\Delta_1 .$$

where $\tilde{\epsilon}$ is the porosity contributed by the joints.

For example, in a small-scale structure with $\Delta_1 = 1$ m and $\Delta_2 =$ 0.1 mm, $k = 0.2$ darcy, very much larger than that of the rock itself at Taupo and more like the *in situ* values. A similar permeability is found for a larger-scale structure with $\Delta_1 = 100$ m if $\Delta_2 = 0.5$ mm, or more impressively with $\Delta_1 = 1$ km if $\Delta_2 = 1$ mm. In these examples the channel Reynold's numbers are 0.5, 50, 500 for the Wairakei flow of 10^{-4} cm s^{-1} and all below the value of order 10^3 at which the channel flow becomes turbulent.

These dimensions are of the same orders of magnitude as revealed by visual inspection of extensive rock exposures. It is certainly not necessary to invoke exclusively the large "fissures" of popular conception (in which Δ_2 is of order 1—10 cm) with their consequent very large permeabilities (of order $10^2 - 10^5$ darcy). Undoubtedly such fissures exist, especially near the surface, and they will have a tremendous effect on the flow near them, an effect of considerable importance in exploiting the area; however, a permeability of 0.1 darcy is more than adequate to permit all the hydrothermal phenomena in a thermal area. Thus the permeability will be more related to the physical condition of the rock, now and also during the time it was laid down, than to its petrology. In particular, the permeability may be expected to be large in volcanic debris and flows, both due to the chaotic laying down and the cracking due to differential cooling. Further, the permeability will decrease with depth due to the increasing collapse of voids under compression.

Darcy's relation in two-phase flow

When the fluid is in two phases the flow of one phase is affected by the presence of the other phase. The flow rates of each phase are correspondingly reduced.

By analogy with Darcy's relation for single-phase flow we define the relative permeabilities of liquid water, F_f and gas (water vapour plus any degassed volatiles), F_g such that:

$$m_f = kF_f(p' - \rho_f g)/\nu_f,$$
$$m_g = kF_g(p' - \rho_g g)/\nu_g,$$

where m_f, m_g are the mass fluxes of the water and gas. For steady co-current vertical flow the total mass flux is:

where
$$m = m_f + m_g = k(\alpha p' - \beta g),$$
$$\alpha \equiv (F_f/\nu_f) + (F_g/\nu_g),$$
$$\beta \equiv \rho_f(F_f/\nu_f) + \rho_g(F_g/\nu_g).$$

Rearranging:

$$p' = (\beta g + m/k)/\alpha.$$

In horizontal flow the role of gravity is zero. Hence we have the above relations with $g = 0$. Thus for example the pressure gradient:

$$p' = m/k\alpha.$$

For steady horizontal radial flow in an annulus of thickness b, we have the constant mass rate

$$M = 2\pi \, rbm.$$

Hence writing

$$y = \log(r/r_2)$$

we have

$$dp/dy = m\alpha.$$

In the case of single-phase flow of liquid water $F_f = 1$ and $\alpha = 1/\nu_f$, so that there is the familiar isothermal solution

$$p - p_2 = [(m/\nu_f) \log(r/r_2)].$$

We notice that the equations of motion involve:

(i) quantities obtained from the properties of water-substance;

(ii) the single independent parameter m/k, a suitable dimension-less form could be $\sigma = m\nu_0/gk\rho_0$, where $\nu_0\rho_0$ are values for liquid water at some standard temperature (say $20°C$).

Clearly F_f, F_g are functions of the void geometry and the relative volumetric proportions of liquid water and gas. For the purpose of discussion, I use the simple relations:

$$F_f = e^3, \qquad F_g = (1 - e)^3,$$

where e is the volumetric liquid fraction.

Properties of the fluid

The most important constraint imposed on the behaviour of the system is the equation of state of water. This is represented in Fig. 10.1. The main feature of this relation is that: above a temperature $T_c = 374°C$ (the *critical* temperature), regardless of the pressure, the fluid is a gas; above a pressure $P_c = 0.215$ kbar, the fluid is either liquid or gas; below P_c, the fluid can be liquid, a mixture of liquid or gas, or a gas (vapour).

First consider a deep system of depth greater than the head equiv-alent to P_c(~ 2 km). A number of possible thermodynamic paths are sketched for an initially cold fluid particle which enters the system at

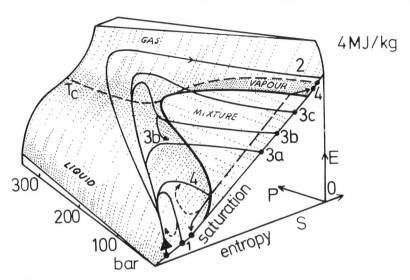

Fig. 10.1. Equation of state of water-substance: pressure, P; enthalpy, E; entropy, S. Various possible thermodynamic paths through a single-pass pipe-like system are indicated.

atmospheric pressure, penetrates the system and is discharged again at atmospheric pressure:

Path 1.   If the maximum temperature reached by the fluid particle T_{max} is small, the fluid particle remains throughout in the liquid state.

Path 2.   If T_{max} is large (rather larger than T_c), the fluid particle remains liquid as it passes through the recharge system and into the heating volume until it is heated to T_c, when it wholly converts to gas. It will be further heated before being discharged as a gas, cooled somewhat by expansion during its ascent through the discharge system.

Path 3.   Between these two extremes the discharge will be a mixture of liquid and vapour:

(a) If the temperature T_{max} is rather low, the fluid will remain a liquid until it begins to flash near the surface.

(b) For a somewhat higher temperature, the fluid may briefly be heated above T_c and become a gas.

(c) If T_{max} is sufficiently high, the fluid may not again pass through the liquid state before changing into a mixture of liquid and vapour as it approaches the surface.

Path 4. Regardless of the water temperature, evaporation can occur at an interface between water and steam. Whereas paths 1—3 refer to a fluid which does not have its distinct phases separated, path 4 does involve the accumulation of the steam in a distinct part of the system from that of the water. The process is different from that of path 2 where at any point the fluid is either entirely water or entirely steam; unless other energy sources are available the water enthalpy is sufficient to evaporate only a small proportion of steam; path 4 is thus more akin to path 1. The consequences of path 4 will be left till the discussion of steaming ground.

The possible paths for a shallow system (depth < 2 km) are similar to those above, except for the simplification that the sequence — liquid, mixture, gas — is maintained.

Finally, this simple picture will be obscured in practice because in a given discharge all the particles have not traversed the same thermodynamic path; in particular, it ignores the mixing of the discharge with near surface ground water.

10.2. FREE CONVECTION

The detailed features of a hydrothermal system can be discussed quantitatively by considering them to arise from the diffuse flow of the water in a medium defined by its conductivity and permeability, both of which are (almost everywhere) continuous functions of position in the medium.

Consider the simple system shown in Fig. 10.2. A "pot" contains

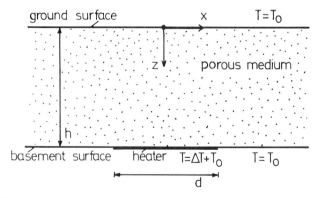

Fig. 10.2. Arrangement for study of free convection in a porous medium.

a horizontal homogeneous slab of large extent: uniform thickness h, permeability k and conductivity of the saturated rock K_m ; it is saturated with water of density ρ, coefficient of expansion γ, specific heat c and kinematic viscosity ν; and is placed in a gravitational field of acceleration g. The temperature is uniform at the ambient temperature T_0 over the boundary except for a portion, the heater, of diameter (or width) d and area $A_1 = \frac{1}{4}\pi d^2$ where it is elevated to $(T_0 + \Delta T)$. It will be possible to express any feature of interest in terms of these quantities. For the moment, let the net mass transfer be zero.

Certain simplifications can be made by assuming the Boussinesq approximation that density variations are only important in producing buoyancy forces.

(1) We take $\rho = \text{constant} = \rho_0$, except in the buoyancy combination (uplift/unit mass) $\gamma g \Delta T \equiv g \Delta\rho/\rho_0$.

(2) The permeability and viscosity occur only through Darcy's equation in the combination k/ν.

(3) The conductivity and specific heat arise in the equation of conservation of energy and occur in the combination $\kappa_m = K_m/\rho c$, where ρc is the thermal capacity of water (heat is advected solely by the water). This is the effective thermal diffusivity of the water-saturated rock.

(4) Except for the buoyancy term these parameters are taken to be independent of temperature; no serious difficulty arises from this in the laboratory, but in the field where variations of ρ and particularly ν are large, these variations must be taken into account.

Hence the system can be specified by: κ_m, k/ν, h, d, $g\Delta\rho/\rho_0$. These five quantities involve only length and time; there will therefore be three dimensionless parameters which are sufficient to specify the system. A convenient set is:

$$d/h, \qquad \nu/\kappa_m, \qquad A_m = g\frac{\Delta\rho}{\rho_0}kh/\kappa_m \nu;$$

a geometrical ratio, the Prandtl number (more or less a constant for liquid water systems) and the Rayleigh number. The Rayleigh number is simply a dimensionless density difference.

Heat transfer

The first aspect of interest is the ability of the system to transfer power. If the power entering the slab through $z = z_1$ is F_1 it is convenient to define a dimensionless heat flux N, the Nusselt number by:

$$F_1 = NK_m A_1 \Delta T/h.$$

In the steady state, as here, the power F_0 leaving the slab through $z = z_0$ is such that $F_0 = F_1$.

The Nusselt number has been studied in the laboratory over an extensive range of each of these parameters; only the broad features concern us here. The experimental results which are summarized in Fig. 10.3 reveal three fairly distinct regions.

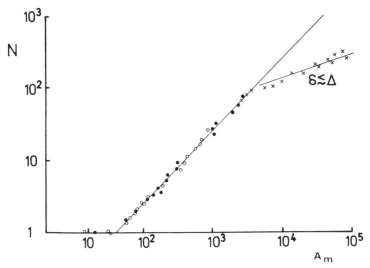

Fig. 10.3. Heat transfer characteristic for free convection in a porous medium. Nusselt number, N, as a function Rayleigh number, A_m. Laboratory model data.

(1)   At small A_m, the heat transfer is dominated by conduction. Thus for $h \ll d$, $N \to N_0 = 1$ as $A_m \to 0$.

(2)   As A_m is increased, there is a smooth but relatively rapid change near a particular value of A_m to a relation

$$N/A_m = 0.020 \pm 10\%.$$

Notice the simple nature of this result. In dimensional form the heat transfer is independent of both h and K_m.

(3)   At higher values of A_m, effects occur which are specific to a laboratory-size model. When the thermal boundary layer thickness on the heater $\delta_1 = h/N$ approaches the grain size

Δ of the porous medium the flow near the heater is no longer dominated by the porous medium — indeed near the heater the fluid behaves as if the porous medium was not present. Since $d/\Delta \gg N$ in geothermal areas this effect is of no importance there.

Stability

For the case $d \to \infty$ there is the possibility that the fluid can remain in stable hydrostatic equilibrium even if $A_m > 0$. Lapwood (1948) has shown that this is the case provided.

$$A_m \leqslant A_{mc} = 4\pi^2 .$$

In general, regardless of the geometry, for $A_m \lesssim 10$, convection is very weak and $N \approx 1$.

Temperature and velocity distribution — the mushroom

Typical experimental distributions of temperature and velocity are shown in Fig. 10.4. Throughout the bulk of the fluid gradients are small; but in the fluid adjacent to the heater, in the hot column which rises above it and near the surface, gradients are large, characteristic of boundary-layer phenomena. It is seen that relatively cold fluid is horizontally incident upon the heater, where it is heated by conduction through the (thin) layer between it and the heater, whence both the convergence of streamlines and the increase in buoyancy due to rising temperature produce a vertical jet or plume of rapidly moving hot fluid. The plume spreads only a little as it rises, since the lateral heat loss will be relatively small. As it approaches the upper surface, and it spreads out over the surface, slowing by increase in extent and cooling by conductive loss to the surface. This produces a near surface maximum in the temperature distribution with depth except near the top of the plume. Near the heater and the surface, and to a lesser extent in the plume (which is largely produced by convergence) it is the simultaneous presence of a moving stream and the relatively weak conductive heat transfer which maintains the large gradients. Outside these regions the cool fluid merely takes up a weak induced motion.

Above the heater, then, rises an increasingly bulbous volume of hot water to be called "the mushroom". It is this mushroom which will be sought by geothermal power engineers and the existence of which may lead to thermal areas being found at the surface.

GEOTHERMAL SYSTEMS

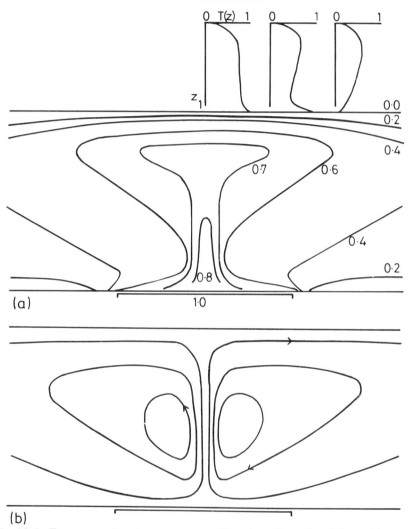

Fig. 10.4. Free convection in a porous medium, two-dimensional flow: (a) temperature distribution, with three temperature profiles; (b) streamlines, $A_m \approx 10^3$.

The relative intensity of the gradients increases monotonically with the Rayleigh number.

Independent changes in depth of fluid, h, do not change the power transfer nor the distributions near the heater, in the lower portion of the plume or near the surface, but rather move the bulb of the mushroom up and down — the greatest change being in the form of the surrounding induced flow.

Discharge

The second aspect of major interest is the ability of the system to produce a net transfer of matter in a surface discharge. Provided the surface is permeable and there is access to water, such a steady discharge will occur if the pressure is lowered over a portion of the surface or throughout a volume with access to the surface. The consequent flow in the medium can be considered as an interaction between the flow due to the discharge alone and the flow due to the free convection.

Fig. 10.5 shows these effects. The effect of the discharge alone is shown in the isothermal velocity distribution for a system in which the discharge is produced by lowering the pressure an amount ΔP inside a volume V and conducting the discharge to the surface in a

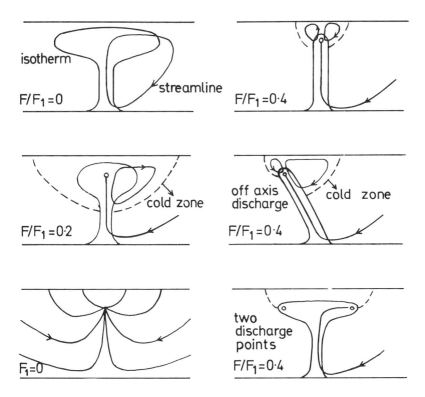

Fig. 10.5. Free convection in a laboratory hydrothermal system with mass discharge and recharge. The dot shows the discharge point. Influence of discharge on the temperature and velocity distribution, for various power inputs, F, and power outputs, F_1.

pipe. For a discharge system the volumetric discharge is

$$Q \propto k V^{1/3} \Delta P / \mu,$$

proportional to ΔP. In nature, ΔP, determined by the local top-ography, will be of order $1-10$ bar. Although the form of the iso-thermal velocity distribution is independent of Q, and although the more rapidly moving fluid is confined to distances of a few length-scales from the discharge volume, as Q is increased the flow velocities become comparable to those of the free convective system and an increasingly extensive interaction occurs.

The most striking effect is the gradual collapse of the mushroom bulb. This is shown at various values of F/F_1, where F is the power transmitted by the discharge and F_1, is the power input, at each value the system being allowed to reach a steady state. At small values of F/F_1 the temperature distribution is hardly altered from that of simple free convection with $F = 0$, but at larger values the flow, especially near the surface and in the vicinity of the mushroom bulb, it becomes more and more dominated by the discharge. Notice especially the increasing zone of near zero surface thermal gradient outside the mushroom. Also, in increasing proportion, colder water from the upper part of the layer enters the discharge and the tem-perature falls. The temperature of the discharge is plotted in Fig. 10.6 as a function of F/F_1 at fixed h, obtained by the model experi-ments with $N = 20$. Thus, although the power output of a thermal area can be temporarily (for the order of 10^1-10^3 year) increased by boring into the store of hot water in the mushroom, so that

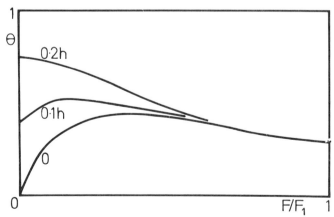

Fig. 10.6. Discharge temperature as a function of F/F_1 the ratio of discharge power to input power, for various discharge point depths 0, $0.1h$, $0.2h$. Model experiment with $A_m \approx 10^3$, $N \approx 20$.

temporarily we may have $F \gg F_1$, as the mushroom collapses and as the system moves to its new steady state the power output will return to the natural value but at a lower discharge temperature.

With suitable F, the mushroom can be collapsed, regardless of the location of the discharge point, indeed in nature it may often be at a considerable distance from the top of the plume. This fact has tremendous importance for the exploitation of a thermal area for power generation, since a power project will operate at values of F/F_1 which will collapse the mushroom, regardless of the location of the bores, which should therefore be located by considerations of local permeability and topography rather than any necessary attempt to locate them directly over the upwelling.

Throughout all these changes (at fixed A_m) the Nusselt number, N, remains constant at $0.02A_m$. This is because, even when the velocities produced by the discharge are sufficient to completely collapse the mushroom, they are still relatively small at the heater, and the fall of temperature near the lower surface has no effect since the recirculated fluid is already cold when it is incident on the heater. This is the most striking demonstration of the independence of the details of the heat transfer processes near the heater and near the upper surface. From this point of view, *the discharge process is merely a feature of the heat transfer process near the upper surface* and does not affect the processes whereby the energy enters the system and when $F \doteqdot F_1$, the system can be regarded, in the pipe-approximation which ignores the details of how the heat enters the system, as a forced convective system. Because of this independence there will be considerable profit in comparing laboratory and geophysical data of the bulb of the mushroom even though details of the heater may be unknown.

10.3. PIPE MODEL OF THE RECHARGE–DISCHARGE SYSTEM

It is very useful to have a simple quantitative method for describing aspects of hydrothermal systems. The commonest and simplest system is that of a liquid filled system in which the bulk of the fluid passes once through the system. This suggests that a "plumber's" model analogous to a domestic hot water system would be appropriate. In a steady state such systems are controlled by pressure differences and resistance to flow of the fluid.

The *gross* features of a quasi-steady hydrothermal system can be discussed quantitatively by replacing it by an equivalent pipe-like

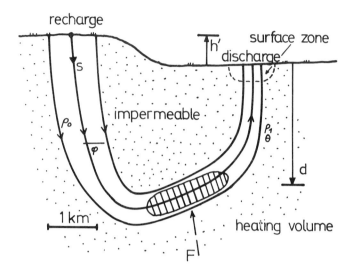

Fig. 10.7. Pipe model of a flow-dominated hydrothermal system.

structure, rather like a thermal syphon (Fig. 10.7) through which the fluid passes only once and by evaluating the equations of mass conservation, energy conservation, momentum and state. For a given physical arrangement of the system and a given energy input, it is desired to derive the consequent mass discharge and its enthalpy.

The mass flow per sec, M, across a particular cross-section of the pipe is the same at every cross-section of the flow and is also equal to the mass discharge per second — it is assumed that fluid neither enters nor leaves the system except at the surface.

If F is the power supplied to, or by, the heating volume to heat this fluid, ignoring for the moment the generally small power loss by conduction from the heating volume and discharge system to the surface, the mean fluid enthalpy, E, can be defined in terms of the energy flow EM at any cross-section in the discharge system by:

$$F = EM.$$

Darcy's relation for the flow of the fluid through the system can be written, allowing for gravity:

$$dP/ds = \rho g \sin \phi - \nu M/kA,$$

where A is the cross-sectional area of the pipe and ϕ is the angle between the horizontal and ds. Integrating completely around the system $(\int_s)$ from recharge surface to discharge surface, so that $\int_s dP = 0$,

$$\Delta P = gMR,$$

where

$$\Delta P = g \int_s \rho \sin \phi \, ds,$$

and

$$R = \int_s \nu \, ds / kAg.$$

R is the *resistance* of the system.

These relations can be re-written,

$$M = \Delta P / gR, \qquad E = gFR / \Delta P.$$

Thus given R, ΔP, F, the mass discharge and enthalpy are defined. However, because both ρ, ν are functions of temperature and pressure it will be necessary in general to use a process of successive approximation to solve for M, E.

A very simple model system is one for which the density is everywhere ρ_0 in the recharge (descending) column and everywhere $\rho_1 (< \rho_0)$ in the discharge (ascending) column. Then where h is the depth of the system and h' is the excess height of the recharge area above the discharge area, except for marginal or low temperature springs, where generally ΔP is close to zero, the term $\rho_0 gh'$ is much smaller than $(\rho_0 - \rho_1)gh$. For example, in a system such as that at Wairakei, $(\rho_0 - \rho_1) \doteqdot 0.2$ ton m^{-3} for water at $250°$C in the discharge system; hence the pressure $(\rho_0 - \rho_1)gh$ for a possible $h = 5$ km is equivalent to a 1 km head of water — much larger than local topographical variations. Hence, in our present model, place $h' = 0$, so that

$$\Delta P = (\rho_0 - \rho_1)gh,$$

$$M = \left(1 - \frac{\rho_1}{\rho_0}\right)\frac{1}{r},$$

$$E = Fr \bigg/ \left(1 - \frac{\rho_1}{\rho_0}\right),$$

where

$$r = R / \rho_0 h.$$

Notice that r is independent of h, M, and E and can be evaluated for any particular flow path using values of ρ given in the thermodynamic tables. This has been done (Fig. 10.8) for the case of a heating volume at a depth corresponding to $2P_c$ (i.e., 4 km) and in which the greatest part of the resistance R is established within the discharge column (i.e., $R_0 \ll R_1$), so that the mean pressure in the discharge column is P_c.

A number of features are apparent. Firstly, with high energy input or large resistance, the discharge is not large enough to transport the

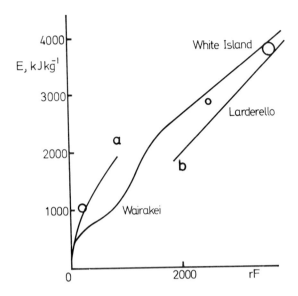

Fig. 10.8. Enthalpy of discharge from a pipe-like system with $h = 4$ km. Data from Wairakei, Larderello, White Island.

energy by hot water and the discharge is gaseous. Thus $\rho_1 \ll \rho_0$ and

$$M \approx 1/r, \qquad E \approx Fr.$$

Notice that these relations are linear in M and E, and that M is not (directly) a function of F but is merely determined by the head of cold recharge water. Secondly, at low energy inputs or small resistance the water will remain in the liquid state. Thus from

$$(\rho_0 - \rho_1) = \gamma \rho_0 E/c$$

we have

$$E^2 = cFr/\gamma, \qquad M^2 = \gamma F/cr.$$

Hence the relations are quadratic in E and M. The relations provide an excellent approximation provided the water is everywhere liquid.

At Wairakei the natural discharge was equivalent to a net 1 ton s^{-1} at 250°C (the net flow which enters the 400 m surface zone). Hence assuming that all the resistance arises in the discharge system of cross-sectional area A, with: $\rho_1 = 0.8 \text{ g cm}^{-3}$, $\nu_1 = 1.4 \times 10^{-7} \text{ m}^2 \text{ s}^{-1}$, evaluated at pressure of 0.2 kbar, $k = 0.1$ darcy, we have $r = 2 \times 10^{-3} \text{ m kg}^{-1} \text{ s}^{-1}$ and hence $A = 0.7 \text{ km}^2$. This area is much smaller than that of the thermal area. The model experiments show it to be a good estimate. The point corresponding to Wairakei lies close to the curve for a wet discharge.

Equivalent circuit

A simple formalism can be created as follows.

Flow rate, Q

Taking a reference density, $\rho_0 = 10^3 \, \text{kg m}^{-3}$ (corresponding to that of cold liquid water near 1 bar and $0°C$) we can define the "equivalent" volumetric flow rate, Q, such that $M = \rho_0 Q$. If the actual flow rate is Q' at fluid density ρ' then

$$\rho_0 Q = \rho' Q'.$$

Head, Δh

If this flow occurs because of a pressure difference Δp between the inlet and outlet of the system, we can similarly define the "equivalent" head, Δh such that $\Delta p = \rho_0 g \Delta h$. If the actual difference of fluid head is $\Delta h'$ at fluid density ρ' then

$$\rho_0 \Delta h = \rho' \Delta h'.$$

Resistance, R

In slow flows through a permeable system the flow rate is proportional to pressure difference so that we can define a flow resistance,

$$R \equiv \Delta h / Q = \Delta h' / Q'.$$

Note that here R is defined in terms of equivalent head and equivalent volumetric flow rate. The expression is analogous to that of Ohm's law for the flow of electricity with Δh in the role of potential difference (volts), Q as electrical current (amperes), R as resistance (ohms). To emphasize this analogy, where appropriate I use the unit symbol $w \equiv s \, m^{-2}$ for units of flow resistance.

The resistance of a flow tube of length l, cross section area a, and permeability k is from, Darcy's relation, $R = \nu l / kga$. As an example consider: $k = 10^{-12} \, m^2$, about 1 darcy, $l = 1 \, km$; $a = 1 \, km^2$; $\nu = 2.94 \times 10^{-7} \, m^2 \, s^{-1}$ at $100°C$ for which $R = 30 \, w$. Notice that R is a function of temperature owing to the variation of the kinematic viscosity with temperature. The viscosity ratio $\nu/\nu(100°C)$ for liquid water is for $100(50)350°C$: 1.00, 0.68, 0.54, 0.46, 0.42, 0.40: it is convenient to remember that this ratio is 0.5 at $225°C$. Thus at $225°C$ the above resistance is only 15 w. The variation of kinematic viscosity with pressure is very small, for example from $2.94 \times 10^{-6} \, m^2 \, s^{-1}$ at 1 bar to $3.04 \times 10^{-6} \, m^2 \, s^{-1}$ at 200 bar.

In most of the models here the variation of resistance with temperature is ignored. This is acceptable when the resistances are phenomenological parameters, otherwise the variation must be taken into account. In a numerical model this can be accomplished by recalculating the resistances at the new temperatures after each iteration.

The pipe-like model can be thought of in terms of an equivalent circuit as shown schematically in Fig. 10.9.

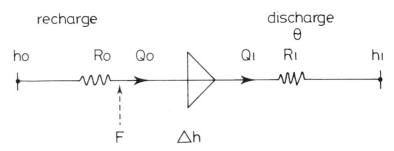

Fig. 10.9. Schematic equivalent circuit of an open liquid-filled hydrothermal system in a steady state.

The total pressure head

$$\Delta P/\rho_0 g = h_0 - h_1 + \Delta h$$

where h_0, h_1 are the heads at the recharge and discharge surfaces and Δh is the head difference produced by the density difference between the recharge and discharge elements. In the case of cold recharge and a nearly vertical hot plume,

$$\Delta h = \int (\rho_0 - \rho) ds/\rho_0$$

where the integral is from the heater to the surface. For example, if the hot column were 5 km deep and of uniform temperature $250°C$, so that $(\rho_0 - \rho)/\rho_0 \approx 0.2$, then $\Delta h \approx 1$ km. We can think of this head difference as being provided by a pump, shown as the triangular symbol in the figure.

Δh is a function of ρ and so of temperature. Hence, for a given geometry, if the power input F changes, as it must during the life of the system, then so will the hot column temperature and Δh. Similarly the resistance R will change. If necessary in model studies these effects on h and R can be allowed for. In the field, while it will not usually be possible to measure the quantities which contribute to R, nevertheless R itself can be measured by a field drawn down test from its macroscopic definition: (equivalent extra discharge rate)/ (change in head).

In order to illustrate working with such an equivalent circuit of this type consider the case, already discussed above, of a liquid system of depth d; $h_0 = h_1$, no topographic effects; and small input power F so that the mean temperature θ of the heat column is sufficiently small to take $\rho = \rho_0(1 - \gamma\theta)$. Then:

$$\Delta h = \gamma g\theta d; \qquad Q = \Delta h/R,$$

where $R = R_1 + R_2$, the sum of the recharge and discharge resistances;

$$F = \rho_0 c_0 Q\theta$$

where $\rho_0 c_0$ is the specific heat capacity of liquid water; hence

$$F \propto \theta^2,$$

as found in laboratory model experiments; and if the hot column is of uniform cross sectional area,

$$R = \nu d/kga.$$

Noting that the conductive power for the same geometry would be proportional to $a\rho_0 c_0 \kappa\theta/d$, then the Nusselt number $N \propto R_m$, the Rayleigh number; and so on.

The equivalent circuit represents the essential physics and emphasizes field-measureable quantities.

11. The Reservoir

Even in the laboratory, detailed studies of time-dependent, three-dimensional convection in a porous medium are a complex and difficult business. In the field the amount of measurable data is very limited in space and time. Numerical techniques are now available for elaborate simulation of such systems and already a number of studies have been made. Nevertheless the field data is quite inadequate to allow calibration of these elaborate models. As we shall see there are much simpler models which merely set out to relate the input and outputs of major parts of the system. Such models have the outstanding advantage of using components whose parameters are directly measurable. Consider for example the hydraulic resistance between the inlet and outlet of a reservoir. The permeability and visocity and their distributions which contribute to the resistance may be completely unknown, but the resistance is obtained from the loss of head required for an increment of flow rate.

There is a further aspect of a hydrothermal system which is important: its temporal behaviour. The gross temporal response of a system to short-term changes in its surroundings is determined by the most voluminous of its components, the reservoir. This chapter considers the role of the reservoir and conditions in it which determine the time-scales of the system. The ideas are then applied in later chapters as a key aspect of the identification of particular systems.

It should be noted that current studies and exploration of active hydrothermal systems only apply directly to the high-level reservoirs simply because the duration of exploitation has been short compared to the overall time-scales of the natural systems. In other words our measurements of response are as yet restricted to that of the reservoirs.

The effects of thermal conduction are small. It is, then, appropriate to ignore conduction altogether and regard hydrothermal systems as controlled by flow. The water-substance and energy can then be envisaged as contained in reservoirs. Further, these reservoirs can be treated as independent bodies except that are interconnected by pipes. We are representing a three-dimensional system as a one-dimensional piece of plumbing.

194

11.1. OPEN LIQUID-FILLED RESERVOIR

The description of the reservoir presented here is of a single "black box" supplied with a single inlet and outlet. Such a model can be expected to be applicable to a well connected natural system (such as Wairakei). Where, however, a particular field has distinct parts a number of simple reservoirs can be run together (such as Wairakei—Tauhara or Ohaki—Broadlands).

Properties of the reservoir

Capacitance, C

If the change in total mass of fluid in the reservoir is ΔM for change of pressure head Δh, I define the capacitance C such that

$$\Delta M / \rho_0 = C \Delta h,$$

where $\Delta M / \rho_0$ is the equivalent volume of fluid. Again, this is by analogy with electricity where the equivalent volume is analogous to the charge (coulombs). In the case of a reservoir which is a vertical prism of depth d and porosity ϵ with an open upper surface of area A filled with fluid of density ρ, the mass of fluid, $\rho \epsilon A d$, and equivalent head, $\rho d / \rho_0$, give a capacitance $C = \epsilon A$.

Other cases, such as gassy reservoirs, can be handled in a similar way except that, in general, the capacitance is a function of the thermodynamic state of the reservoir.

Thermal volume, Φ

The thermal capacity of a reservoir is conveniently defined such that the total thermal energy of the reservoir (relative to $0°C$) is:

$$U = \rho_0 c_0 \Phi \theta,$$

where $\rho_0 c_0$ is the specific heat capacity per unit volume of reference liquid water and $\theta(°C)$ is the spatial mean reservoir temperature. For a reservoir of volume V, the energy is

$$U = \rho_* c_* V \theta,$$

where $\rho_* c_*$ is the specific heat capacity per unit volume of the water-substance-saturated rock. Thence,

$$\Phi = V(\rho_* c_* / \rho_0 c_0).$$

Provided the porosity of the rock is small $\rho_* c_*$ will be close to that of the rock itself. In any event the ratio $\rho_* c_* / \rho_0 c_0$ is typically 0.5— 0.6. Without further ado, I use the value 0.5 for this ratio.

The dimensions of Φ are those of volume and lacking an appropriate term I refer to Φ as the thermal volume (or thermal capacitance).

For the simple case of an open liquid reservoir of cross-sectional area A, depth d and porosity ϵ we have

$$\Phi \approx \tfrac{1}{2} Ad = \tfrac{1}{2} Cd/\epsilon.$$

For example with $A = 10\,\text{km}^2$, $d = 5\,\text{km}$, $\epsilon = 0.1$ we have $\Phi = 50\,\text{km}^3$, $C = 2\,\text{km}^2$.

Constancy of reservoir parameters: R, C, Φ

Strictly the resistances, capacitances and thermal volumes are parameters defined in the vicinity of a particular operating point of the system. Consider for example a liquid reservoir which is an open vertical homogeneous prism. Then C will be independent of the fluid level, but Φ will not — it can only be considered a constant if $\Delta h \ll d$, i.e. the fluid range is small compared with the prism depth.

If the reservoir is of an irregular shape, then both C and Φ will be functions of the level of fluid in the reservoir. Furthermore, since the reservoir itself is not a fixed structure, the parameters are strictly functions of the state of the system.

Units used in reservoir models

h:	pressure head	m
C:	capacitance	m^2
R:	resistance	$\text{m}^{-2}\,\text{s} \equiv \text{w}$
Φ:	thermal volume	m^3
Q:	flow rate	$\text{m}^3\,\text{s}^{-1}$

Note that flow rate is stated in $\text{m}^3\,\text{s}^{-1}$ of equivalent liquid water so that normally $1\,\text{m}^3\,\text{s}^{-1}$ is equivalent to $1\,\text{ton}\,\text{s}^{-1}$.

11.2. BEHAVIOUR OF AN OPEN-SURFACE LIQUID RESERVOIR

A conceptual model of a hydrothermal system similar to a domestic hot water system is sketched in Fig. 11.1 A large cold water reservoir is supplied with meteoric water. Some of the cold water passes

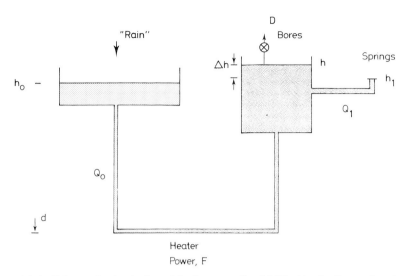

Fig. 11.1. Schematic physical model of an open liquid-filled hydrothermal system.

through a pipe past a heater and the heated water supplies a small hot water reservoir. Fluid leaves the system as a natural overflow. By means of an adustable siphon an "artificial" discharge of hot water is also taken from the system. For the moment, we consider the case in which the water remains liquid. Later, the case in which some of the fluid is gas or steam will be considered. We are interested in the possible steady states of this system and, if there is a change in the conditions, how the system moves from its initial to its new steady state.

Simple drainage

Suppose an open surface liquid reservoir of fluid level y which has just been filled to a level y_1 and drains to a level y_0, as sketched in Fig. 11.2. The equivalent discharge rate is:

$$Q = (y - y_0)/R,$$

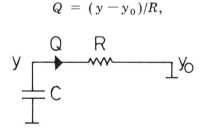

Fig. 11.2. Simple drainage of a reservoir, schema and nomenclature.

and conservation of mass requires:

$$\rho_0 Q = -\frac{d}{dt}\rho_0 Cy = -\frac{d}{dt}\rho_0 C(y - y_0).$$

Provided C is independent of y (as in a vertical prism, or where $\Delta y \ll d$), the system is linear. Writing $\tau = RC$ then we have:

$$y = y_0 + (y_1 - y_0)\exp(-t/\tau).$$

Thus the system time-scale is RC.

The case of simple refilling of an open surface reservoir through a recharge resistance will be similar.

Simple reservoir with recharge and discharge

Consider now a similar system but with both recharge and discharge as sketched in Fig. 11.3. In addition let there be an artifical discharge at equivalent rate D directly from the reservoir. Then,

$$Q_1 = (h_1 - y)/R_1, \qquad Q_2 = (y - h_2)/R_2,$$

and the net flow into the reservoir is:

$$Q = Q_1 - Q_2 - D.$$

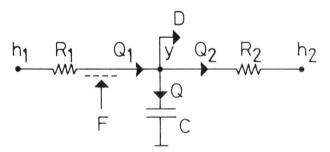

Fig. 11.3. Reservoir with drainage and recharge, schema and nomenclature.

For the moment we restrict $h_1 > y > h_2$, in order to avoid the complication of reverse flow of cold fluid ($Q_2 < 0$).

(i) *Steady state.* In the steady state, $Q = 0$ which on rearrangement of terms gives:

$$y = h_2 + R[-D + (h_1 - h_2)/R_1],$$

where

$$1/R = 1/R_1 + 1/R_2.$$

Define y_0, y_∞ as the steady state values for $D = 0$ and steady discharge D. Then measuring y relative to level y_0, writing $y' = y - y_0$, we have:

$$y'_0 \equiv 0, \qquad y'_\infty = -RD.$$

Thus, relative to the original state of no artifical discharge, the ultimate drawdown for a steady discharge is RD.

(ii) *Non-steady state*. Consider a system in equilibrium with no artificial discharge which has imposed an artifical discharge D at time $t = 0$. conservation of mass requires:

$$\frac{d}{dt} Cy' = Q = -D - y'/R.$$

Hence,

$$y' = y'_\infty [1 - \exp(-t/\tau)],$$

where

$$\tau \equiv RC.$$

Several points should be noticed.

(1) The system has a single time-scale RC. The net resistance R is obtained from R_1, R_2, in parallel. This arises because, seen from the reservoir, flow enters or leaves in both paths to fixed levels. Note that both R_1 and R_2 need to be non-zero, otherwise the system is fixed.

(2) After the transient response has passed, the ultimate level y_∞ is independent of C. The steady state is independent of the capacitance.

(3) The time-scale of the system is independent of the artificial discharge D. The role of D is confined to the amplitude of the changes in level y. The ultimate drawdown, $Y \equiv y_\infty(D)$ is such that $dY/dD = -R$, again the level being controlled by the resistances in parallel.

(4) The relations for the steady state are readily obtained by the conventional methods of electrical circuit theory. For example, we note that, for discharge $D = 0$, the head difference $(h_1 - h_2)$ is distributed in proportion to the resistances, that is:

$$Q_1 = Q_2 = (h_1 - h_2)/(R_1 + R_2), \qquad y - h_2 = Q_2 R_2, \quad \text{etc.}$$

(5) The reservoir parameters R, C can be determined solely from the measurement of $y'(t)$.

(6) The rate of recharge of fluid is:

$$Q_1 = D(1 - \exp(-t/\tau)),$$

so that near $t = 0$ the discharge comes largely from fluid stored in the reservoir, but as $t \to \infty$ the discharge comes from recharge. When $D > 0$, the total mass in the system diminishes to an ultimately smaller amount.

Simple reservoir with heat supply

Let the above system now be provided with a heat supply of power F applied to the recharge system as sketched in Fig. 11.3. Ignoring the effects of temperature on the flow through changes of density and viscosity, so that the mass flow and energy equations are uncoupled, we then have a flow given as a function of time as determined in the previous section. The conservation of energy requires

$$d(\rho_0 c_0 \Phi \theta)/dt = F - \rho_0 c_0 (Q_2 + D)\theta,$$

where θ is the reservoir temperature (relative to $0°C$), the original recharge fluid is assumed to have zero energy, the discharge is at the temperature of the reservoir and the specific heat of the discharge fluid $c \approx c_0$, and $\Delta y \ll d$ so that Φ is nearly a constant.

Noting that $Q_2 = (y - h_2)/R_2$, and using the above form of y, after some rearrangement we have:

$$\frac{d\theta}{dt} + \frac{1}{T}\theta = f - \beta,$$

where

$$1/T = [D + (y_\infty - h_0)/R_2]/\Phi,$$

$$f = F/\rho_0 c_0 \Phi,$$

$$\beta = (y - y_\infty)\exp(-t/\tau)\theta/\Phi R_2.$$

We notice at the outset that the thermal system has two time-scales: τ, determined by the hydrological parameters; and T, a thermal time-scale. The factors which affect the transient flow do not influence T. Note that:

$$Q_{1\infty} = [(h_1 - h_2) + R_2 D]/(R_1 + R_2),$$

$$Q_{2\infty} = Q_{1\infty} - D,$$

$$T = \Phi/Q_{1\infty},$$

so that T is determined by the thermal volume and the ultimate flow rate into the reservoir.

Rather than solve the linear equation formally it is of greater interest to consider two extreme cases.

(i) $T \ll \tau$. The term $d\theta/dt$ is negligible and

$$\theta \approx T(f - \beta) \sim Tf = F/\rho_0 c_0 Q_\infty$$

Here the thermal time-scale is so short that rapid temperature adjustment occurs, the temperature is simply a function of the flow and in effect the thermal system is independent of Φ. This case is not of geothermal interest.

(ii) $T \gg \tau$. The role of the transient term β is negligible, since in times of order τ the temperature has hardly changed. The thermal behaviour is determined solely by the time-scale T and the ultimate flow rates. We have:

$$\theta \approx \theta_\infty + (\theta_0 - \theta_\infty) \exp(- t/T), \qquad \theta_\infty = fT.$$

This is the case of geothermal interest.

As a guide to possible values of τ/T consider a hypothetical case: natural system discharge, $1 \, \text{ton s}^{-1}$; system depth, $5 \, \text{km}$; system head $\approx 0.2 \times 5 \, \text{km} = 1 \, \text{km}$ (for $250°C$ reservoir fluid at $0.8 \, \text{g cm}^{-3}$); then $R = 1000 \, \text{m}/1 \, \text{m}^3 \, \text{s}^{-1} = 10^3 \, \text{w}$, and porosity, $\epsilon = 0.1$. For a prismatic reservoir

$$\tau/T = RCQ_\infty/\Phi \approx 2eRQ_\infty/d \approx 0.04.$$

For example if $\tau = 10 \, \text{yr}$ then $T = 250 \, \text{yr}$.

The system behaviour is illustrated in Fig. 11.4 for a system originally cold and with no flow or heat supply which at $t \geqslant 0$ is provided

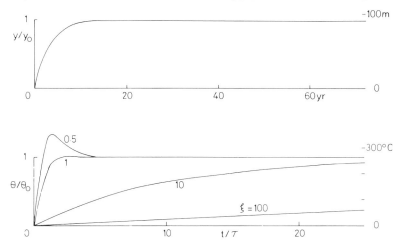

Fig. 11.4. Reservoir characteristic following a sudden change in discharge. System pressure, y and temperature for various values of time-scale ratio, $\xi = T/\tau$.

with a power supply and flow. The data are given in dimensionless form together with actual values for the case: $F = 10^9$ W, $D = 0$, $h_1 = 1000$ m, $h_2 = 0$, $R_1 = 1000$ w, $R_2 = 100$ w, $C = 10^6$ m^2 giving $\tau \approx 2.9$ yr, and various values of Φ chosen to cover the range $T/\tau = 0.5$–10.0.

(i)   For $T/\tau \lesssim 1$, the system response is affected by the flow changes. For progressively smaller values of T/τ there is an increasing temperature overshoot — already pronounced in the data shown at $T/\tau = 0.5$. This is an artifact of this particular case where the flow starts from zero with a fixed power input and the balancing discharge from the reservoir lags the thermal response.

(ii)   For $T/\tau \gtrsim 10$, the thermal response is unaffected by the details of the flow change, only by its ultimate value — the response being that of a simple exponential of time-scale T.

11.3. TWO-PHASE RESERVOIR: STEADY STATE†

If a liquid-filled reservoir is sufficiently depressurized, part of the reservoir will contain a two-phase mixture of liquid water and steam. This tendency will be enhanced if the liquid contains a soluble gas. When a substantial part of the reservoir is filled with two-phase fluid, its behaviour is very different from that of a simple open liquid-filled reservoir. We now turn our attention to multi-component multi-phase reservoirs by considering the case of reservoirs containing two-phase mixtures of water and carbon dioxide, a fairly common situation and one for which the case study of the Ohaki—Broadlands system, described later, is typical. Note that the simplest two-phase situation of the fluid being only water plus steam is a special case in which the gas content is very small.

Steady state behaviour is described first and then time-dependent behaviour.

Whether a system is in a steady state or not its behaviour is dominated by its thermodynamic behaviour. This is particularly the case for two-phase systems and especially so for two-phase two-component systems. Such a system, having more degrees of freedom, is able to respond to changes in external contraints by a simple rearrangement of the proportions of its phases. A knowledge of the thermodynamic properties of the fluid together with a few simple dynamical principles is sufficient to describe a wide range of phenomena. We start with a description of the thermodynamic properties and apply them to steady state sampling of a gassy reservoir.

† Except for readers particularly interested in the details of two-phase phenomena, the remainder of this chapter could be skipped on a first reading.

Properties of a mixture of water and a soluble gas, CO_2

For the moment use the subscripts l, liquid phase = liquid water + dissolved gas; v, vapour phase = steam + exsolved gas; and retain, f, pure liquid water; g, pure steam. At a given pressure P and temperature T for a given mixture of H_2O and CO_2 let the mass fraction of CO_2 in the mixture be λ and the partial pressure of CO_2 in the vapour P_c. The data below are for CO_2 and are taken from a compilation by Sutton (1976). The main interest here is to be able to determine the state of the mixture and in particular its density.

The mass fraction, n_l, of CO_2 in the liquid phase.

In the ideal system of Henry's relation the molar fraction $x_l \propto P_c$. Hence, $n_l = 44x_l/(26x_l + 18) \approx (44/18)x_l$ for $x_l \ll 1$, where the molecular weights of H_2O, CO_2 are 18, 44. Thus for the relative amount of CO_2, with $\lambda \ll 1$, we have:

$$n_l = \alpha P_c, \qquad \alpha = \alpha(T).$$

Experimental data for $\alpha(T)$ are shown in Fig. 11.5. Over the range $T \approx 150\text{--}300°C$, the scattered data can be fitted by a second-order polynomial (slightly different from that of Sutton, 1976):

$$\alpha = \sum_{i=0}^{2} a_i (T/100°C)^i,$$

with $a = (2.8, -0.85, 0.55) \times 10^{-4}\ \text{bar}^{-1} \pm 25\%$.

For application over a restricted range of temperature it is sufficient to take $\alpha = $ constant, for example: $4 \times 10^{-4}\ \text{bar}^{-1}$ over 200–280°C; $3 \times 10^{-4}\ \text{bar}^{-1}$ over 100–200°C. The value $\alpha = 4 \times 10^{-4}\ \text{bar}^{-1}$ has been used here unless noted to the contrary.

The mass fraction, n_v, of CO_2 in the vapour phase

In the ideal system of Dalton's relation the molar fraction is proportional to the partial pressure. Experimental data, given by the above authors, fits better to

$$n_v = P_c/P$$

over the range 100–300 bar to $\pm 5\%$ up to 280°C, and to $\pm 25\%$ to 300°C.

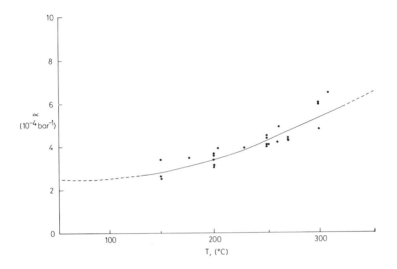

Fig. 11.5. Solubility coefficient, α, for carbon dioxide as a function of temperature (After Ellis and Golding, 1963; Malinin, 1974; and others). Second order polynomial curve used in this work.

The enthalpy, h_c, of CO_2 gas

Expermimental data for h_c, relative to the triple point of water, can be represented by a third-order polynomial in the independent variable $(T/100^\circ C)$ with:

$$a = (0.205, 86.46, 2.257, -0.019)\,\text{kJ kg}^{-1}.$$

Values obtained from this polynomial are given in Table 11.1.

Table 11.1 *Enthalpy of CO_2.*

$T(^\circ C)$	$h_c(\text{kJ kg}^{-1})$	$h_{soln}(\text{kJ kg}^{-1})$	$h_{lc}(\text{kJ kg}^{-1})$
0	0.2		(-280)
50	44.0		
100	88.9	-170	-80
150	134.9	-65	70
200	182.0	70	250
250	230.1	220	450
300	279.4	490	770
350	329.6		(1220)
400	380.9		

The enthalpy, h_{lc}, of dissolved CO_2

In solution $h_{lc} = h_{\text{soln}} + h_{gc}$. Experimental values of h_{soln} are given by Ellis and Golding (1963). Some of these are included in Table 11.1. A third-order polynomial fit to h_{lc} with independent variable $(T/100°C)$ gives mean error $5\,\text{kJ}\,\text{kf}^{-1}$ with $a = (-558, 673, -257, 60)\,\text{kJ}\,\text{kg}$.

In most cases, where the total CO_2 content is small, the contribution of h_{lc} to the energy balance can be ignored. For example in liquid at $250°C$ with $\lambda = 0.01$ in 1 kg liquid we have water energy, $1.08\,\text{MJ}$ and dissolved gas energy, $0.0045\,\text{MJ}$, about 0.4% of the total. In all calculations here the role of h_{lc}, is ignored — in any event, in the cases of interest, the bulk of the CO_2 is in the vapour phase.

The specific volume, v_l, of the liquid phase

Since the relative amount, n_l, of CO_2 in the liquid phase will usually be rather small it is assumed to be sufficiently accurate to take $v_l = v_f$, the specific volume of pure liquid water.

The specific volume, v_c, of the pure gas phase

Over the range of temperature of interest here CO_2 gas behaves nearly as an ideal gas so that (at absolute temperature $\tilde{T}$):

$$v_c = R\tilde{T}/P_c, \qquad R = 8.3143 \times 10^3 /44 \times 10^5$$
$$= 1.9 \times 10^{-3}\,\text{m}^3\,\text{kg}^{-1}\,(\text{K}^{-1}\,\text{bar})$$

(or written $\rho_c = \beta_c P_c /\tilde{T}$, with $\beta_c \approx 530\,\text{kg}\,\text{m}^{-3}\,(\text{K}\,\text{bar}^{-1})$).

Experimental data over the range $T = 100$–$330°C$ and pressures up to 20 bar indicate this is accurate to about $\pm 2\%$.

Thermodynamic state of a mixture of water and soluble gas, CO_2

Consider unit mass of fluid, as shown schematically in Fig. 11.6 which at depth is entirely liquid water at temperature, T_0, with dissolved gas (CO_2) of mass proportion, λ.

In studies of the discharge from a reservoir we wish to know the mass proportions of liquid, $(1 - \xi)$ and vapour, ξ (water vapour plus degassed CO_2), for the liquid the mass proportion, x, of dissolved CO_2 and for vapour the mass proportion, y, of degassed CO_2. This

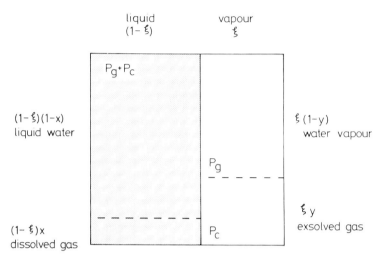

liquid   vapour
$(1-\xi)$ ξ

$P_g + P_c$

$(1-\xi)(1-x)$ $\xi(1-y)$
liquid water   water vapour

P_g

ξy

$(1-\xi)x$ P_c exsolved gas
dissolved gas

Fig. 11.6. Conceptual arrangement of unit mass of a two-phase, two-component system in thermodynamic equilibrium.

is obtained from the properties of the constituents together with the relations for the conservation of mass of water-substance, mass of CO_2 and energy. Then given the equations of state of the constituents we obtain the other quantities of interest.

Mass of CO_2

With CO_2 concentrations n_l, n_g in the liquid and vapour phase the mass of CO_2 is:

$$\lambda = (1-\xi)n_l + \xi n_g$$

Noting that

$$x \equiv n_l = \alpha P_c = \alpha y P, \qquad \text{where } y \equiv n_g = P_c/P,$$

we have

$$\lambda = [(1-\xi)\alpha P + \xi]\,y.$$

It is worth noting immediately that since in most cases where there is a vapour phase we are dealing with $P \sim 40\,\text{bar}$, $\xi \sim 0$ and $\alpha \approx 4 \times 10^{-4}\,\text{bar}^{-1}$, then $\lambda \approx \xi y$, giving the value y as if all the CO_2 were in the gaseous phase. Under these circumstances the system is insensitive to α.

Energy

The total enthalpy, h_0, is presumed given (for example if $\lambda = 0$, $h_{f0} = h_f(T_0)$). The energy of the various constituents is:

$$(1 - \xi)(1 - x)h_f, \qquad \text{liquid water}$$

$$(1 - \xi)xh_{lc}, \qquad \text{dissolved } CO_2$$

$$\xi(1 - y)h_g, \qquad \text{steam}$$

$$\xi y h_{gc}, \qquad CO_2 \text{ gas}$$

so that

$$h_0 = h_f + \xi h_{fg} - \xi y(h_g - h_{gc}) - x(1 - \xi)(h_f - h_{lc})$$

where

$$h_{fg} = h_f - h_f,$$

and

the specific enthalpy of the liquid is: $h_l = (1 - x)h_f + xh_{lc}$;

the specific enthalpy of the vapour is: $h_v = (1 - y)h_g + yh_{gc}$.

At a sufficiently high pressure all the gas will be in solution, so that $x = \lambda$; no vapour will be present, so that $\xi = 0$, and the total enthalpy is solely that of the liquid:

$$h_0 = h_f - \lambda(h_f - h_{lc}).$$

Thus, the total specific enthalpy can be specified directly by stating h_0. Otherwise it can be obtained indirectly from the equivalent temperature, T_0, and the CO_2 mass fraction, λ, of the equivalent entirely liquid substance. Note that then h_0 is for a liquid composed of a fraction $(1 - \lambda)$ of H_2O, λ of CO_2 at temperature T_0 and pressure $P_s(T_0) + \lambda/\alpha(T_0)$, namely, pressure $P = P_* = P_s^* + P_c^*$, where we write

$$P_s^* \equiv P_s(T), \qquad P_s^* \equiv \lambda/\alpha(T).$$

Total mass

The volumes of the two phases are: liquid, $(1 - \xi)/\rho_l$; vapour, ξ/ρ_v; so that the density, ρ, of the fluid mixture is:

$$\rho = 1 \Big/ \left[\frac{(1 - \xi)}{\rho_l} + \frac{\xi}{\rho_v} \right].$$

Here

$$\rho_v = \rho_g + \rho_c, \qquad \rho_g = \beta_g P_s/\tilde{T}, \qquad \rho_c = \beta_c P_c/\tilde{T}.$$

so that

$$\rho_v = [\beta_g(1 - y) + \beta_c y] P/\tilde{T}.$$

The quantities are not strict constants·but over the range of interest, $\beta_g \approx 260 \text{ K kg m}^{-3} \text{ bar}^{-1}$, $\beta_c \approx 530 \text{ K kg m}^{-3} \text{ bar}^{-1}$, approximately in the ratio β_c/β_g of the molecular weights of H_2O and CO_2.

The above form of ρ is adequate for most calculations. Otherwise polynomial representations of β_c, β_g, can be used. Here ρ_g has been obtained from the polynomial representation of $v_g = 1/\rho_g$.

Methods of solution

Of the numerous possibilities there are two situations which occur frequently.

(i) Given pressure, P, specific enthalpy, h_0, and gas content, λ_0. In, for example, a study of the flow of fluid through the ground and up a bore, the pressure can be obtained step by step. Each new step requires among other things knowledge of the mixture density. This case of given pressure (etc.) will be solved by an iteration, here called the "P-iteration".

(ii) Given temperature, T, mixture density, ρ, and gas content, λ_0. In, for example, a study of a non-steady two-phase reservoir, the temperature is controlled by the thermal capacity of the rock of the reservoir and is obtained explicitly, while gross conservation of mass of H_2O and CO_2 give the reservoir mixture density, ρ and gas content, λ. This case of given temperature (etc.) will be solved by an interation here called the "T-iteration".

The basis of the method is the same for both cases. It should be noted in practice that, in certain ranges of the variables, the new iterate needs some digital filtering (a simple low-pass filter) to promote (rapid) convergence.

Iterative solutions

The "T-iteration". Noting that the given total pressure (of the vapour) is
$$P = P_g + P_c$$
where
$$P_c = yP, \qquad P_g = (1 - y)P,$$
we can arrange the above expressions to
$$\xi = [(h_0 - h_f) + \xi y(h_g - h_{gc}) + x(1 - \xi)(h_f - h_{lc})]/h_{fg},$$
$$y = \lambda/[(1 - \xi)\alpha P + \xi],$$
$$x = \alpha yP,$$
$$P_s = (1 - y)P,$$
$$T = T(P_s).$$

For a given P and an initial estimate $\xi = 0$, $x = 0$, $y = 0$, this arrangement rapidly converges: for example, about five iterations to obtain T to the nearest $0.05°C$.

Note that if $\lambda = 0$, that is no CO_2, then $P_s = P$ and no iteration is necessary since

$$T = T(P_s), \qquad \xi = \xi(T).$$

The "P-iteration". Rearranging the above expressions we have:

$$\xi = (v - v_l)/(v_v - v_l),$$

$$y = \lambda/[(1 - \xi)\alpha P + \xi],$$

$$P = P_s/(1 - y); \qquad P_s = P_s(T).$$

For a given T and v_v and initial estimate $\xi = 0$, $y = 0$, $P = P_s$, this arrangement rapidly converges. Thereafter we obtain x, h_0, etc., directly.

Acceptable approximations

(1) $\alpha = $ constant. This is referred to in the discussion of n_l. For example, with $T_0 = 250°C$, $\lambda = 0.05$, over the range $0{-}100$ bar, the difference between values for $\alpha = \alpha(T)$ and $\alpha = 4 \times 10^{-4}$ bar^{-1} are less than: T, $0.1°C$; P_s and P_c, 0.1 bar; ξ, x, y, e, 0.001.

(2) Zero enthalpy contribution of dissolved gas. This is setting $x = 0$ in the expression for the total enthalpy, h_0. The liquid is considered as pure water. This is an acceptable approximation for λ as large as about 0.05. For example, with $t_0 = 250°C$, $\lambda = 0.05$ over the range $0{-}100$ bar, errors reach: T, $5°C$; P_s and P_c, 3 bar; ξ, x, y, e, 0.03. For $\lambda = 0.02$ the figures are: T, $0.5°C$; P_s and P_c, 0.4 bar; ξ, x, y, e, 0.003. For $\lambda \gtrsim 0.1$ the errors are quite unacceptable.

It is of physical interest and numerical convenience to note that if a vapour phase is present, except very near the boiling—degassing point, that $x \ll 1$, i.e. the bulk of the CO_2 is in the vapour phase. There is negligible error in deleting the terms in x in the iteration, and unless noted otherwise this has been done here. For example if we consider the case: $T_0 = 250°C$; $\lambda = 0{-}0.2$, over the range $P = 0{-}40$ bar, the error in neglecting the x-terms is at most $0.3°C$ (too low).

Illustrative example. The behaviour of the thermodynamic system is demonstrated in the following figures obtained for $T_0 = 250°C$ and various λ.

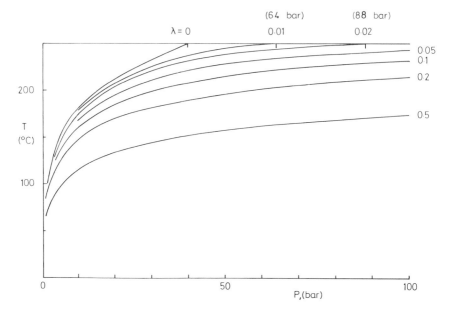

Fig. 11.7. Boiling—degassing curves $T(P)$ for a water—carbon dioxide mixture; of CO_2 mass fraction λ_0, $T_0 = 250°C$.

(1) $T(P)$, for various λ, is shown in Fig. 11.7. The effect of the presence of a soluble gas is pronounced even for λ of only a few per cent. At a given pressure the temperature is lowered, or at a given temperature the pressure is increased. The boiling—degassing range is increased; for example up to 40 bar at $\lambda = 0$; up to 160 bar at $\lambda = 0.05$.

(2) The relative amounts x, y, of CO_2 in the liquid, and in the vapour, as shown in Fig. 11.8, increase with pressure. As $P \rightarrow 0$ we have $x \rightarrow 0$ and $y \rightarrow \lambda/\xi$, a non-zero value. As $P \rightarrow P^*$ we have $x \rightarrow \lambda$ and $y \rightarrow \lambda/\alpha P^*$.

(3) The proportions of CO_2 in the liquid and vapour, $(1 - \xi)x/\lambda$, $\xi y/\lambda$, are shown in Fig. 11.9. At low pressures, the CO_2 is predominantly in the vapour phase. As $P \approx P^*$, the proportion in the liquid approaches unity while that in the vapour goes to zero ($\xi y \rightarrow 0$ since $\xi \rightarrow 0$). Also shown are the mass proportion of vapour, and e, the volumetric proportion of liquid (the liquid saturation).

(4) The partial pressures of steam and CO_2 gas, P_s, P_c, are shown in Fig. 11.10(a), where $P_s > P_c$ except near P^*. For comparison, (b) shows a case where $P_s < P_c$ except near $P = 0$.

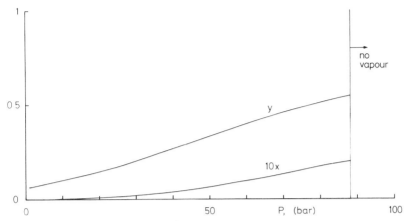

Fig. 11.8. The relative amount of soluble gas, CO_2, in the liquid, x, and vapour, y, of a mixture of water and carbon dioxide: $T_0 = 250°C$, $\lambda_0 = 0.02$.

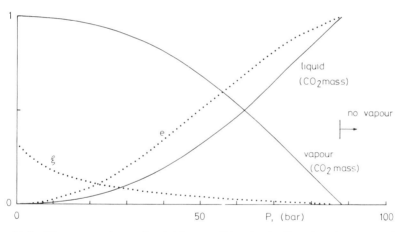

Fig. 11.9. The proportions of soluble gas, CO_2, in the liquid and vapour of a mixture of water and carbon dioxide. $T_0 = 250°C$, $\lambda_0 = 0.02$. The vapour mass fraction, ξ, and the saturation, the liquid volumetric fraction, e, are also shown.

11.4. RESERVOIR SAMPLING THROUGH A PERMEABLE THROTTLE

Consider the thermodynamic state of fluid withdrawn from a reservoir. If the pressures are sufficiently high for the fluid to be in and remain in the liquid state, the discharge will have the same thermodynamic state as that in the reservoir. If, however, (i) the

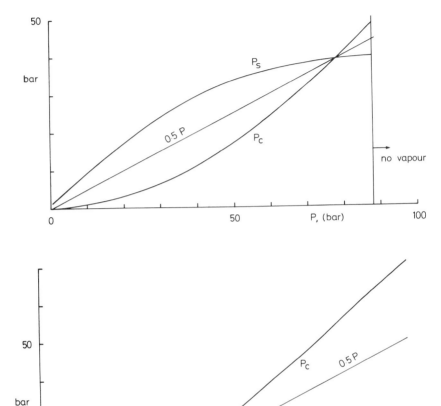

Fig. 11.10. The partial pressures of steam and CO_2 gas in a mixture. (*a*) $T_0 = 250°C$, $\lambda_0 = 0.02$. For comparison the line, partial pressure $= 0.5P$ ($P =$ the total pressure) is shown. (*b*) $T_0 = 300°C$, $\lambda_0 = 0.05$.

reservoir fluid has a vapour phase, and (ii) the discharge occurs through a porous permeable medium, then the thermodynamic state of the discharge will be different from that of the reservoir. This arises because of the differential throttling effect of the two phases on each other in flow through a permeable medium.

Simple sampling

Consider a large homogeneous reservoir maintained by an external system so that the fluid has specific enthalpy, h_0 and total gas content, λ_0. Withdraw fluid at a steady rate through a short section of permeable medium. Arrange the flow rate to be sufficiently small so that $\Delta P/P \ll 1$, where P is the reservoir pressures and ΔP the pressure drop across the throttle. Then the state of the separate liquid and vapour phases will be the same in the throttle as in the reservoir. The arrangement is sketched in Fig. 11.11.

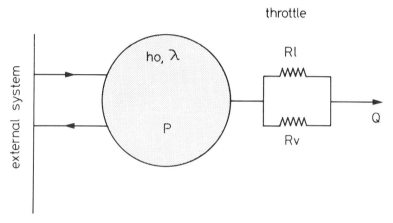

Fig. 11.11. Schematic arrangement for simple sampling of a two-phase reservoir through a permeable throttle.

The permeable throttle

The role of the discharge resistances, R_l, R_v, the key feature here, is obtained as follows. From the Darcy relationship for two-phase flow we can take:

$$R_l = R_*(\nu_l/\nu_{l0})/F_l, \qquad R_v = R_*(\nu_v/\nu_{l0})/F_v,$$

where R_* is the resistance for single-phase cold liquid water for which $\nu_l = \nu_{l0}$ and $F_l = 1$; the arbitrary cold reference temperature is taken as $20°C$; F_l, F_v are the relative permeabilities of liquid and vapour, and are functions of the fluid saturation, $e = $ (liquid volume)/ (liquid volume + vapour volume). Purely for illustration I will take:

$$F_l = e^3, \qquad F_v = (1-e)^3 \quad \text{with } 0 \leqslant e \leqslant 1.$$

The net resistance, R is such that $1/R = 1/R_l + 1/R_v$. Note that in practice the throttle would be extensive and the effects would need to be integrated through the throttle volume. The throttle is here made of a thin permeable zone in which the fluid components are the same as those in the reservoir.

Hence, if the total equivalent flow rate through the throttle is Q, the rates of flow of the separate phases are:

$$Q_l = (1 - \zeta)Q, \qquad Q_v = \zeta Q,$$

where

$$\zeta = R/R_v,$$

and the discharge specific enthalpy and total gas fraction made up from the contributions carried in each phase are:

$$h = (1 - \zeta)h_l + \zeta h_v,$$

$$\lambda = (1 - \zeta)x + \zeta y.$$

The quantities h_l, h_v, x, y are obtained as described above for a closed reservoir.

No gas, $\lambda_0 = 0$

The sample output for a particular case is illustrated in Fig. 11.12. The enthalpy of the sample rises rapidly, as P/P_* is reduced from unity, reaching a broad plateau, in this case with $h/h_0 \approx 2.5$. This behaviour arises from the differential effects of the permeable throttle on the discharge of liquid and vapour. The relative flow proportion of vapour ζ, rises rapidly as P/P_* is lowered, in this case for $P/P^* \lesssim 0.6$, $\zeta \approx 1$ and the discharge is largely vapour. The contribution to ζ which arises from the variation of the kinematic viscosity ratio is small. The dominant contribution to variation of ζ comes from the effect of saturation on the relative permeability ratio.

Role of soluble gas, $\lambda_0 > 0$

The sample output of enthalpy ratio, h/h_0 and total gas ratio, λ/λ_0, is illustrated in Fig. 11.13 for $\lambda = 0.01, 0.02, 0.05$.

The enthalpy ratio is a nearly monotonic function of P/P^* for all values of λ_0, including zero as discussed above. In contrast to the case $\lambda_0 = 0$, there is a region near P/P^* where $h/h_0 \approx 1$, and the extent of this region increases with λ_0. For example at $\lambda_0 = 0.05$, we find $h \approx h_0$ for $P/P^* \gtrsim 0.6$. As for the case $\lambda_0 = 0$, the behaviour of the sample enthalpy is dominated by the effects of saturation on the permeable throttle.

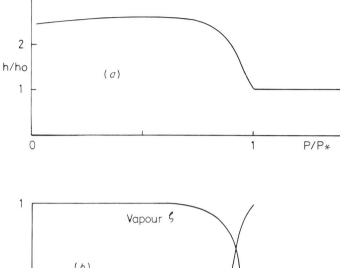

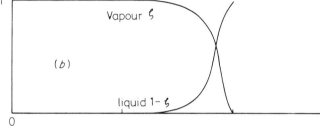

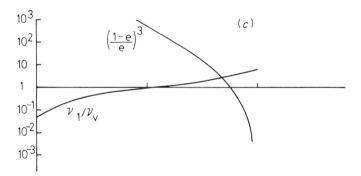

Fig. 11.12. Properties of a reservoir sample obtained through a permeable throttle as a function of reservoir pressure: (a) Specific enthalpy ratio, h/h_0; (b) mass fraction of the sample as vapour, ζ and liquid $(1 - \zeta)$; (c) factors contributing to the throttle operation. $T_0 = 250°C$, $\lambda_0 = 0$ (no gas).

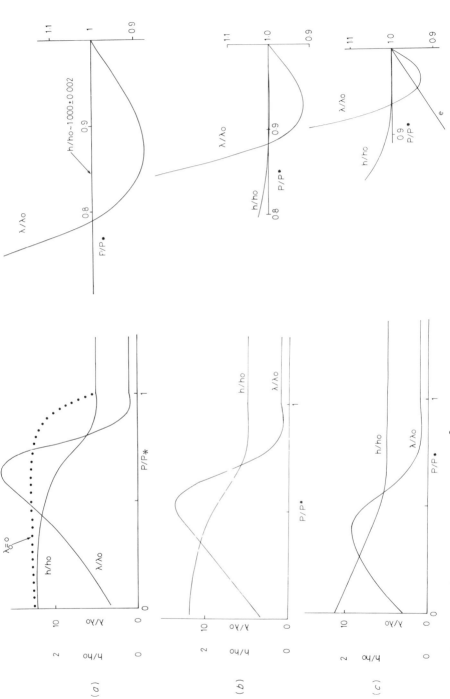

Fig. 11.13. Sample of a gassy reservoir, $T_0 = 250°C$, at gas contents $\lambda_0 = 0.01, 0.02, 0.05$. Curves of sample specific enthalpy ratio, h/h_0 and sample CO_2 mass fraction ratio, λ/λ_0.

The total gas ratio is a strongly varying function of P/P^* with a weak minimum near $P/P^* = 1$, an asymmetrical peak of large amplitude in mid-range, and a roughly linear decrease at lower pressures. The peak amplitude is notable. For example with $\lambda_0 = 0.01$ the total gas ratio reaches 16.7 at $P/P^* \approx 0.65$. The amplitude of the maximum decreases and that of the minimum increases with λ_0.

Behaviour near $P/P^* = 1$

This is a distinct region in which $\lambda/\lambda_0 < 1$ (and $h/h_0 \approx 1$). Near P/P^* there is a range in which ζ is very small; the extent of this region increases with λ_0. Hence as $P/P^* \to 1$, $\lambda/\lambda_0 \to x/\lambda_0$. If, for example, P/P^* is lowered from unity, the initial behaviour is determined by x, the concentration of soluble gas in the liquid phase, which falls rather slowly from its initial value, $x/\lambda_0 = 1$. Once however P/P^* has been lowered sufficiently for ζ to be significantly greater than zero, and furthermore noting that $y \gg x$, then ζy rapidly becomes the dominant term and the bulk of the soluble gas is flowing in the vapour phase.

The behaviour is thus dominated by that of ζ. This resistance ratio is itself derived from the two effects of kinematic viscosity, ν_l/ν_v and saturation through the relative permeability ratio, F_l/F_v taken here as $e^3/(1-e)^3$. The role of variation of the viscosity ratio is small. The form of ζ is dominated by the saturation ratio.

Behaviour near $e \approx 0.5$

There is a pronounced peak in sampled total gas content ratio, λ/λ_0. As we have seen, except near $P/P^* = 1$, $\lambda \sim \zeta y$. Below $P/P^* = 1$, ζ increases rapidly owing to the effects of saturation on the relative permeabilities but ultimately as the pressure falls $\zeta \to 1$, and $\lambda \sim y$. Throughout, y falls with P, so that once this condition is reached the total output gas falls.

Complete sampling

Consider now the case in which all the fluid supplied to the reservoir is withdrawn through a permeable throttle. This situation is at the other extreme from the case just considered. There, a negligibly small fraction of the reservoir fluid was withdrawn. In view of the discussion of the rather elaborate behaviour above, it may seem at first sight to be a paradox that, in this extreme case, the fluid discharge has precisely the same specific enthalpy and total gas concentration as that of the reservoir. But, of course, under the circumstances, it cannot be otherwise. Both the total enthalpy and gas content are conserved, so that in the steady state, the mass discharge rate being equal to the reservoir input rate, the output specific enthalpy and gas concentration must equal that in the reservoir.

This apparently very simple situation reveals a "self-limiting" feature of the behaviour of a permeable throttle. The feature of interest is its ability to adjust its total and relative resistances over a very wide range in such a manner as to provide a "thermodynamic buffer" between its input and output. Suppose that, at a particular moment, the total flow rate through the system is Q, and the flow rates in the throttle are Q_l, Q_v. The quantity

$$\Delta H = Q_0 h_0 - (Q_l h_l + Q_v h_v)$$

is a measure of the excess enthalpy of the reservoir arising from the difference between the rate of input and output to the reservoir. Suppose we find $\Delta H > 0$, the enthalpy rate through the throttle is too low. The throttle rate can be increased if the reservoir pressure falls, as seen in the simple sampling case above. Thus, the system will find a reservoir pressure such that $\Delta H = 0$.

Although it is obvious, it is of interest to note that in equilibrium the resistance ratio, $\zeta = \xi$, the vapour mass fraction. All the vapour generated in the reservoir must flow "through" R_v, all the liquid "through" R_l; where else can it go? Whereas for a closed reservoir, ξ is determined solely from the given specific enthalpy, here it is determined from a dynamical equilibrium.

It is of some interest to simulate numerically the above process of the system adjusting its reservoir pressure until $\Delta H = 0$. It is particularly striking to try this for given h_0, λ_0, but a variety of values of R^*, the throttle "single-phase" resistance. The consequent values of P, Q_0, are all the same regardless of R^*.

Thus for a two-phase reservoir with a single discharge and recharge there is a "characteristic" pressure which is a function solely of h_0 and λ_0.

Withdrawal from a reservoir

The two situations just described are extreme examples of withdrawal from a reservoir. Consider now the system shown in Fig. 11.14. Fluid is circulated through a reservoir by means of a pump and recharge—discharge system to an infinite source of specific

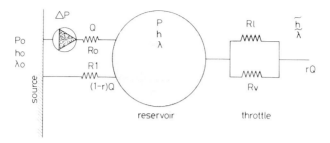

Fig. 11.14. Schematic arrangement for partial sampling, through a permeable throttle, of a two-phase reservoir with recirculation to the source.

enthalpy h_0, gas content λ_0, and pressure, P_0. A proportion , r, of the recharge flow rate, Q, is withdrawn through a permeable throttle. Let the reservoir fluid have specific enthalpy, h, and gas content, λ, while the corresponding mean quantities discharged through the throttle be $\tilde{h}$ and $\tilde{\lambda}$. If any possible two-phase effects in the recirculation system can be ignored, then:

$$P = P_0 + (1 - r)R_l\Delta P/[R_0 + (1 - r)R_l] \, ,$$

$$Q = (P_0 + \Delta P - P)/R_0 \, .$$

Conservation of enthalpy and soluble gas require

$$h = (h_0 - r\tilde{h})/(1 - r),$$

$$\lambda = (\lambda_0 - r\tilde{\lambda})/(1 - r).$$

Hence, for given P for presumed values of h and λ, the method described above for the case of a closed reservoir allows calculation of T, x, y, etc., and corresponding values of $\tilde{h}$, $\tilde{\lambda}$. New estimates of h and λ can be then found and the iteration continued.

Comment on sampling in practice

It is clear that interpretation of measurements of the enthalpy and gas content of a bore discharge, in order to determine the state of the undisturbed reservoir fluid, is not straightforward. Data for which the down hole pressure $P \approx P^*$ will give a more or less direct reading of h_0, λ_0. Otherwise, spot readings could be most misleading. Only a set of measurements over the fullest range of P/P^* are entirely satisfactory.

Of course if $P/P^* > 1$ there is no problem on this account.

11.5. TWO-PHASE RESERVOIR: UNSTEADY SYSTEMS

For a simple discharge experiment of short duration, the behaviour is largely determined by a single (though non-constant) time-scale controlled by the combined effect of the system resistance as seen by the discharge volume and the capacitance of that volume. We therefore look a little more closely into the role of these effects.

Resistance of a permeable throttle

Consider the two-phase flow through a conduit of cross-sectional area, A, and length, l. Since we have

$$p^l = \Delta P/l = m/k\alpha$$

and by definition

$$\Delta P = \rho_0 g \Delta h, \qquad \Delta h = QR,$$

then:

$$R = R_* u,$$

where

$$R_* = \nu_f/kgA, \qquad u = 1/[F_f + F_g(\nu_f/\nu_g)].$$

The resistance, R is related to the resistance for one-phase liquid flow, R_*, and the mutual resistance ratio, u, a function of e and $n = \nu_f/\nu_g$. This ratio $u = 1$ when the liquid saturation $e = 1$, otherwise for $0 < e < 1$, $u > 1$, having its extreme value when $e = 0$ with $u = 1/n$. For $F_f = e^3$, $F_g = (1 - e)^3$, values of R/R_* are shown in Fig. 11.15 at various temperatures as a function of e, appropriate

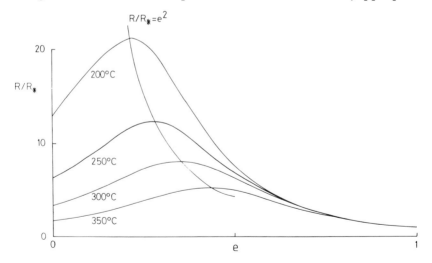

Fig. 11.15. Relative resistance, R/R_*, of a permeable throttle as a function of liquid volumetric fraction, e, at various temperatures.

to the draining of an isothermal reservoir. (i) $e \approx 1$: in this region, $u \sim 1/F_f$ and the role of n is negligible. (ii) $e^2 = R/R_*$: maxima of u occur on this line, the amplitude increasing rapidly with decreasing temperature. (iii) $e \approx 0$: in this region, $u \sim 1/n$, the role of n being dominant.

Thus, for a reservoir being progressively drained from an initial $e = 1$, the associated recharge and discharge resistances will at first rise and if draining becomes sufficiently advanced will fall again, but not to the original value. This effect is greater at lower temperatures.

Capacitance of a two-phase reservoir

Consider a reservoir of volume V, porosity ϵ, containing fluid at presssure P. By definition the total mass of fluid $M = \rho_0 Ch$ where $P = \rho_0 gh$. Hence if the volumetric proportion of liquid is e we have:

$$C/C_l = [e + (1-e)\rho_v/\rho_l] P_l/P,$$

$$C_l = \epsilon V\rho_l g/P_l,$$

where $P = P_l$ for $e = 1$, all liquid. Near $e = 1$, $C \approx C_l$ and near $e = 0$, $C \approx C_0 \equiv C_l\rho_v P_l/\rho_l P$ which is generally less than C_l since except near the critical point $\rho_v/\rho_l \ll 1$.

In nearly isothermal discharge from a two-phase reservoir, the role of capacitance is dominated by e. The effects of e on the capacitance for various temperatures and gas contents are shown in Fig. 11.16.

(a) For a given gas content, the variation of C/C_l is greater at lower temperatures, owing to the increasing predominance of the gas pressure over the steam pressure. With $\lambda = 0.02$, as shown: at $300°C$ the effect of the gas is minor and C/C_l is a monotonic function at e; at $250°C$ there is a substantial interval, $0.6 \lesssim e \lesssim 1$ in which C/C_l is nearly constant before falling nearly linearly to a low value; at lower temperatures there is a pronounced maximum in C/C_l.

Thus, for a reservoir being progressibely depressurized, provided λ is sufficiently large, the capacitance first increases before decreasing — of course, the total mass in the reservoir falls in proportion to CP. Hence if such a reservoir were recharged through a fixed supply resistance the time-scale of the reservoir would first increase and then decrease. Under some circumstances, for example $T_0 = 250°C$, $\lambda = 0.02$, the initial phase would have a constant time-scale and behave as if it were a one-phase liquid reservoir.

(b) For a given temperature the effect of various gas contents follows a similar pattern. In the example shown, with $T_0 = 250°$ for

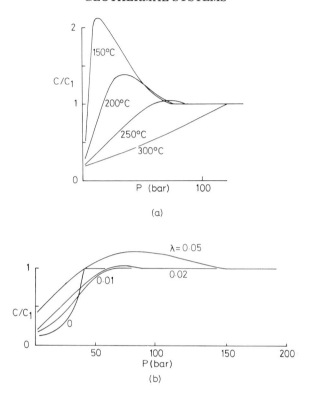

Fig. 11.16. Capacitance of a two-phase gassy reservoir as a function of pressure. (a) $\lambda_0 = 0.02$, $T_0 = 150, 200, 250, 300°$C; (b) $T_0 = 250°$C, $\lambda_0 = 0, 0.01, 0.02, 0.05$.

$\lambda \gtrsim 0.01$, there is an interval of nearly constant C/C_l and this interval becomes substantial for $\lambda \gtrsim 0.05$. The dramatic effect of even a small amount of gas is emphasized by comparison with C/C_l for the case $\lambda = 0$, no gas. Then the initial drop as saturation falls below $e = 1$ is steep.

Combined two-phase reservoir and permeable throttle

The behaviour of a reservoir will be determined not only by its internal conditions but by the effects of the recharge—discharge system. This interaction is particularly strong for a two-phase reservoir for which either or both of the supply or discharge resistances are permeable throttles. A quantity of great practical interest is the pressure response time-scale, $\tau = RC$, where here both $R = R(e)$ and $C = C(e)$.

Using the forms of R, C just described above, the time-scale for a variety of circumstances is shown in Fig. 11.17. Suppose the system is being depressurized and for the moment concentrate attention on

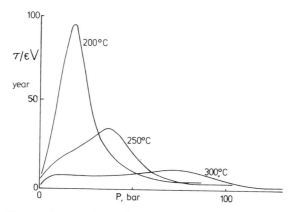

Fig. 11.17. Time-scale, $\tau = RC$, of a combined two-phase reservoir and permeable throttle as a function of reservoir pressure. $R_* = 1000$ w, $V = 1\,\text{km}^3$, $\epsilon = 0.1$, $\lambda_0 = 0.02$ at $T_0 = 200, 250, 300°\text{C}$. Note that τ is proportional to ϵV.

the data for $T_0 = 250°\text{C}$, $\lambda_0 = 0.02$. Above the pressure $P_* = 88$ bar, only one phase is present and $\tau = R_* C_l$. As P falls below P_*, and the liquid saturation $e < 1$ the time-scale begins to rise — initially dominated by increased R, since C remains nearly constant until $P \lesssim 60$ bar. At $P \approx 37$ bar the time-scale is a maximum, largely determined by the maximum in R. Thereafter the time-scale falls steadily. The range of values of τ is striking: $P = P_*$, 3 years; $P = 37$ bar, 32 years; $P = 1$ bar, 4.5 years. For a given gas mass fraction these effects are muted at higher temperatures and enhanced at lower temperatures.

The time-scale is proportional to ϵV. The quoted figures are for $V = 1\,\text{km}^3$, $\epsilon = 0.1$, $R_* = 1000$ w a rather small reservoir of moderate permeability. For larger reservoirs with V of order $10\,\text{km}^3$ typical time-scales for a strongly depressurized system will be of order $10^2 - 10^3$ yr.

Response of a two-component, two-phase reservoir

The mass of fluid in a single-phase liquid-filled reservoir is determined by the "level" of liquid in the reservoir. If the reservoir is closed except for localized discharge and recharge it will not manifest a capacitance. However, in a reservoir containing a two-phase fluid the mass of fluid in the reservoir will depend strongly on its thermodynamic state. The behaviour of such a system is of great interest not only because some natural systems are of this type but that with intense exploitation originally liquid reservoirs may be sufficiently depressurized to become two-phase systems.

Consider the system sketched in Fig. 11.18, a reservoir supplied from an effectively infinite source. This situation will apply for

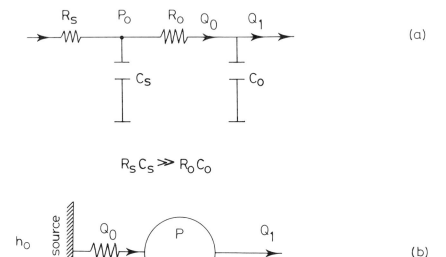

Fig. 11.18. Schematic arrangement for a simple model of a two-phase, two-component reservoir: (a) circuit; (b) cartoon.

processes of time-scale $R_0 C_0$ provided $R_0 C_0 \ll R_s C_s$ where $R_s C_s$ is the time-scale of the deep recharge system. The system is presumed to have been undisturbed for a sufficiently long time to be in a steady state with conditions in the high-level reservoir the same as those of the source. Then for an interval of time $0 \leqslant t \leqslant t_l$ an artificial discharge of equivalent volumetric rate $Q(t)$ is imposed. We wish to describe the response of the system for all $t \geqslant 0$.

A considerable simplification of the representation is possible under certain circumstances. Provided the fluid is in two phases and the thermal volume is sufficiently large (always so in practice) then the reservoir temperature hardly changes and the behaviour is dominated by changes in the saturation, e. This is justified *a posteriori*. Thus, in the analysis we treat quantities like ρ_l, ρ_v as constants.

Conservation of mass of water-substance

The mass of water-substance, M in a reservoir of volume V and mean connected porosity ϵ is:

$$M = \epsilon V[e\rho_l + (1-e)\rho_v],$$

where e is the volumetric proportion of liquid. Conservation of mass requires.

$$dM/dt = m_0 - m_1 = \rho_0(Q_0 - Q_1),$$

where m_0, m_1 are the net mass rates of input and output to the reservoir. The input $m_0 = \rho_0 Q_0$, where Q_0 is the equivalent volumetric flow rate (and ρ_0 the reference density), is determined by conditions in the source and reservoir. If the input is as a single-phase,

$$Q_0 = (P_0 - P)/\rho_0 g R_0.$$

The output $m_1 = \rho_0 Q_1$ is presumed determined by adjustment of the bore field.

It is convenient to define:

$$\rho = M/\epsilon V; \qquad q_0 = Q_0/\epsilon V; \qquad q_1 = Q_1/\epsilon V,$$

so that q_0, q_1 are the equivalent volumetric flow rates per unit fluid volume, quantities with dimensions $(\text{time})^{-1}$, and ρ is the mean fluid density. Then

$$d\rho/dt = A, \qquad \text{where} \quad A = \rho_0(q_0 - q_1).$$

Conservation of energy

The energy, U, of the fluid-saturated rock is:

$$U = V[\rho_* c_* T + \epsilon\{e\rho_l h_l + (1-e)\rho_v h_v\}].$$

Conservation of energy requires:

$$dU/dt = \rho_0(h_0 Q_0 - h_1 Q_1).$$

In practice, to a close approximation, $U = \rho_* c_* VT$ so that

$$dT/dt = Y,$$

where

$$Y = \epsilon\rho_0(h_0 q_0 - h_1 q_1)/\rho_* c_*.$$

Conservation of mass of carbon dioxide

The mass of carbon dioxide, N, in the reservoir is:

$$N = \epsilon V[e\rho_l n_l + (1 - e)\rho_v n_v].$$

Conservation of mass requires:

$$dN/dt = \rho_0(n_0 Q_0 - n_1 Q_1),$$

writing $n = N/\epsilon V$. We have

$$dn/dt = C,$$

where

$$C = \rho_0(n_0 q_0 - n_1 q_1).$$

Solution of time-dependent models

The model equations can be integrated forward in time from an initial state: $\rho = \rho(0)$, $T = T(0)$, $n = n(0)$ defined by conditions in the source, in the order:

$$q_1 = Q_1(t)/\epsilon V;$$

$$q_0 = (P_0 - P)/gR_0 \epsilon V;$$

$$d\rho/dt = A; \quad dT/dt = Y; \quad dn/dt = C;$$

$$v = 1/\rho; \quad \lambda = n/\rho;$$

$$P = P\text{-iteration}(v, T, \lambda).$$

There are no essential numerical approximations in this method. The only complication is the time-consuming iteration.

Notice that ϵ occurs in combination with V except in the factor $\rho_* c_* /\epsilon$ of the energy equation. Hence, in so far as $T \approx$ constant, the behaviour is independent separately of ϵ, except through its contribution to the fluid volume ϵV.

Simple nearly isothermal model

A simplification of the representation is possible under certain circumstances. Provided the fluid is in two phases, and the thermal volume is sufficiently large or the duration of discharge sufficiently short, then the reservoir temperature hardly changes and the behaviour is dominated by changes in the saturation, e. This is justified *a posteriori*. Thus in the following analysis (Grant, 1977) we treat quantities like ρ_l, ρ_v, as constants.

Hence $M = M(e)$, and after differentiating we have:

$$de/dt = X,$$

where

$$X = \rho_0(q_0 - q_1)/(\rho_l - \rho_v).$$

The energy equation is as before:

$$dT/dt = Y.$$

In the expression for N, recalling that:

(i) $n_l = \alpha P_c$, with $\alpha \approx$ constant;

(ii) $n_v = P_c/P$, where $P = P_s + P_c$, so that at nearly constant T, and hence constant P_s, $dn_v = -P_s dP_c/P^2$;

we have:

$$dN = \epsilon V[\{\rho_l n_l - \rho_v n_v\}de + \{e\rho_l\alpha + (1-e)\rho_v P_s/P^2\}dP_c].$$

Hence, rearranging this as an expression for dP_c:

$$dP_c/dt = Z,$$

where

$$Z = \frac{\rho_0(n_0 q_0 - n_1 q_1) - (\rho_l n_l - \rho_v n_v)X}{[e\rho_l\alpha + (1-e)\rho_v P_s/P^2]}.$$

The three conservation relations for M, U, N provide corresponding rate equations for e, T, P_c. They have been derived as a perturbation about the state $T = $ constant. In other words, this model allows investigation of the role of gas content in virtual isolation from the effects of temperature. As we shall see, this role manifests itself largely through the effects of the gas pressure, P_c.

The model equations can be integrated forward in time from an initial state: $e = e_0$, $T = T_0$, $P = P_0$, defined by conditions in the source in the order:

$$q_1 = Q_1(t)/\epsilon V, \text{ given};$$

$$q_0 = (P_0 - P)/gR_0\epsilon V;$$

$$de/dt = X, \quad dT/dt = Y; \quad dP_c/dt = Z;$$

$$P_s = P_s(T); \quad P = P_s + P_c.$$

In order to illustrate the behaviour of a system of this type, Fig. 11.19 shows the response of a reservoir with: $V = 1 \text{ km}^3$, $\epsilon = 0.1$, source pressure 80 bar at 250°C and various fixed supply resistances for the two cases of $\lambda_0 = 0.02, 0.05$ for a field drawdown experiment

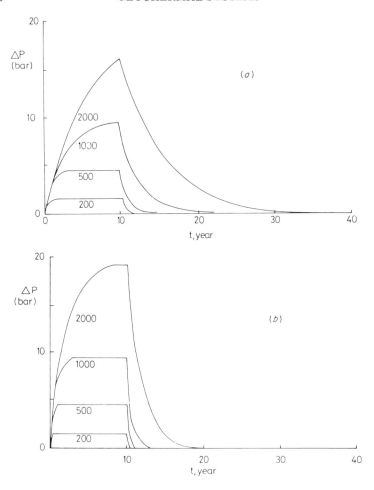

Fig. 11.19. Pressure response of a two-phase two-component reservoir with an artificial discharge of $100\,\mathrm{kg\,s^{-1}}$ for 10 years. $T_0 = 250°\mathrm{C}$, $P_0 = 80\,\mathrm{bar}$, $V = 1\,\mathrm{km^3}$, $\epsilon = 0.1$, $R_* = 200,\ 500,\ 1000,\ 2000\,\mathrm{w}$. (a) $\lambda_0 = 0.02$ ($e_0 = 0.92$, at $R_* = 1000\,\mathrm{w}$, $\tau = 2.90\,\mathrm{yr}$); (b) $\lambda_0 = 0.05$ ($e_0 = 0.57$, at $R_* = 1000\,\mathrm{w}$, $\tau = 1.93\,\mathrm{yr}$).

in which an artificial discharge of $100\,\mathrm{kg\,s^{-1}}$ is maintained for $10\,\mathrm{yr}$ and then the wells are closed. (a) For $\lambda_0 = 0.02$ (and hence $e_0 = 0.92$), the drawdown reaches an equilibrium (steady state) for $R \lesssim 1000\,\mathrm{w}$, for which $\tau = RC = 2.9\,\mathrm{yr}$. (b) For $\lambda_0 = 0.05$, with $e_0 = 0.57$, equilibrium occurs for $R \lesssim 2000\,\mathrm{w}$, for which $\tau = 3.8\,\mathrm{yr}$.

Apart from the non-linearity produced by $R(e)$, $C(e)$, this system behaves similarly to a single-phase system.

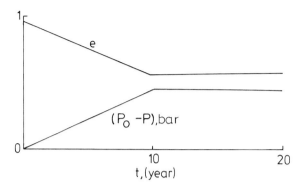

Fig. 11.20. Response, for a discharge of $100 \, \mathrm{kg \, s^{-1}}$ for 10 years, as in Fig. 11.19, for $T_0 = 250°C$ and $\lambda_0 = 0$ (no gas). Curves are of pressure drop, $(P_0 - P)$, and liquid volumetric saturation, e.

The behaviour of a reservoir with zero or minor gas content is quantitatively different. Fig. 11.20 shows the pressure response for $\lambda_0 = 0$, $T_0 = 250°C$, for a similar hypothetical drawdown experiment. Although the change in e is considerable, the change in system pressure is very small — at most about 0.45 bar in the example shown. The temperature, during this short interval, is buffered by the large thermal capacity of the (rock) reservoir and as a consequence the pressure is nearly constant. The extra pressure of a soluble gas is not available.

12. Wet Systems: Taupo

One of the best-known liquid dominated systems, that at Wairakei, New Zealand lies in a distinct zone of intense hydrothermal systems located as shown in Fig. 12.1.

The North Island volcanic belt extends from Mt. Ruapehu to White Island in the Bay of Plenty. White Island itself, and the three peaks Ruapehu, Ngauruhoe, and Tongariro are active volcanoes. Between Mt. Ruapehu and White Island, the 1886 eruption of Mt. Tarawera and a series of explosions at Frying Pan Flat are the only historic eruptions; but there are many youthful cones and domes (Kear, 1957), and carbon dating of volcanic ash and hydrothermal explosion deposits have given ages ranging from 4000 to 800 years ago (Baumgart and Healy, 1956; Lloyd, 1959).

The general geology of the Taupo District is based on the work of Grange (1937). Grange identifies the volcanic belt as occupying an extensive structural depression, which he named the Rotorua–Taupo graben. The oldest rocks are Mesozoic greywackes, which make up the ranges east and west of the depression. It is assumed that the basement in the central depression is formed by the same Mesozoic greywackes that are seen in the ranges; the exposed rocks are almost exclusively Pliocene to Recent acid volcanics. Extensive ignimbrites predominate, together with rhyolitic tuffs, domes, and flows. Dacite and andesite, though well distributed, are subsidiary, and there are a few isolated basalts, usually associated with large rhyolite masses.

12.1. GROSS FEATURES

Gravity survey and general structure

Fig. 12.2 shows a residual Bouguer anomaly map of the Taupo district. Where the basement topography is not too steep, the slab formula can be used to give a rough indication of basement depth, h, where $g' = 41.8\,h\Delta\rho$ milligal per $km\,g\,cm^{-3}$. Thus for example with $\Delta\rho = 0.24\,g\,cm^{-3}$ we have 1 km per 10 milligal. A suitable value for

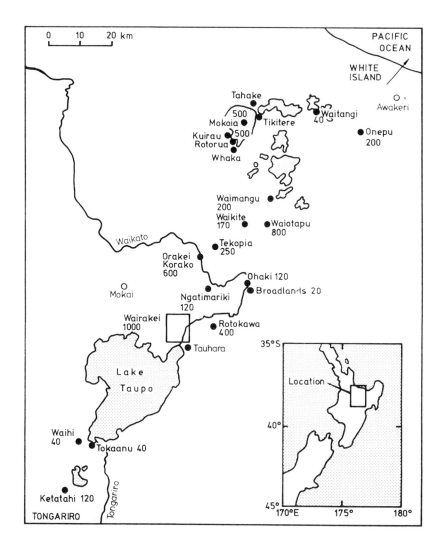

Fig. 12.1 Natural heat discharge in MW of the Taupo hydrothermal systems. Based on various measurements by: J. Healy (Grange, 1955); Ellis and Wilson (1955); Gregg (1958); Studt (1958); Lloyd (1959); Wilson (1960) and others. Round figures by the author.

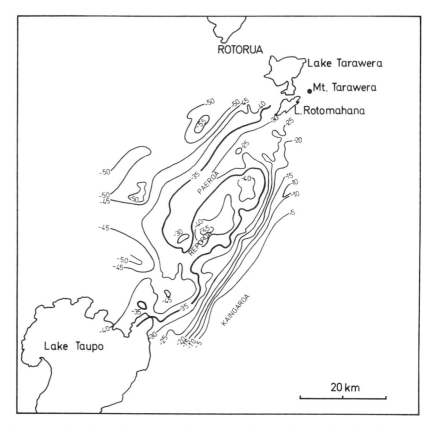

Fig. 12.2. Gravity pattern, in milligal, in the Taupo district. Residual Bouguer anomaly obtained from ordinary sea-level Bouguer values (adjusted for altitude, terrain and latitude, with $2.0\,g\,cm^{-3}$ for the assumed density of material above sea level) less the regional Bouguer values extrapolated from surrounding Mesozoic greywacke basement using $2.62\,g\,cm^{-3}$ in reducing to sea-level. Note that since most of the area is 200—400 m above sea-level, a density change of 0.1 g cm^{-3} produces about 1 milligal change in residual Bouguer value (From Modriniak and Studt, 1959).

$\Delta\rho$ is uncertain. A value of $0.5\,g\,cm^{-3}$ was used in the original analysis, in which depths to 4 km were found (Modriniak and Studt, 1959, their Fig. 5). A mean value assuming porosity 0.1 and equal parts of saturated breccia, ignimbrite, rhyolite and surface data suggest $0.3\,g\,cm^{-3}$. The effects of compaction at depth are unknown. Nevertheless, typical depths of around 5 km in the deeper parts of the depression are indicated.

Magnetic survey and the volcanic rocks

Whereas the gravity data are related primarily to basement structure, the magnetic survey shown in Fig. 12.3 gives the distribution
of the volcanic rocks within the cover, since the magnetization of
most of these is more than ten times that of the basement rocks.
The ratio of remanent to induced polarization is high in many of the
outcropping rhyolites, lower in the ignimbrites and the more basic
volcanics, and very small in the non-welded fragmental rocks. Detailed
mapping depends largely on this ratio, for the contrast in suscepti-
bility between the massive and fragmental rocks is quite small, many
of the so-called pumice breccias being tuffs, differing from the
ignimbrites only in the absence of welding. All these rocks are
normally magnetized.

Three areas can be distinguished. Over the Mesozoic rocks in the
east, the field is practically undisturbed, the normal latitude variation being the only significant feature, as is to be expected owing
to the weak magnetization of the greywacke. Over most of the
Kaingaroa plateau, there is still no very great departure from the
latitudinal variation, despite the high magnetization of many of the
ignimbrites. The reason for this is that such rocks occur in continuous,
uniformly magnetized sheets, which do not greatly disturb the regular magnetic pattern except at their edges. In addition, they are
underlain at depths of 100 m or so by virtually non-magnetic greywacke. Steep magnetic gradients coincide with the position of the
Kaingaroa margin, and in the depressions farther west the field
becomes complex. Although there are many anomalies which cannot
be directly correlated with the surface geology, this complexity
appears to be chiefly due to the rhyolitic rocks. Young andesite and
dacite masses like Maungaongaonga, although highly magnetized, do
not, as a rule, greatly influence the field, showing that they are
surface features with quite narrow roots.

Detailed stratigraphy

Ordinary field mapping together with numerous drill cores allows
an elaborate description of the stratigraphy. The following summary
is from Grindley (1965). Some typical sections are shown in Figs.
12.4—12.6. The principal formations shown in these sections are as
follows, starting at the surface:

(a) *Wairakei pumic breccias.* A material with relatively small particle size and high porosity (up to about 0.3), and a high permeability.

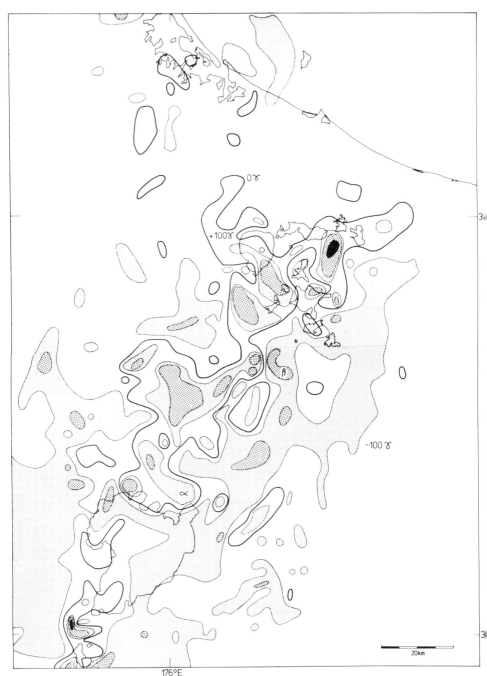

Fig. 12.3. Magnetic field amplitude variation in the Taupo district measured at 1.6 km above sea level (Gerard and Lawrie, 1955) with regional effects removed (Whiteford, 1976).

Heavy line: 0 gamma.   Light shading: amplitude > 100 gamma.
No shading: amplitude < 100 gamma. Heavy shading: amplitude > 200 gamma.

Known rhyolite domes, analysed in detail are: α, target anomaly, diameter 7.3 km, thickness 2 km; β, Waiotapu rhyolite, diameter 3 km, thickness 0.8 km. Assumed total magnetic intensity 200 gamma (Modriniak and Studt, 1959).

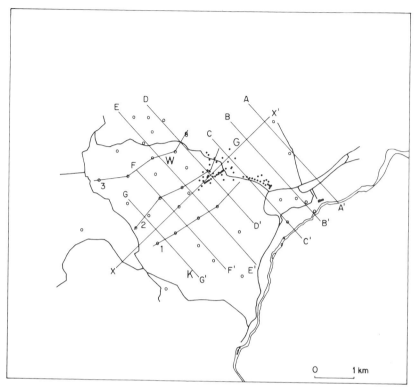

Fig. 12.4. Locality map for the Wairakei area. Various section and profile lines, referred to in the text, are indicated. Dots and small circles are bores. K = Karapiti; G = Wairakei Geyser Valley.

In some local thermal areas deposition of silica by hot spring waters may greatly reduce porosity and permeability, and in others acid alteration by steam and gases converts the breccia to impermeable clay minerals.

(b) *Huka formation.* This is a siltstone with a low permeability. It constitutes a partial cap formation over a large part of the area, and appears to separate the upper ground water system from that beneath, except where it is breached by permeable fault zones. Whilst the ground water table in the permeable formations above the siltstone has shown little regular change since the beginning of exploitation, the hydrostatic pressures in the formations beneath have fallen.

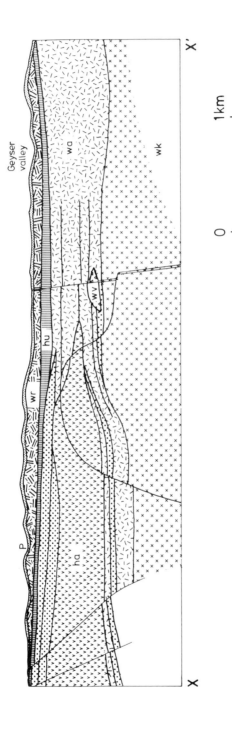

Fig. 12.5. Stratigraphic section at Wairakei (through bore holes: 220, 203, 216, 24, 15, etc.). No vertical exageration. After Grindley (1965).

p: pumice
wr: Wairakei breccia
hu: Huka formation (mudstones, early Hawera age)
ha: Haparangi rhyolite
wa: Waiora formation
wv: Waiora valley andesite
wk: Wairakei ignimbrites (Nukumaruan age)

(c) *Waiora formation*. This consists of permeable pumic breccias. It forms an important feature and many of the drillholes feeding the power station produce from some level in this formation, often from near the bottom.

(d) *Wairakei ignimbrites*. Ignimbrite is a relatively dense rock with low porosity and low permeability when unfractured. In large masses, it is generally traversed by many permeable joints and these, together with brecciated fault zones, result in a high average permeability. Also, the upper and lower surfaces of the formation may be brecciated. One section of the Wairakei ignimbrites forms the base of the present production area.

(e) *Rhyolite*. This appears as lens-shaped masses. In its physical properties, rhyolite has some similarity to ignimbrite. Permeability and porosity of the rock are low except where the structure is broken by faults, jointing, or surface brecciation. In any large mass, both the upper and lower surfaces are liable to be broken up by thermal and mechanical stresses during deposition, and considerable further shrinkage jointing will occur as the mass cools, so that overall permeability can be high while average porosity remains low.

(f) *Ohakuri formation*. This is a pumic breccia, non-welded lapilli crystal and vitric tuff. The assumed average thickness is about 400 m. The volcanic material below this formation is virtually unknown.

(g) *Greywacke basement*. This is alternating dark grey argillite and redeposited sandstone with minor tuffs, complex deformation, slaty cleavage, rodding and boudinage structures. Its thickness here is estimated to be about 7 km. The porosity is low but the macroscopic permeability is high.

The age of Wairakei thermal water

Ordinary springs are widely scattered over the earth's surface. Hot springs are also common but tend to be grouped within tectonic structures. The prominent hot springs of New Zealand fall into three groups: alpine margin, in the south and east; older volcanic district, in the north; younger volcanic district of the Taupo zone. Only the latter are our present concern.

Measurements of oxygen and hydrogen isotopic ratios at Wairakei, shown in Fig. 12.7, follow the pattern discussed above, of neutral

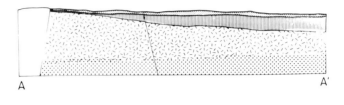

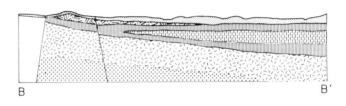

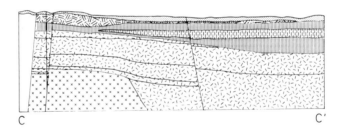

Fig. 12.6. (Above and facing.) Stratigraphic sections at Wairakei (After Grindley, 1965). Ornament as in previous figure.

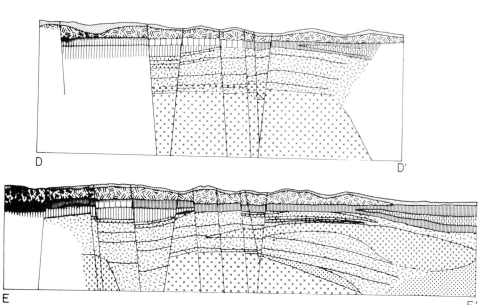

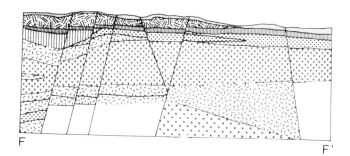

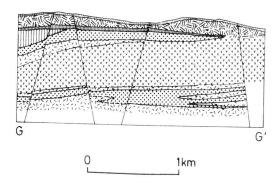

D   D'

E   E'

F   F'

G   G'

0   1km

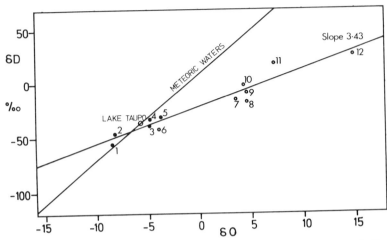

Fig. 12.7. Deuterium—oxygen isotope variation, in parts per thousand, diagram for New Zealand thermal waters. After Banwell (1963) from an original diagram by H. Craig. Discharge and pH: (1), Fumarole, Geyser Valley, 4.6; (2), Karapiti fumarole, 6.4; (3), Champagne pool, geyser valley, 6.5; (4), Bore 20, Wairakei, 7.7; (5), Black geyser, Spa, 4.8; (6), Crater lake, Ruapehu, 1.2; (7), Black geyser, Waiotapu, 1.7; (8), Perpetual spouter, Ketatahi, 1.7; (9), Terminal pool, Tikitere, 2.9; (10), Black pool, Ruahine, 1.7; (11), pool near Big Donald, White Island, 0; (12), Spouter, middle crater, White Island, 0.4.

and acid waters. This is meteoric water. The isotopic relationship of the acid springs is similar to that of other acid spring areas. The neutral waters are somewhat unusual. The near neutral deep chloride water at Wairakei shows no oxygen or hydrogen isotope shift. This has two extreme interpretations: (i) rapid passage of the meteoric water through the hydrothermal system; or (ii) a system of sufficiently long life that the ambient rocks are in isotopic equilibrium with the water. The isotopic evidence alone cannot resolve these possibilities. In the case of the Taupo systems with reservoir time-scale of about 250 yr and a total system life in excess of 0.5 Myr, so that the system has been flushed more than 10^3 times, isotopic equilibrium is likely.

Direct estimates of the ages of the thermal waters have been made using the naturally occurring radioactive isotopes of carbon, ^{14}C and hydrogen, ^3H ($=$ T, tritium).

Simple reservoir model

Consider the system shown in Fig. 12.8. Label the recharge, discharge systems with 0, 1. Define the quantities: M, volumetric flow

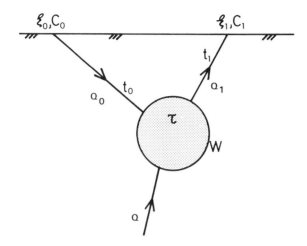

Fig. 12.8. Model and nomenclature for water age analysis.

rate: t, the time in an arm of the recharge—discharge system; C, concentration of the radioactive decay product; ξ, the ratio of the decaying isotope concentration to the stable decay product; λ, the radioactive decay constant; W, the reservoir volume. Let an amount $Q_1 - Q_0$ of dead water be added from depth, the fluid in the reservoir be well mixed, and the concentration be small. Then, simply by evaluating the amounts of the active constituent in its passage trhough the system we find:

$$\frac{\xi_1}{\xi_0} = \frac{C_0 Q_0}{C_1 Q_1} \frac{\exp[-\lambda(t_0 + t_1)]}{(1 + \lambda\tau)}, \qquad Q_1 = Q_0 + Q,$$

where $\tau = W/Q_1$. Here ξ_1/ξ_0 is the isotopic ratio relative to local meteoric water and τ is the mean residence time of fluid in the reservoir. Notice these cases:

(i) $\tau \approx 0$: a small reservoir, the system behaves like a pipe with a residence time $t = t_0 + t_1$;

(ii) $t \approx 0$, we have a sort of "wine-barrel" problem, the simple mixing of active meteoric water and dead water in the reservoir;

(iii) $\tau \approx 0$, $t \approx 0$ is the case of simple dilution.

The age of the deep water

The small amount of radiocarbon, with half-life of 5800 yr, in

CO_2 from Wairakei (Fergusson and Knox, 1959) indicates a residence time of order 10^4 yr for the bulk of the deep water.

The age of the water of the surface zone

Tritium with a half-life of 12.5 yr, usually measured in units such that 1 TU (tritium unit) corresponds to $T/H = 10^{-18}$ atom/atom, is a convenient marker for surface zone systems with residence time of 0–100 yr (Lal and Peters, 1962). Results obtained with this method are rather poor. Interpretation is made uncertain by the role of several reservoirs of various residence times and by the variable man made atmospheric tritium deposits from thermonuclear bombs. Various data are shown in Fig. 12.9. The natural background level has been estimated at about $\alpha = 1.5 \pm 0.5$ TU. Bomb tritium appears from 1956 onward. I have provided a time-scale of apparent age, t, from tritium value, $\xi = \alpha \exp(-\lambda t) + \beta$, which will apply for water older than that containing bomb tritium. Some water is close to the detection limit, of apparent age about 50 yr; some water is definitely of age of order 10 yr.

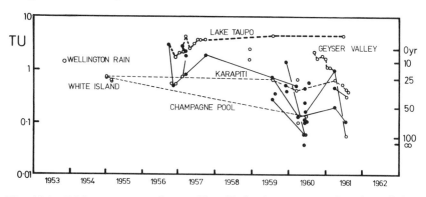

Fig. 12.9. Tritium content of some New Zealand waters as a function of time. *Note*: 1 tritium unit (TU) is such that $T/H = 10^{-18}$ atom/atom. A nominal age-scale for a natural level of 3 TU and instrumental background level of 0.05 TU is shown. (Nuclear bomb tritium detected in Hutt river mid 1959.) From a compilation by Banwell (1963).

Clearly these apparent ages arise from the discharge sampling water over a range of depths. It is interesting to note that early in its life a new bore has rising tritium values followed by a fall. This is to be expected since the initial drawdown requires a greater proportion of high-level fluid to enter the bore until the readjustment of the surface layers is complete.

About all that can be said from data of this type is that the surface zone at Wairakei has waters of ages from 0—100 yr.

Heat flow

Regional surface conductive heat flux.

Measurements of the conductive heat flux over the central part of the North Island are shown in Fig. 12.10. Five distinct regions can be recognized:

(1) In the NW, values are normal, about $50 \, mWm^{-2}$. Note that this includes the area of Mt. Egmont.

(2) In the SE, values are normal, but rather low, about $25 \, mWm^{-2}$.

(3) Between these two broad regions, in the south, a transition zone of width about 200 km, with intermediate values.

(4) Between these two broad regions, in the hydrothermal zone an abrupt transition region, of width about 30—40 km, in which the conductive heat flow is *zero*.

(5) Within zone 4, but not shown in the diagram, are small hot patches of high conductive and very high convective heat flux.

Heat flow by discharge

The surface heat flow for each of the thermal areas in the Taupo district has been determined by Healy (Grange, 1955). More recent work indicates the need to multiply his figures by about 2. This rather arbitrary figure is used here. The revised figures are shown alongside each area in Fig. 12.10. The total surface heat flow is 5000 MW. Included in the area of $2500 \, km^2$ are two intense groups, Waiotapu and Wairakei, with flows of the order of 1000 MW; four moderate groups, Rotorua, Tikitere, Rotokawa, and Orakei Korako, with flows of the order of 300 MW; seven small groups, with flows of the order of 100 MW; and four very small groups, with flows of the order 30 MW. The regional average heat flux is $2 \, W \, m^{-2}$, with values averaged over the more intense parts of each area of the order of $500 \, W \, m^{-2}$.

The Taupo thermal area needs, for a life-time of a million years, a supply of energy equivalent to that in a column of rock of depth 100 km and temperature excess 500 K, together with a transport

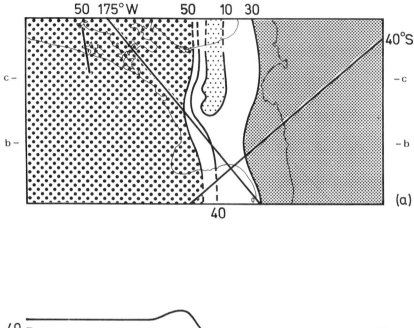

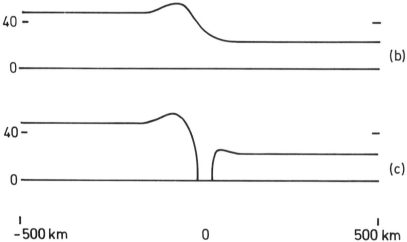

Fig. 12.10. Regional conductive heat flux in $mW\,m^{-2}$ near the Taupo hydrothermal systems. (*a*) An area 100 km × 500 km across the thermal zone. (*b*) Transverse profile, in the south-west and outside the hydrothermal zone. (*c*) Transverse
profile across the hydrothermal zone. Adapted from Studt and Thompson (1969)
and the author.

mechanism to transfer the energy from the rock at depth to the base of the Taupo hydrothermal systems.

Near surface temperature distribution

The location and form of individual systems in a liquid-dominated zone are most simply found from shallow temperature surveys. Data for the middle portion of the Taupo zone are shown in Fig. 12.11. The outline of these warm areas can also be found from ground or aerial surveys of plant distribution patterns.

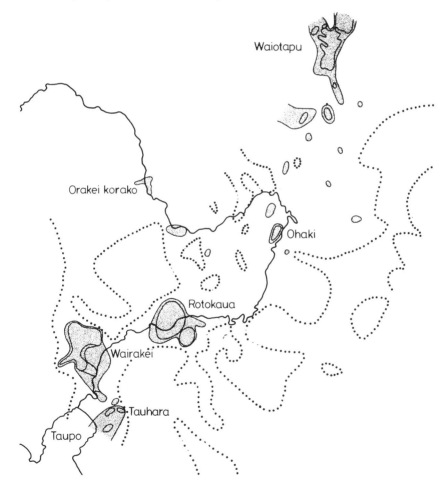

Fig. 12.11. Regional distribution of temperature at 1 m depth in part of the Taupo region. Shaded areas about $2°C$ above normal, dotted margin measurably above ambient. (After Thompson, 1960.)

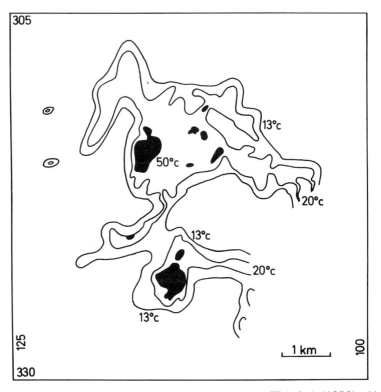

Fig. 12.12. Temperature distribution at 1 m depth, Wairakei (1958). (After Thompson *et al.*, 1961).

A detailed survey is shown for the Wairakei area in Fig. 12.12. The pattern is very complex. It is controlled by the variety of local discharge mechanisms.

12.2. TEMPERATURE AT DEPTH

Figure 12.13 shows vertical temperature sections. Figure 12.14 gives horizontal sections showing the isothermal patterns at various levels. These diagrams give a nearly complete piciture on the entire local hot area from the surface to a depth of about 1 km. There is a general similarity between the horizontal sections at all levels, but the vertical sections show a rather complex pattern of interpenetrating hot and cold zones, especially towards the margins.

The deepest set of isotherms shows a single inner area of maximum temperature (270°C) which might be regarded as the common

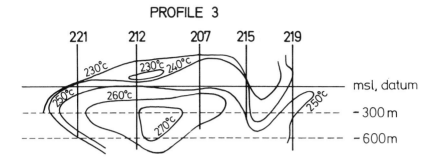

PROFILE 3

221   212   207   215   219

230°c   230°c 240°c

260°c

250°c   250°c

270°c

msl, datum

- 300 m

- 600 m

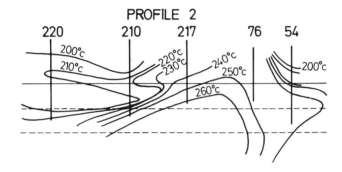

PROFILE 2

220   210   217   76   54

200°c

210°c

220°c
230°c

240°c
250°c   200°c

260°c

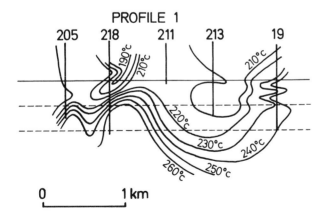

PROFILE 1

205   218   211   213   19

190°c   210°c   210°c

220°c

230°c

240°c

260°c   250°c

0   1 km

Fig. 12.13. Temperature distributions, vertical sections at Wairakei. Profile lines shown on Fig. 12.4. (After Banwell, 1963.)

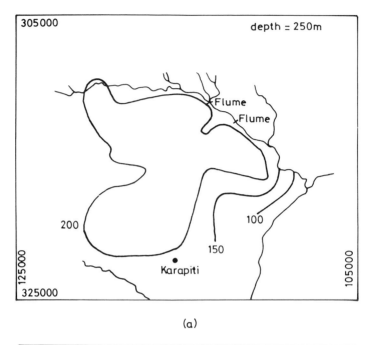

(a)

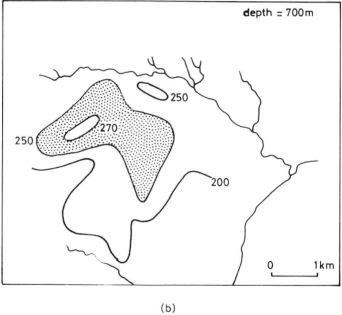

(b)

Fig. 12.14. (a), (b). See caption facing page.

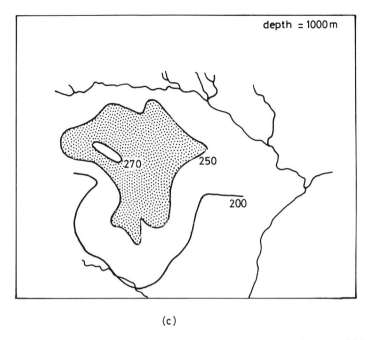

(c)

Fig. 12.14. Temperature distribution at various depths above datum at Wairakei: (a) 150 m; (b) −300 m; (c) −800 m. Datum is sea-level, about 400 m below mean ground level. (After Banwell, 1963.)

hot water source for the whole area. Reference to the sections, however, shows that this maximum is underlain by lower temperatures, so that the feed cannot be vertically beneath. The profiles suggest that the feed path dips towards the south and east, towards Bore 218.

The mushroom, its lobe structure

The temperature distribution at Wairakei is well known to a depth of 1000 m, poorly known from 1000 m to 1500 m and completely unknown below 1500 m. Nevertheless there are enough data to be able to identify a rather distorted "mushroom" of hot water. It should be noted that the raw temperature data are accurate only to ±2°C and that in the drawing of isotherms there is considerable freedom of choice when data points are widely spaced and the temperature distribution is more or less uniform. Fortunately the mushroom is located above 1500 m so that quite a good identification is possible.

A summary of the data presented above is shown in Fig. 12.15.

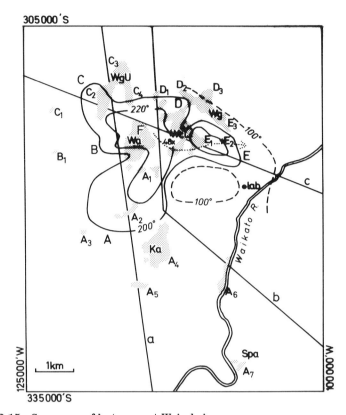

Fig. 12.15. Summary of hot areas at Wairakei:
  (i)   Areas of surface heat flow greater than $400 \, W \, m^{-2}$; after Thompson *et al.*, (1961). Total heat flow of area shown is about 1000 MW.
  (ii) Portion of $100°C$ isotherm at 250 m amsl (depth about 150 m); after Banwell (1957).
  (iii) $200°C$, $220°C$ isotherm at 150 m amsl (depth about 250 m); after Banwell (1957).
  (iv) Section lines a, b, c of Fig. 12.16.

The isotherms are centred below the surface thermal areas and are considerably simpler in form than the surface distribution. This pattern is just what would be expected from a section across the upper portion of a mushroom which is distorted by outward flowing lobes of hot water. There are five lobes, labelled A—E. Further details are revealed by three vertical sections, (a), (b), (c), shown in Fig. 12.16.

Only a system dominated by the flow of water could produce such distributions. These are now discussed in detail:

(a) Section (a) reveals, ignoring the details for a moment, a body of water at 200°C or more, 4 km wide and 1.4 km deep. The 200°C isotherm is closed, so that the hot water cannot enter the 4 km × 1.4 km volume by flow in plane (a), rather the flow must be inclined to plane (a). Since Fig. 12.15 reveals that the fluid to the west is colder and that to the east is hotter, the flow must be from the east (towards the reader).

Close inspection of the temperature distribution reveals three local maxima A, B, F; zones where the temperature exceeds 210°C, 220°C, 250°C respectively:

Local maximum A, corresponds to a flow of hot water from the east, more rapid than the general flow which produces the 200°C isotherm. This flow is part of lobe A which spreads to the west and in particular will feed thermal areas A_1, A_2, A_3.

Local maximum B, rather small, similarly corresponds to lobe B, feeding hot water to the west, in particular to thermal area B_1.

Local intense maximum F, also feeds hot water to the west; the upper portion of F can be identified as corresponding to fluid flowing from the east into lobe C.

(b) Section (b) similarly reveals a closed body of water at 200°C or more, 3.5 km wide by 1.4 km deep, but in this case since the hotter fluid is to the west, the flow of hot fluid is from the west (away from the reader). It may be argued, however, that the flow could be from the vicinity of lobe E toward the west; indeed in the early stages of exploration, when temperature data was available only down to 500 m, lobe E appeared to be possibly distinct from the zone of lobes B, C, D, but data to 1000 m reveals that there is no separation.

Whereas in section (a) the 200°C isotherm is nearly an oval, the 200°C isotherm of section (b) reveals a prounced angularity to the south, clearly produced by a flow from north to south in the plane of section (b), a flow of hot water over colder water at depth and in which the hot water loses heat to the ground surface. This is merely another view of part of lobe A as it moves south spreading as it goes, the portion shown in section (b) in particular feeding thermal areas $A_4 - A_7$.

There is a single local maximum F which corresponds to hot fluid moving to the east to feed lobes E, D. (These lobes are seen not to be as distinct as would appear from the large indentation DE shown in Fig. 12.15 — this indentation is a shallow superficial feature.)

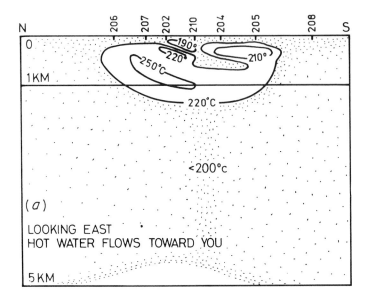

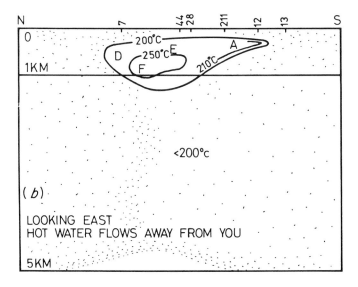

Fig. 12.16. The "mushroom" of the Wairakei hydrothermal system. Temperature distributions for vertical sections on lines a, b, c indicated on Fig. 12.15. Originally adapted by the author (Elder, 1966) from data presented by Banwell (1957, 1961). Figure continued on facing page.

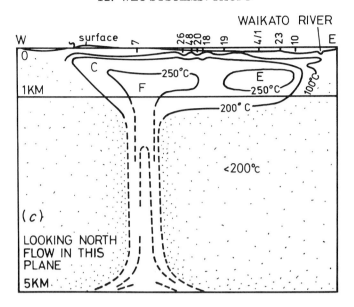

(c) Section (c) also reveals a body of water at 200°C, 4.5 km wide and extending at least to 1.5 km deep, but in this section it does not appear to be closed. The stem of the mushroom has been indicated (rather arbitrarily) in this section. There are local maxima E, F. Inspection of the isotherms shown in Fig. 12.15 suggests that in section (c) the fluid moves almost entirely in plane (c). The hot fluid which flows to the east supplies lobe E, the flow to the west supplies lobe C — the lobe C was poorly revealed in section (a) but here it is quite clear. Notice how the uplifted tips of F associated with lobes C, E could suggest, if only data to 500 m was available, the existence of two distinctly fed zones. The break shown in section (c) between F, E arises because the fluid which supplies E flows first slightly south and then into lobe E, so that while the fluid in local maximum E is moving largely to the east there is a small component to the north (away from the reader). The depth of this flow is 500–1000 m, and therefore below the indentation shown between lobes D, E.

It must be pointed out that the location of the mushroom stem with the present data is rather arbitrary and a more reliable location must await data to greater depths. Certainly it cannot be below the deep bore 48 which completely passes through the 250°C zone —

bore 48 also passes nearly through the 200°C zone since at 1.1 km depth the temperature has fallen to 210°C. The temperature distribution for bores 18, 19 (drilled to 950 m, 1050 m below datum) are similar to that of bore 48, so that between these bores the deep 200°C isotherm will be nearly horizontal: the stem will therefore not be located here. The location of the stem north and south is controlled by sections (a), (b): possibly as far north as bore 207, possibly as far south as bore 28; in section (a) the extent to which F extends to the south is not reliably known, it could extend possibly 1 km further; however, in section (b) the temperature distribution is quite well defined except at depth between bores 28, 211 but even with extreme distortion the stem could not be located further south than 0.3 km south of bore 28. A possible location compatible with all the available data is shown in Fig. 12.15 (under letter F).

Unfortunately, the temperature data are insufficient to completely discount the possibility of a second stem located below E, or the possibility that if there is a single stem it is under E feeding into F and thence to A, B, C, D. Ellis and Wilson (1960), nevertheless, present striking evidence for a common source of the water beneath E and F by a comparison of the atomic ratios of the chemical constituents of the waters. Further, they note that potassium is depleted in thermal water by deposition in the rock and that depletion is more advanced in E than F. (F is the region of smallest depletion.) A flow from F to E would produce this result since the fluid in E will have been longer in contact with the rock. Unfortunately, their data rely on bore discharges largely from depths of 300–600 m, so that there is a break between E and F as in the temperature section (this region is also very sparsely sampled). However, while their data do not preclude two separate but chemically identical sources for E and F they do preclude a flow from E to F. This conclusion is strongly supported by the pressure distributions evaluated by Studt (1957, 1958). They reveal beneath E a flow from west to east as dotted on Fig. 12.15.

The re-entrant zones AB, DE, EA do not necessarily represent inward-moving colder water, the temperature distribution could be produced by the hot lobes moving out into nearly stagnant colder water (possibly slow-moving recent meteoric water). This is certainly the case for zone EA which is a shallow feature extending down to 400 m below datum. The 100°C isotherms at 250 m below datum in the vicinity of zone AE are also shown in Fig. 12.15. Above this level the feature is established: on the southern side by the south-east spreading portion of lobe A, a thin shallow flow at 350 m below datum; on the northern side by the south-east flow of lobe E, a thicker and deeper flow which 500 m below datum, where lobe A does not exist, spreads nearly completely under the upper stagnant zone. Zone X of section (a) is probably also a nearly stagnant zone.

On the other hand, zone Y of section (a) probably corresponds to a weak shallow inflow of colder water from the west associated with the Wairoa thermal area Wa. Similarly the re-entrant zone DE appears to be a shallow inflow of colder water towards the south-west associated with the Lower Waiora thermal area WaL. It is 300 m wide and extends from 250 m to 400 m below datum. Below this depth the flow from the east is a single uninterrupted stream passing from D to E and spreading out under the stagnant zone EA.

There are thus three major lobes: A, which supplies thermal areas $A_1 - A_7$; C, which supplies B_1, $C_1 - C_4$; E, which supplies $D_1 - D_3$, $E_1 - E_3$. Lobe A, though rather shallow is the most extensive; the flow runs horizontally 5 km from F to thermal area A_6; it is quite possible that lobe A also supplies the small thermal area A_7 at the Spa, and may even run under Taupo township. Lobe E supplies the thermal area WaL by a largely vertical flow, but the Wairakei Geyser Valley Wg is supplied by fluid moving to the north-east.

The most complicated part of the system is in the vicinity of bore 48 where a strong vertical flow above the deeper lobe E is responsible for the Lower Waiora thermal area. The region is strongly affected by the colder inflow of DE from the north-east; a very complex pressure distribution is found, and during exploitation falls of water level have been largest here, as have changes in the surface heat flow.

There is little obvious correlation of the lobe structure of the mushroom with geological or topographical features:

(1) However, we notice that below 200 m the temperature distribution does not resemble the detailed pattern of the surface heat flow: this must be because below 200 m the flow is largely horizontal. Put another way, we see that the details of the surface heat flow arise in the upper 200 m of the system. In this upper zone there is an increasing tendency for the fluid to move vertically below the thermal areas, and associated with these vertical flows are inward flows of colder fluid. The depth of 200 m coincides with the upper surface of the so-called Huka beds, lake siltstones about 100 m thick. It is quite possible that these beds impede vertical flows except in fractured zones. The whole area is indeed heavily fractured and apart from the Waiora and Karapiti areas the thermal areas at Wairakei are rather elongated.

There has, however, been a considerable over-emphasis of the importance of fractures especially when there is no apparent correlation between reported fractures and the thermal areas; there is a strong possibility that one factor in the location of a thermal area is the existence of a fractured zone in an otherwise impermeable bed, but below 200 m depth, on the contrary, the evidence reveals that while the volume is heavily jointed and fractured there is no particular fracture of special importance.

(2) At depth, the existence of ignimbrite or rhyolite sheets does not appear to affect the temperature distribution as would be the case if they were impermeable.

(3) All the features of the temperature distribution can be produced by hydrodynamic factors provided suitable surface discharge areas are given. These surface discharge areas will be determined by the same topographical and structural factors that determine the location of the water table.

12.3. CONVECTIVE SYSTEM

The large heat flows *can* be supported by a free convective system. But can the quantitative results on heat flow found in the model experiments be applied to the geophysical system? There is an immediate difficulty; whereas in the laboratory, model experiments using only small temperature differences, γ/ν is nearly constant, with the large temperature differences found in the geophysical system, γ/ν is a rapidly varying function of temperature.

The best we can do is to evaluate γ/ν at ΔT.

This is unsatisfactory. Estimates based on this procedure may be in error by an order of magnitude. The quantity $\gamma/\nu \approx 0.1 \, \text{s} \, \text{m}^{-2} \, {}^{\circ}\text{C}^{-3}$ ± 25%. Hence

$$N = A_m / A_{mc}$$

becomes

$$F_1 = SkA_1 (\Delta T)^4 ;$$

$$S \approx \text{constant} \approx 3 \times 10^4 \, \text{W} \, \text{m}^{-4} \, {}^{\circ}\text{C}^{-4} .$$

Let us apply this expression to Wairakei. It is not clear to what extent the Taupo systems are separate; so firstly we consider a local system, say that at Wairakei with a heat flow of 1000 MW. Take $T_1 = 350°\text{C}$ and $k = 0.05$ darcy; substituting we find $A_1 = 40 \, \text{km}^2$. This value is reasonable, it corresponds to a circle of diameter 7.5 km, so that assuming $h = 5\,\text{km}$, $d/h = 1.5$, the Rayleigh number $A_m = 1.5 \times 10^4$ and the Nusselt number $N = 300$ — thus with $d/h = 1.5$, it would be necessary to reduce Ra by a factor of 200, to prevent the convective flow. In other words, the heat flow is 200 times equivalent conductive heat flow.

If, secondly, we regard A_1 as the whole area of the Taupo basement ($2.5 \times 10^3\,\text{km}^2$), the known value of $F_0 = 5000\,\text{MW}$ requires $T_1 = 215°\text{C}$. It is hardly likely that the basement temperature is uniform and extrapolation indicates basement temperatures in the range $0-400°\text{C}$; rather $T_1 = 215°\text{C}$ could be regarded as a mean basement temperature. It is more likely that the basement temperature is extremely non-uniform. Assuming the basement temperature is zero except in hot patches at $374°\text{C}$, an area of $250\,\text{km}^2$, only 10% of the basement, is necessary for the hot patches. Indeed if A_1 is regarded as the area of the intruded masses accessible to the circulating fluid, because these masses are not flat — more likely greatly contorted — the basement area affected may be much less than $250\,\text{km}^2$.

Collapse

The mushroom discovered at Wairakei is collapsed. At Wairakei, Thompson *et al.* (1961) give $F/F_1 = 0.96$. But the mushroom does not give the appearance of a collapsed system, since there are several discharge points.

Temperature variation with depth

It is remarkable that the temperature distribution with depth over a large part of the Wairakei area is below but close to the BPD in the upper $400\,\text{m}$. At first sight this appears to be fortuitous and of no special significance. We notice, however, that with the Nusselt number $N = 300$ and $d = 7.5\,\text{km}$ the heater boundary layer thickness $\delta_1 = d/N = 25\,\text{m}$. The surface boundary layer thickness will also be of this order. Unless other mechanisms appear, small values of δ_1 will require the temperature to exceed the BPD over portion of the boundary layer. At Wairakei, the flow in the upper 1 km is largely horizontal with cooling by conduction to the surface (as in the model experiments) sufficient to maintain temperatures below the BPD; the possibility of the Huka beds partially confining the fluid at depth will help this process since then the upper boundary layer will be attached to the lower surface of the Huka beds. If conduction to the surface is, however, not able to transfer all the heat supplied and no water discharge is possible, the water table temperature will rise till the heat can be discharged by the mechanism of steaming ground.

Stored energy

Accepting the point of view that a hydrothermal system can be compared to a mushroom of hot water (and rock) surrounded by cold water (and rock) the total energy stored in the Wairakei system can be estimated to be 10^{19} J (equivalent to 20 km^3 of rock at 200°C). This would be used up at the discharge rate of 1000 MW in 250 years.

If all the Taupo systems are treated in this way, and assuming that the ground is everywhere cold except in the mushrooms, the total energy stored in the Taupo systems is of order 5×10^{19} J. The total volume of the Taupo depression is of order 10^4 km^3, and if it were a normal region without convection, the energy content based on a mean temperature of 100°C would be 2×10^{21} J. Thus due to the convective systems the Taupo depression contains only of the order of 2% of the energy of a corresponding normal area. This is because most of the volume is filled with cold recharge water.

12.4. A PIPE APPROXIMATION FOR WAIRAKEI IN THE NATURAL STATE

Chloride discharge

During 1954, discharge from Lake Taupo to the Waikato river averaged about 140 ton s^{-1}. Between the lake outflow and Aratiatia rapids, a site downstream of the Wairakei area, the chloride concentration increased from 9.6 PPM to 15.3 PPM — the increase of 5.7 PPM presumably arising from the discharge of thermal chloride water of amount $5.7 \times 10^{-6} \times 140$ ton s^{-1} = 0.8 kg s^{-1}. The individual sources of this chloride water have been estimated similarly as shown in Table 12.1.

This chloride discharge is comparable with that of other thermal regions (Ellis and Wilson, 1955). With discharge of chloride in kg s^{-1}: Valley of Ten Thousand Smokes, Alaska, 1919 only, 40; Kilauea, Hawaii, 1; Yellowstone, 0.8; Californian hot springs, 0.5; Umnak Island, Aleutians, 0.2; Iceland, 0.2; Steamboat Springs, Nevada, 0.1; Arima, Japan, 0.05.

Table 12.1. *Chloride output, Taupo area.*

Region	Output (kg s^{-1})
Spa area	0.16
Waipouwerawera stream	0.02
Waiora valley	0.01
Wairakei: Bores	0.34
Geyser valley	0.16
Lower stream	0.02
Unaccounted	0.09
Total	0.80

Taking the mean chloride content of the deep water, as measured in Wairakei bores as 1320 PPM and assuming the chloride discharge comes solely and directly from this deep water, the thermal water discharge is $0.8/(1320 \times 10^{-6})$ kg s^{-1}, that is 0.6 ton s^{-1}. An early estimate of the corresponding heat flow, although erroneous, is of interest. Assume that the mean deep-water temperature is 255°C corresponding to the measured mean steam mass proportion of the bore discharges of 0.309 at the ambient boiling point of 99°C, and that all the heat arises from this water by cooling it to the ambient temperature of 12°C: then the energy available in the water fraction is 1063 KJ kg^{-1}. The liquid-water-carried heat flow is then 600×1063 KW = 640 MW.

Heat flow

Estimates of the heat flow from the Wairakei thermal area have been given by Ellis and Wilson (1955) and by Thompson *et al.* (1961). Another evaluation is given here as a preliminary step to establishing a pipe-approximation for Wairakei. A scheme of this is shown in Fig. 12.17 and should be studied as the argument proceeds.

Reliable data for the discharge from the Wairakei area is restricted to only two quantities:

(1) There is a surface discharge of $M_S = 300$ kg s^{-1} $\pm$ 10% of steam. This figure is based on the data of Thompson *et al.* (1961). Because of large systematic instrumental defects, early measurements underestimated the steam flow, so that we only consider the most recent data of 1960; an estimated 30 kg s^{-1} for the Spa area is included. Systematic errors may

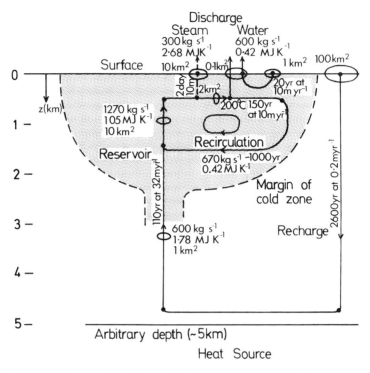

Fig. 12.17. Diagram of a pipe model of the Wairakei hydrothermal system.

still exist, since due to greater care and thoroughness, a further $40\,kg\,s^{-1}$ was "found" between 1958 and 1960.

(2) There is a natural discharge of $(415 \pm 45)\,g\,s^{-1}$ of chloride to the only drainage channel which leaves the area, the Waikato River (Ellis and Wilson, 1955).

The steam is considered to arise entirely by the evaporative mechanism of steaming ground by evaporation at the water table. A small percentage of flashed steam will occur in geysers but this is included in the water discharge. The defect in the calculation of Ellis and Wilson (1955) was their assumption that all the steam was flashed, in which case the $255°C$ water they considered would yield 31% steam; whereas evaporation at the water table does not restrict the ratio of steam/water. No restriction can be placed on the temperature of the evaporative surface; it can be in excess of $100°C$ or on the other hand much less; in the extreme case, all the energy flowing into a lobe could be lost from steaming ground. The small percentage of heat lost by conduction alone is included in the steam total: it is 4%

at Wairakei. It must be emphasized that cooling by loss of steam at depth is purposely ignored here.

The chloride discharge is considered to arise from the flushing of a body of "chloride water" at depth. If this body had a uniform chloride concentration equal to that of the mean chloride content of bore water, $1320\,\text{PPM}$, the discharge would be $M_W = 315\,\text{kg s}^{-1} \pm 10\%$ of "chloride water", here taken as $300\,\text{kg s}^{-1}$. (It is fortuitous that the discharges of "chloride water" and steam are equal.)

The mean temperature at which the chloride water discharge enters the surface systems is difficult to estimate; it could be as high as $250°\text{C}$. There is little modification of isotherms below the $200°\text{C}$ level by the surface systems (possibly influenced by the presence of the bottom of the Huka beds at the $200°\text{C}$ level); the temperature at the base of the surface systems is taken here to be $200°\text{C}$.

The heat flow is then $240\,\text{MW}$ as chloride water, and $760\,\text{MW}$ as steam, with a total of $F_0 = 1000\,\text{MW} \pm 10\%$. The possible error arises about equally from possibly errors in the field data and the uncertainty in the proportion of the energy of the chloride water discharge which is given up as steam.

Surface zone

Temperatures measured in boreholes reveal a relatively homogeneous body of $250°\text{C}$ water at depths between 0.5 and 1 km. A high proportion of the energy lost from this water passes through the surface zone as evaporated steam. There is more than sufficient loss of steam to lower the water from $250°\text{C}$ to the $200°\text{C}$ of the chloride water. However, the chloride water, since it is discharged at the surface, cannot exceed $100°\text{C}$. This loss of enthalpy could also be due to loss of steam but, if this is to be the case, since surface water discharges have average salinity no more than $1000\,\text{PPM}$, it is necessary to invoke some "mixing" with cold ground water. Thus if all the chloride water discharge is cooled by "mixing", $300\,\text{kg s}^{-1}$ of cold ground water must flow in the surface zone to reduce the $300\,\text{kg s}^{-1}$ flow of chloride water at $200°\text{C}$ to $100°\text{C}$. Until simultaneous measurements of both temperatue and salinity in the surface zone are made the "mixing" hypothesis remains a speculation. Thompson *et al.* (1961) have only been able to identify a water discharge of $200\,\text{kg s}^{-1}$, but this must be regarded as a minimum value, since loss by seepage is difficult to detect.

Deep circulation

The net surface loss of $1000\,\text{MW} \pm 10\%$ and $600\,\text{kg s}^{-1} \pm 10\%$ of fluid requires at depth a flow of $600\,\text{kg s}^{-1}$ of fluid of enthalpy $1760\,\text{kJ kg}^{-1} \pm 20\%$. This enthalpy corresponds to a temperature of $355 \pm 30°\text{C}$ if the fluid is at its boiling point (depth 2.5 km) or $410°\text{C}$ at depth of 5 km. These values are close to the critical temperature $T_c = 374°\text{C}$ and at the present stage are quite acceptable.

No restriction can be placed on the depth of the $600\,\text{kg s}^{-1}$ recharge except that it is below 2.5 km.

Recirculation

The model studies revealed a fundamental feature of the circulation of a free convective system undergoing discharge — "exhausted" water is recirculated at moderate depth, there being a clear separation between the recirculating water and the recharge water which flows to great depths. Indeed, the surface thermal areas will be entirely within the curve where this boundary meets the surface.

If the flow rate of recirculated water is $Q'\,\text{kg s}^{-1}$ and its temperature is arbitrarily taken as $100°\text{C}$, since the recirculated water together with the water from depth must produce water at $250°\text{C}$, we have the energy equation in units of kg K s^{-1}:

$$2.5 \times 10^5 + 100\,Q' = 250\,(600 + Q').$$

Hence, $Q' = 670\,\text{kg s}^{-1}$ and the flow of $250°\text{C}$ water is $1270\,\text{kg s}^{-1}$.

Notice that the total enthalpy flow of the $250°\text{C}$ water is $1340\,\text{MW}$, greater than the net flow by the recirculation flow of $280\,\text{MW}$.

Circulation times

To complete the approximation of the flow by that in a set of pipes it remains to specify the cross-sectional areas of each pipe. The values shown are rather arbitrary. The surface areas used are based on Fig. 12.15, the cross-sectional area of the rising column is evaluated from the results of the model experiments and the cross-section area of the upper horizontal flow is estimated from lobe cross-sections shown in Fig. 12.16. The corresponding velocities can then be calculated and consequently the time to flow along each branch. The times are surprising both for their magnitude and the large range of scales: of order 10^4 years for the deep recharge, 10^3 years for the immediate recirculation and recharge, 10 years for the surface discharges.

12.5. EFFECTS OF EXPLOITATION

Drilling commenced in 1950 and continued till 1968 with about 8 new bores per year. The majority of bores reach depth 600—700 m. The production area is roughly 3 km^2, but the bulk of the output is drawn from a much smaller area — 80% production from 0.5 km^2. The discharge history is summarized in Table 12.2 and in Fig. 12.18.(a) The mass discharge rate is primarily determined by external constraints (number of bores, management policy, etc.). It increased more or less uniformly to more than 2 ton s^{-1}. The subsequent decline is, however, largely owing to the progressive degradation of the reservoir (more about this below). (b and c) Throughout the reservoir

Table 12.2. *Annual mean quantities for Wairakei*, 1953—1976. Discharge rate, M; mean specific enthalpy of bore discharge, h; total borefield discharge power, F; changes in mean pressure, ΔP at 274 m below sea level (NZMWD data base); and values from the simple model.

	M (ton s^{-1})	h (MJ kg^{-1})	F (10^3 MW)	ΔP (bar)	Model ΔP (bar)
1953	0.16	1.035	0.17	0	0.1
54	0.23	1.051	0.24	0	0.4
55	0.34	1.075	0.36	0	0.9
56	0.67	1.106	0.74	0.1	1.6
57	0.64	1.121	0.72	0.5	2.5
58	0.68	1.076	0.73	1.7	3.3
59	1.19	1.072	1.27	3.1	4.5
60	1.59	1.122	1.79	5.2	6.3
61	1.34	1.121	1.50	7.6	8.0
62	1.64	1.105	1.81	9.7	9.6
63	2.35	1.142	2.69	13.3	11.8
64	2.29	1.140	2.62	15.5	14.4
65	2.05	1.150	2.36	17.6	16.4
66	1.98	1.142	2.26	18.6	18.0
67	1.89	1.126	2.13	19.2	19.2
68	1.52	1.148	1.75	20.0	20.0
69	1.76	1.144	2.00	20.6	20.6
70	1.70	1.126	1.92	21.4	21.3
71	1.73	1.105	1.91	21.7	21.9
72	1.64	1.109	1.82	22.0	22.3
73	1.54	1.116	1.71	22.3	22.6
74	1.49	1.113	1.66	22.6	22.7
75	1.47	1.064	1.56	22.8	22.8
76	1.49	1.090	1.63	22.8	22.8
Average	1.39	1.17	1.63		

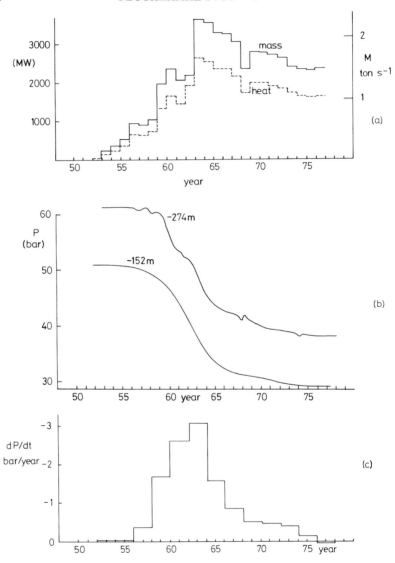

Fig. 12.18. Wairakei field characteristics during exploitation. (*a*) Discharge rate of mass, m, in ton s^{-1}; heat F, in MW. (*b*) Reservoir pressure, in bar, at depths 152 m, 274 m below sea-level. (*c*) Annual mean rate of change of pressure, in bar yr^{-1}. (Taken from a compilation by Allis, 1978).

pressures fell uniformly. After an interval of about 20 years a new (only apparent) hydrological equilibrium has been reached with net falls of about 23 bar.

The reservoir fluid

The main change in the (apparent) state of the reservoir fluid has been in its specific enthalpy; see Fig. 12.19. In the natural state at production depth the fluid was entirely liquid, with liquid volume fraction $e = 1$.

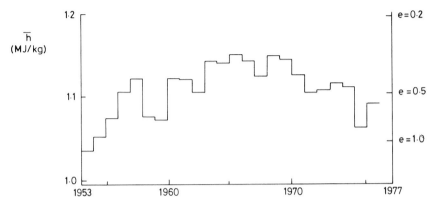

Fig. 12.19. Annual mean discharge specific enthalpy, $\bar{h}$, of all Wairakei bores as a function of time. Also shown, lines at specific enthalpy for water—steam mixture of temperature $245°C$ volumetric liquid fraction, e.

(i) Already, however, by 1954, the mean enthalpy indicates that boiling is occurring in the ground. Since the gross field pressure change is only 0.1 bar, this must be a local effect confined to the vicinity of individual bores, as a consequence of local drawdown.

(ii) During the interval of near maximum discharge the liquid fraction, $e \approx 0.4$ and some of this possibly arises because of the overall pressure fall.

(iii) During the current interval of diminishing discharge the enthalpy is falling again. Since the field remains depressurized the enthalpy fall must come from effects localized to bores.

We are forced to the conclusion that the bulk of the variation of specific enthalpy arises through local effects.

Ground subsidence

During the period of exploitation the ground surface has fallen by as much as 5–10 m. The average rate of subsidence is shown in Fig. 12.20. The greatest rate is near the production area but somewhat displaced, about 1 km east, from the area of the bulk of the withdrawal of water.

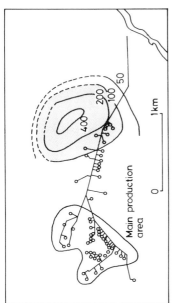

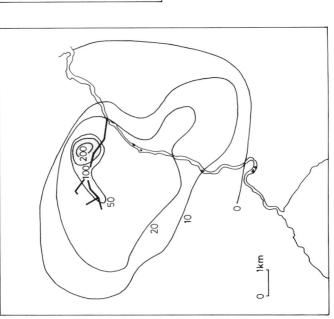

(a)

(b)

Fig. 12.20. Ground subsidence at Wairakei average rate in mm/year: (*a*) 1956—1971; (*b*) detail for production area, 1964—1974.

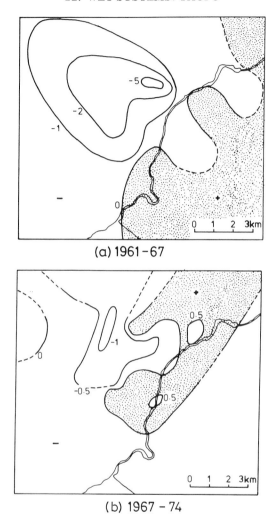

(a) 1961–67

(b) 1967 – 74

Fig. 12.21. Gravity changes at Wairakei, μN kg^{-1} corrected for elevation changes: (a) August 1961—April 1967; (b) April 1967—December 1974. (After Hunt, 1977).

Net reservoir mass loss

Gravity surveys of the area were made from 1950. Of these a consistent reoccupation was made in 1961, 1967, 1974. The distribution of gravity change is shown in Fig. 12.21. This data is corrected for subsidence. From measurements at a bench mark (A97) on the eastern side of the production area, there is (only) a spot value for the change during 1950–1961. The gross changes are summarized in Table 12.3, which gives the integral of the gravity change over the anomaly area, and the mass changes ΔM expressed as equivalent volumes of cold fluid $\Delta V = \Delta M/\rho_0$. These data are interpreted as follows.

Table 12.3. *Total mass changes at Wairakei.*

	$\int g' \mathrm{d}A$ ($\mu N\,kg^{-1}$) km^2	Net loss	ΔV (km^3) Withdrawn	Replaced
1950–61	(-82)†	(0.196)	(0.170)	(0.026)
1961–67	-98	0.234	0.360	0.126
1967–74	-15	0.036	0.400	0.364
Total 1950–74	-195	0.466	0.930	0.516
	(a)	(b)	(c)	(d)

Notes: (a) Area integral of gravity change.
(b) Net loss of equivalent volume $= \int g' \mathrm{d}A /2\pi G \rho_0 ; \rho_0 = 1\,t\,m^{-3}$.
(c) Measured discharge.
(d) Replaced $= (c) - (d)$
† Based on a single station (A97): with g' (1950–1961) $= -4.2\,\mu N$ kg^{-1}; g' (1961–1967) $= -5.0\,\mu N\,kg^{-1}$ and assumed similar distribution of change as for 1961–67 – a very dubious assumption.
Original data from Hunt (1977).

Suppose a net mass change ΔM is produced in a finite volume because of a density change, by rearrangement of the matter in the volume and its surroundings. Then, from Gauss's Theorem, the integral of the gravity anomaly over the area of the anomaly is $2\pi G \Delta M$. Consider the average value of the gravity anomaly, g' over an anomaly area, A. Then

$$g' = 2\pi G \Delta M/A.$$

In the interval 1950–1974, of the $0.93\,km^3$ of equivalent liquid withdraw, $0.52\,km^3$ has been replaced and $0.47\,km^3$ has been withdrawn from that previously stored in the system. The amount taken from "storage" appears to be close to its final amount by 1974. Qualitatively this behaviour follows that of the hydrological system.

12.6. RESERVOIR PARAMETERS

Let us see if the Wairakei system can be represented by a single open liquid-filled reservoir of resistance R and capacitance C (as described in section 11.2).

Simple model

The model system for the fall in head h is:

$$\mathrm{d}h/\mathrm{d}t = (D/C) - h/\tau,$$

where: the borefield discharge rate $D = D(t)$ is given; $\tau = RC$; the pressure fall, $\Delta P \equiv \rho_0 g h$; and at some time $t = t_0$ we have $h = 0$. Values of $h(t)$ are readily obtained by integration and the corresponding ΔP can be compared with the measured data. For the data of Table 12.2, the best fit is for: $R = 160$ w, $C = 1.86$ km^2 with mean absolute error of 0.6 bar. The time-scale $\tau = 9.4$ yr. Note that for these values I have taken the annual average pressure drop as being the mid-year values and fitted them to mid-year values of ΔP.

It is worth noting that the determination of R, C, are predominantly: for R from the data as $t \rightarrow \infty$, since then $h \sim D/R$; for C from the data as $t \rightarrow 0$, since then $h \sim Dt/C \sim$ (total discharge)$/C$.

There is a certain amount of flexibility in the data and the fit. The above fit is poorest in the early stages. The average difference, (model $-$ data) for 1953–1962 is 1.0 bar. This raises the question of the accuracy of the data. An error of 0.5 bar is possible in the early data. Thus if we add 0.5 bar to the measured data of the pressure loss, corresponding to a slightly more rapid early drawdown, the model values are $R = 163$ w, $C = 1.81$ km^2 and mean absolute error 0.5 bar. The fit is not really any better. (Adding 1 bar gives: 165 w, 1.74 km^2, error $= 0.5$ bar).

The simple model is a good representation of the Wairakei reservoir and presumably of other open liquid systems.

Additional estimate of reservoir parameters from large-scale shut-in test

From mid-December 1967 for 109 days the field mass output was reduced from 1.9 to 0.7 ton s^{-1}, that is to about 40% of its previous and subsequent level, by closing and reopening lower enthalpy bores (Bolton, 1969). Some of the data are shown in Fig. 12.22. The field pressure began to recover, apparently immediately, at a rate $\mathrm{d}\bar{P}/\mathrm{d}t \approx +3.3$ bar yr.$^{-1}$ This rate is similar in amplitude to the rate in 1962–1963 which was -3.0 bar yr.$^{-1}$

The behaviour is a most valuable clue to the state of the reservoir in its exploited state.

A qualitative explanation of this recovery is straightforward in terms of our simple open liquid reservoir model. At the time of the partial shutdown the hydrological system is approaching apparent

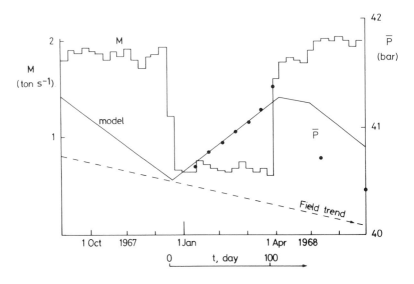

Fig. 12.22. Pressure response during a partial shutdown Dec 1967–Mar 1968. Mass discharge rate, M, in ton s^{-1} and average pressure, P, in bar; 274 m below sea-level.

equilibrium: at the end of 1967 the mean pressure drop was 19.6 bar to be compared with the equilibrium value of apparently 22.8 bar (Table 12.2). Thus, a change in discharge rate produces an effect as if it started from scratch. Furthermore, with a lowered discharge rate the new ultimate drawdown, proportional to discharge, will be smaller, so that the pressure begins to rise.

A quantitative representation using the simple model with $R = 160$ w, $C = 1.86$ km^2, taking the discharge rate as: 1.9 ton^{-1} before the partial shut down; 0.6 ton^{-1} during the shut down of nominal 105 days; 1.5 ton^{-1} for 2 months after the shut down; thereafter at the original rate of 1.9 ton^{-1}, leads to a pressure change during shut down of $\delta P = 0.29$ bar. The actual change is about 0.8 ± 0.1 bar.

Can something be deduced from this discrepancy? Since the final state is determined by R and not by C, with this model there is no flexibility in the choice of R. Indeed the discharge, pressure measurement model-fitting is sensitive to the choice of R and dominated by data in the near equilibrium range. Therefore the partial shut down can be regarded as a measurement of the capacitance near equilibrium, C^1. With $C^1 = 0.7$ km^2 the model gives $\delta P = 0.77$ bar. (Note that $\delta P \propto 1/C^1$). The curve shown on Fig. 12.22 is for these values.

How is this much smaller capacitance to be reconciled with that previously estimated? The effect of the capacitance is dominant in the early stage of a draw down experiment but as a new equilibrium is approached its role falls to zero. Thus the value of $C = 1.86 \text{ km}^2$ is valid for the earlier, more or less unexploited field; the value $C^1 = 0.7 \text{ km}^2$ applies in the exploited state.

The production zone has been sufficiently depressurized to create a two-phase zone. Then the effective capacitance is a function of the fluid state and in particular its liquid fraction, e. Near $e = 1$ we have $C/C_1 \approx eP_1/P$, where $P = P_1$ and $C = C_1$ when $e = 1$. The specific enthalpy of the discharge shows that during 1967–68, $e \approx 0.4$, which suggests $C_1 \approx 0.4 \times 1.86 \approx 0.7 \text{ km}^2$.

The test provides a measurement of the capacitance and thereby the liquid fraction. The partial shut down test behaviour is, therefore, dominated by the reservoir mean liquid fraction.

The model is, however, clearly oversimplified since the value $C^1 = 0.7 \text{ km}^2$ gives a much too rapid field trend prior to the parital shut down. We have ignored the local effects near bores. Prior to the partial shut down local conditions would be close to quasi-equilibrium since the discharge rate had been more or less constant for several years. Following the test, however, it is apparent that the local ground has been disturbed and is not in local equilibrium and immediately after the partial shut down responds to the current disturbed conditions.

A more complete analysis of this test is not given. It is sufficient here to demonstrate that the apparent capacitance of the reservoir diminishes as exploitation proceeds.

The reservoir and the behaviour of peripheral bores

The discharge area is quite a small fraction of the area of the whole reservoir. Nevertheless, gross pressure drawdown spreads rapidly and nearly uniformly throughout the reservoir. It is this observation which allows us to represent the Wairakei reservoir as a single open box. Peripheral bores, however, have a muted trend as shown in Fig. 12.23.

Comment on the adjacent Tauhara sub-system

The Tauhara region is clearly linked hydrologically to the Wairakei region since pressures follow those at Wairakei with a lag of a few months and a reduced amplitude of change. An incomplete very rough identification of the system can be made. A possible arrangement

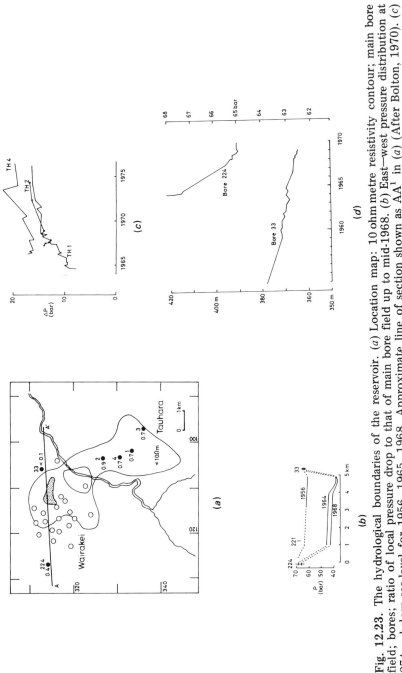

Fig. 12.23. The hydrological boundaries of the reservoir. (a) Location map: 10 ohm metre resistivity contour; main bore field; bores; ratio of local pressure drop to that of main bore field up to mid-1968. (b) East–west pressure distribution at 274 m below sea-level for 1956, 1965, 1968. Approximate line of section shown as AA[1] in (a) (After Bolton, 1970). (c) Pressure drop relative to estimated local undisturbed pressure as a function of time in Tauhara bores TH 1, 2, 4 (NZMWD data). (d) Water level changes in bores 33, 224 and estimated absolute pressure at 274 m below sea level. Note expanded pressure scale (NZMWD data).

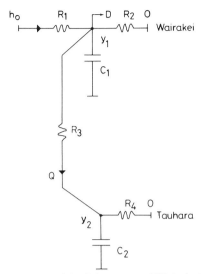

Fig. 12.24. Schematic model of the coupled Wairakei—Tauhara system.

is sketched in Fig. 12.24, which shows the Tauhara region as an area of the Wairakei hydrothermal system. For a steady state, or the hydrological time-scale of the Tauhara area somewhat less than that of the main system, the ratio of pressures changes is:

$$\Delta y_2 / \Delta y_1 \;=\; R_4 / (R_3 + R_4).$$

This has been measured to be about 0.7 (and assuming that the effect is not because of a time lag), so that $R_3 / R_4 \approx 0.4$. If we take the lag in response as about 3 months, and $C_2 \sim 1 \text{ km}^2$, then $R_3 \sim 10\text{ w}$, $R_4 \sim 25\text{ w}$.

The simple model and the gravity measurements

In the simple RC liquid reservoir model the total equivalent volume of fluid permanently removed from the system is $Ch(t)$. This amount, as a function of time, is compared in Fig. 12.25 with the amount removed as estimated from the areally integrated changes in gravity. The two estimates are the same within the presumed limits of accuracy ± 10—20%.

The fit is questionable because of the method of estimating the integrated gravity change in 1950—1961. To presume that the areal distribution of g' is the same in the interval 1950—1961 as that of 1961—1967 (which ignores any time lags in the peripheral parts of the field, both because of hydrological processes and the continued

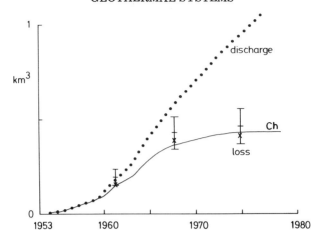

Fig. 12.25. Mass loss from Wairakei, 1950—1976, expressed as equivalent liquid volume. (i) Heavy curve: model estimate, Ch for $R = 160$ w, $C = 1.86$ km^2. (ii) Large crosses: areally integrated gravity change estimate. Vertical lines indicate $\pm 20\%$ error range (adapted from Hunt, 1977 — see text.) (iii) Small crosses: as for (ii) but with 1950—1961 gravity change estimate adjusted to give best fit (see text). (iv) Dotted curve: total discharge, for comparison. Note that total recharge = (discharge — loss) (NZMWD data).

installation of new wells) probably gives an upper bound for the integrated gravity change for 1950—1961. In order to estimate the magnitude of a possible mis-estimate, let us find the value of the integrated gravity for 1950—1961 which gives the best fit to the simple model. The required value corresponds to a net loss for 1950—61 of 0.148 km^3 instead of 0.196 km^3. The mean error from the three gravity measurements is then 0.014 km^3. The required value for the integrated gravity change would be -62 instead of $-82 \, \mu$N kg^{-1} km^2; the gravity change at station A97 would be -3.13 instead of $-4.15 \, \mu$N kg (all changed in the ratio 0.755). These values are quite acceptable but we do not know what they actually were.

In spite of the uncertainty in the data and the model this comparison is of the first importance in the identification of the hydrothermal system. It is the only direct identification of the mass change in a hydrothermal system known (to me at least) that is so unequivocal. The fluid withdrawn from Wairakei affects the system as if it came from a simple open liquid reservoir. In my view, this set of observations — total discharge, system pressure, integrated gravity change as functions of time — together with their representation by means of a simple open liquid reservoir model, is the outstanding piece of verifiable information about hydrothermal systems.

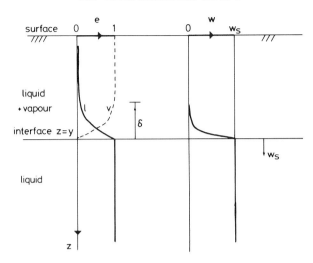

Fig. 12.26. Schematic diagram of the surface zone during exploitation. Qualitative vertical profiles of: (i) liquid volumetric fraction, e; (ii) vertical volumetric velocity, w.

12.7. THE SURFACE ZONE DURING EXPLOITATION

The consistency of the gravity and hydrological data suggest that the relative amount of matter in the surface zone, namely that above the equivalent liquid level, h, of the reservoir capacitance, is small. This suggests the arrangement of the surface zone sketched in Fig. 12.26. There is a sharp interface below which all the fluid is liquid and above which the fluid is in two phases. The interface at depth $z = y(t)$ is descending at a rate $W_s = \mathrm{d}y/\mathrm{d}t$, determined by the artificially imposed discharge and the hydrological parameters of the reservoir. Above the interface there is a net downward drainage of liquid. The vapour will have a net vertically upward velocity or could be nearly stagnant. The key point from the gravity data is that the vertical length scale, δ, of the drainage zone is such that $\delta \ll \Delta y$ where Δy is the change of interface level under exploitation. This requires that the potential rate of drainage of liquid is very much greater than W_s.

The rate at which drainage could occur can be estimated as follows. Consider for the moment a slug of liquid in a permeable vertical column. If the slug falls as a connected body its limiting vertical velocity $W_1 = kg/\nu$, independent of its vertical extent. For example, with water at $250°C$ so that $\nu = 1.34 \times 10^{-7}\,\mathrm{m^2\,s^{-1}}$ in a medium of permeability $k = 10^{-13}\,\mathrm{m^2}$ (0.1 darcy), then $W_1 = 231\,\mathrm{m\,yr.^{-1}}$. If the liquid occupies only a volumetric fraction e of the void space its velocity is $W = W_1 F_l \sim W_1 e^3$. Thus, as the layer is progressively depressurized so that e falls from unity to zero, the drainage velocity will fall from of order 10^2 m yr^{-1} to a very small value. For a total loss of head of about 220 m over about 20 years the zone will be nearly fully drained throughout except for a tiny residual liquid fraction.

The draining liquid will presumably move not as a single connected mass but as isolated slugs of liquid progressively coalescing as the fully liquid level is approached. It should be noted that this is a counter-current slug flow of liquid and vapour.

Considerable support for this picture is provided by data on chemical trends during exploitation. These trends can be described simply by assuming that solutes are carried down through the drainage zone into the liquid zone below (Ellis and Mahon, 1977; Glover, 1970).

13. Gassy Systems: Ohaki

The dominant feature of the mechanism of a liquid system such as that at Wairakei is the hydrology. Where, however, there is sufficient soluble gas in the fluid, the system is dominated by the thermodynamics of the fluid. Such a system is that at Ohaki—Broadlands.

In the course of developing the geothermal resources of New Zealand, the Ohaki—Broadlands system was developed to near full production prior to building an electric power station of proposed output of about 100 MW(e) when, in 1971, further development was suspended. This has provided a unique opportunity to study the partial drawdown *and recovery* of an isolated system. Since the high-level reservoir is gassy, the system is of exceptional interest. Thus, inadvertently, we have had in effect a large-scale experiment. I might add in passing that such experiments, in my view, should be an essential prerequisite before the final production phase is planned, the discharge during the experiment being used in an appropriate pilot plant.

13.1. THE OHAKI—BROADLANDS SYSTEM

The Ohaki—Broadlands system has been studied in detail (NZMWD Report 1977, from which this summary is drawn). As shown in Fig. 13.1, it straddles the Waikato river, NE of Wairakei.

Surface manifestation

Surface activity is confined to the area of warm ground, about $10\,km^2$, shown in Fig. 13.2. The ambient mean temperature is $12.5°C$. The major discharge, of about $20\,kg\,s^{-1}$, is in the Ohaki area in the NW and comes from a few hot pools, a little steaming ground, mud pools and seepages. A prominent feature is the Ohaki Pool, of area $850\,m^2$, which has discharged up to $10\,kg\,s^{-1}$ of $98°C$ chloride water. The small Broadlands area to the east with a few warm springs, gas discharges and little steaming ground discharged about $2\,kg\,s^{-1}$.

277

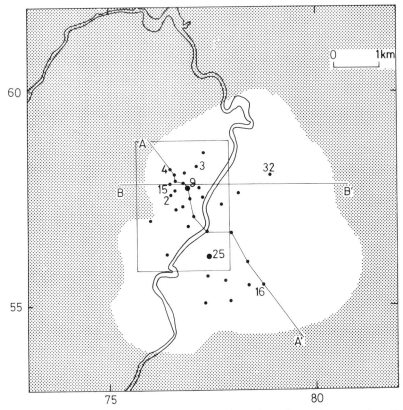

Fig. 13.1. Ohaki—Broadlands hydrothermal system. Locality map, borefield distribution and section lines.

Various estimates of the heat discharge rate have been made. For example Dickinson (1968) gives the following data for a total of 73 MW:

Flow from springs	6.25 kg s⁻¹	1.7 MW	(70—80°C)
Evaporation from pools		18.7 MW	
Steaming ground		1.2 MW	
Seepage to river		51.5 MW	(65°C)

In passing through the region the Waikato river temperature rises $0.2°C$. At a mean flow rate of $140\,m^3\,s^{-1}$, this is an added power of about 120 MW. A power output of about 100 MW is fairly certain, but it could be considerably higher.

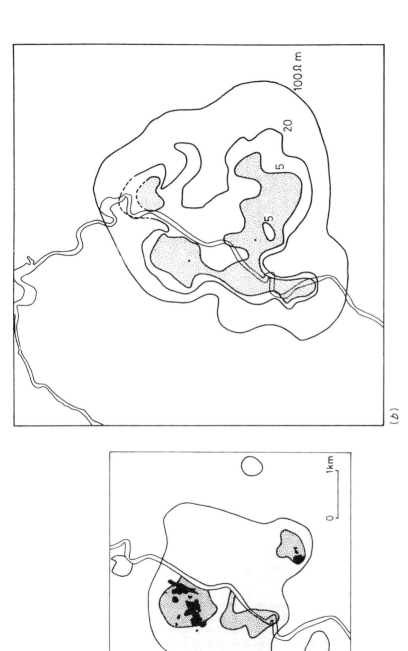

(b)

(a)

Fig. 13.2. Near surface observations at Ohaki—Broadlands. (a): (i) Distribution of modern surface activity; (ii) average near surface temperature gradient; light shading, less than 0.07°Cm⁻¹; heavy lines, 5°Cm⁻¹ enclosing regions of higher gradient (After G. E. K. Thompson and others). (b): Distribution of apparent resistivity in ohm metre (Wenner type survey with 600 foot electrode spread. After Dawson and Rayner (1968).)

If all this power were discharged as water at $80°$C, say, then (relative to $12°$C) the flow would be $0.35 \, m^3 \, s^{-1}$. If all this power were derived from water at $260°$C, typical of the deeper bores, the flow would be $0.1 \, m^3 \, s^{-1}$.

The distribution of apparent resistivity, at depths of order 100 m, reveals a more or less circular region of low resistivity in an ambient of resistivity of more than 100 ohm metre — typical of cold, wet, low-chloride volcanics. The resistivity data can be interpreted simply with a three-layer model (after Dawson and Rayner, 1968): an 8 m surface layer of dry pumice, 5000 ohm m; a wet pumice layer, 265 ohm m; on an infinitely deep hot reservoir, 1 ohm m. This gives the depths of the top of the hot reservoir for the 5, 20, 100 ohm m contours as 50, 80, 140 m. Wider spaced electrodes, various dipole techniques and electromagnetic soundings indicate a very similar though smoothed pattern.

The reservoir fluid

The discharge found in springs and bores is a near neutral chloride —bicarbonate water with, at higher levels, free gas mainly carbon dioxide. The gross distribution of chloride and soluble gas is shown in Fig. 13.3. The average concentrations are Cl^-, 1975 PPM; CO_2, 0.023.

The chloride distribution shows a region of highest concentration, about 1200 PPM, in the Ohaki region to the north with a gradual fall-off into the Broadlands region to the south. At first sight this suggests either: (i) a single upwelling of high concentration, in the north, with progressive dilution by non-chloride surface water as the fluid moves outward; or (ii) a uniform body of moderate concentration, say about 800 PPM, with localized boiling in pockets which is at a maximum in the north.

The carbon dioxide distribution has a broadly west—east pattern, reaching concentrations of greater than 0.04 in the east, and less than 0.005 in the west. There is a localized high in the Ohaki region (near BR21), where the concentration exceeds 0.03. At first sight this suggests, on the presumption of a subsurface outflow from the more active region to the west, that carbon dioxide has accumulated in the part of the field to the east owing to minor surface discharge there.

All this is rather deceptive, however, since the carbon dioxide discharge of bores is extremely variable both in space and time and the pattern may be just a result of inadequate sampling. In other words, local effects may be predominant. Nevertheless a typical carbon

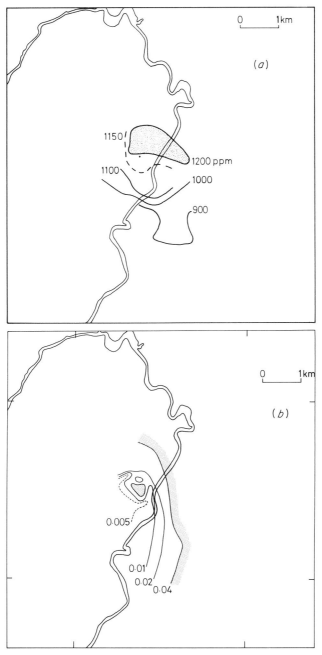

Fig. 13.3. Principal solutes in bore fluid at Ohaki—Broadlands. (a) Chloride concentration in deep fluid, PPM, (b) Carbon dioxide concentration in bore discharges, mass fraction.

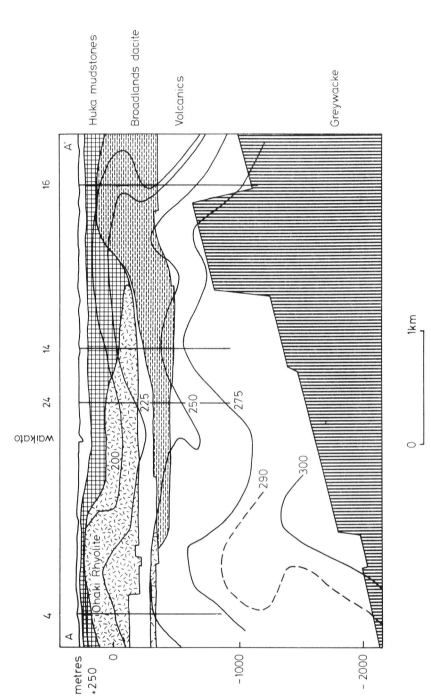

Fig. 13.4. Vertical section (line AA¹ of Fig. 13.1) showing isotherms and some of the geological formation: Huka mudstone, Ohaki rhyolite, Broadlands dacite, greywacke basement. Other fill is recent pumice, various ignimbrites and volcanic breccias.

dioxide concentration of $\lambda = 0.02$ and a corresponding possible gas pressure of about $\lambda/\alpha \sim 50\,\text{bar}$, indicates that in the upper levels of the system, to depths of order 1 km or so, pressures will be dominated by the role of the soluble gas.

Geology

A sequence of acid volcanics, probably all younger than 10^5 years, overlays an old greywacke basement; see Fig. 13.4. The rocks are similar to those found throughout the Taupo—Rotorua district. They are extremely variable in their gross properties of local porosity and permeability. Permeability is extremely inhomogeneous, the Ohaki portion to the NW having fair permeability, the Broadlands portion to the SE having low permeability — typical values for the production volume are $5 \times 10^{-13}\,\text{m}^2$ (0.05 darcy).

The rocks have wet densities of about $2.2 \pm 0.3\,\text{t}\,\text{m}^{-3}$ with porosity 0.25 ± 0.25. An estimate of the connected porosity is not known. We use a nominal value of 0.1 in the model studies.

For the model studies below we take the volumetric specific heat capacity as $\rho_* c_* = 2500\,\text{J}\,\text{m}^{-3}\,\text{K}^{-1}$.

The bore field

The location of the bores has already been shown in Fig. 13.1. They are 25 cm diameter wells with average cased depth of 615 m, uncased depth 450 m and total depth 1055 m. Of all the wells drilled, 18 (about half the total) are good producers with an average output of $54\,\text{kg}\,\text{s}^{-1}$ and specific enthalpy $1.3\,\text{MJ}\,\text{kg}^{-1}$ at WHP $\approx 10\,\text{bar}$. The existing field allows production at about $100\,\text{kg}\,\text{s}^{-1}\,\text{km}^{-2}$. (The corresponding figures for Wairakei are $46\,\text{kg}\,\text{s}^{-1}$ at WHP $\approx 14\,\text{bar}$, but for average cased depth, 450 m and uncased depth, 200 m.)

From a compilation (after Hitchock and Bixley, 1972) a reference pressure, p, is obtained on the assumption that pressures are nearly hydrostatic, where $p = 37.2 - 0.0769(y/\text{metre})\,\text{bar}$, and y is the elevation above mean sea-level. Thus at the mean open depth of bores of about $y = -580\,\text{m}$, $p = 82\,\text{bar}$. A nominal production depth undisturbed pressure of 80 bar will be used in some of the model calculations below. (This gives an *apparent* mean column density of $785\,\text{kg}\,\text{m}^{-3}$, the same as that of pure liquid water at $260°\text{C}$.) The pressure changes quoted below are relative to p. Mean pressures and temperatures for the production bores are: 150 m depth, $164°\text{C}$, 18 bar; 750 m depth, $266°\text{C}$, 63 bar. The range of values being about $\pm 10\%$.

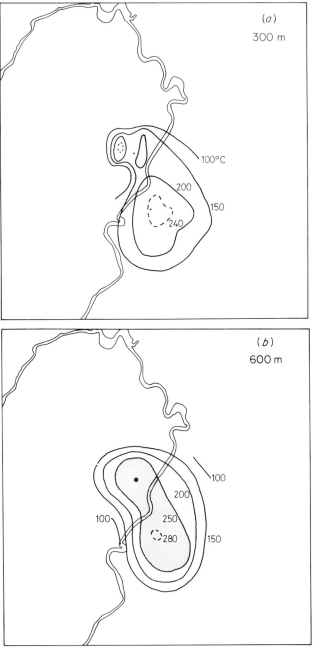

Fig. 13.5. Horizontal sections showing temperatue distribution at nominal depths: (above) (*a*) 300 m (sea level); (*b*) 600 m; (facing page) (*c*) 900 m; (*d*) 1200 m. (Ground surface at typical elevation 300 m above sea level).

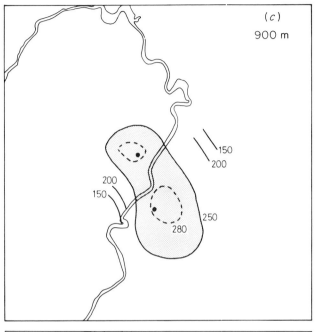

(c)
900 m

150
200

200
150

250
280

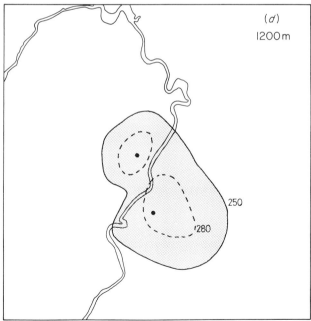

(d)
1200 m

250
280

Temperature distribution

Temperatures reach $300°C$ at depths of $1.5\,km$ or so. The temperature distribution is shown in horizontal sections in Fig. 13.5 and in vertical section in Fig. 13.4. The isotherms are domed upwards in two distinct bumps suggestive of two upwelling regions. The margins of the system are not revealed by the existing data. In the production volume, $10\,km^2 \times 2\,km$, the total energy is about $10^{19}\,J \approx 3.10^5\,MW\,yr$.

The thermodynamic state of the reservoir fluid

In spite of numerous measurements the state of the fluid in the reservoir is poorly known. The striking feature is the extreme inhomogeneity of the system. For example, Fig. 13.6 shows the pressure excess over that of pure saturated water for a variety of downhole

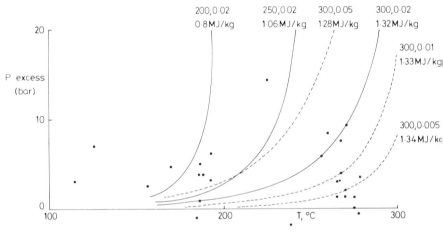

Fig. 13.6. Pressure excess over the pressure of pure saturated water as a function of temperature from measurements in closed in bores, compared with that of an equilibrium water—carbon dioxide mixutre. The curves are for constant specific enthalpy. The pairs of values are $(T_0, °C; \lambda)$.

measurements of non-discharging bores. There is little pattern to be seen. Many of the measurements are suspect owing to accumulation of gas in the well and to localized field disturbances in the vicinity of the well. About the best that can be done (other than a very detailed study) is to obtain a mean state for the reservoir currently sampled. The mean values for the total borefield discharge are: specific enthalpy, $h = 1.3\,MJ\,kg^{-1}$; gas content, $\lambda = 0.02$. There is some evidence that prior to intensive exploitation these values were $h = 1.25\,MJ\,kg^{-1}$, $\lambda = 0.015$. Most of the wells bottom at temperatures near $275°C$, for which $h \approx 1.2\,MJ\,kg^{-1}$. There will be little specific enthalpy enhancement due to input to the bores through permeable throttles unless the pressure is dropped subtantially (see, for example, Fig. 11.13). Thus for a deep fluid of constant specific enthalpy $1.2\,MJ\,kg^{-1}$, and gas content $\lambda = 0.02$ percolating upwards, in the production zone with ambient pressures of $60-80$ bar, the volumetric liquid fraction, $e = 0.41-0.71$. (The corresponding saturation for $\lambda = 0.01$ would be $0.53-0.997$.) A loss of pressure of, for example, 20 bar, under production, would lower the saturation to $e = 0.19-0.41$.

Substantial changes in the amounts of water and carbon dioxide occur in this system for only moderate changes in system pressure.

In the numerical models below I use the crude assumption that $n_0 = n_1 = \lambda_0$ and $h_0 = h_1$, recharge and discharge gas contents and specific enthalpies are the same. For the purpose here of simulating a short discharge experiment, these are minor assumptions.

13.2. THE 1966–1971 OHAKI DISCHARGE EXPERIMENT

From 1966–1971 the field was discharged and then left to recover. A total of 35×10^6 ton was discharged from the borefield — a mere $0.035\,km^3$ equivalent liquid volume. The discharge and the pressure response are shown in Fig. 13.7.

The discharge gas content and specific enthalpy were more or less constant at 0.02 and $1.3\,MJ\,kg^{-1}$, respectively. The pressure drop however is large and rapid and apart from minor details is adequate for the determination of the two key parameters R, C.

Output data are given in Table 13.1. The model simulations referred to later used the monthly average discharge data, although it makes very little difference to use the yearly average data. Over the main phase of the discharge, the four years 1968–1971, the mean discharge as equivalent volumetric discharge was close to $0.25\,m^3\,s^{-1}$.

Pressure distribution

The areal pressure change distribution is shown in Fig. 13.8. The distribution deepened and spread throughout the discharge phase, but following shut down, although the pressures in the central part of the region recovered nearly as fast as they were produced, the outer parts of the field contracted very slowly. This pressure field

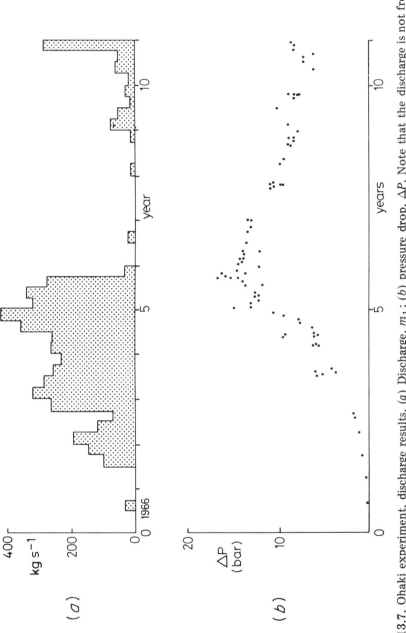

Fig. 13.7. Ohaki experiment, discharge results. (*a*) Discharge, m_1; (*b*) pressure drop, ΔP. Note that the discharge is not from a fixed bore field.

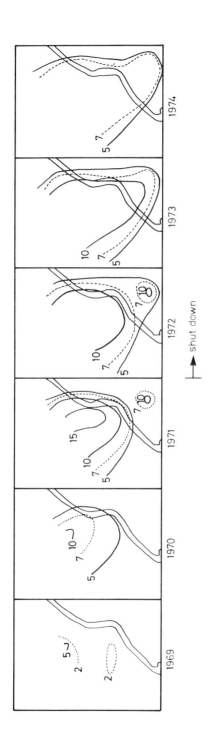

Fig. 13.8. Ohaki experiment, distribution of local pressure change in bar. (Redrawn from a compilation by Ian Donaldson, shown in Grant (1977)). The area is that shown as a rectangle in Fig. 13.1.

Table 13.1. *Ohaki–Broadlands discharge, annual values.*

Year	Mass flow rate $(\mathrm{kg\,s^{-1}})$	Specific enthalpy $(\mathrm{MJ\,kg^{-1}})$	Heat (MW)	ΔP (bar)
1966	11.9	1.237	14.7	0
1967	68.9	1.294	89.1	1
1968	166.6	1.465	244.1	2
1969	268.9	1.434	385.7	5
1970	326.2	1.352	441.2	8
1971	245.9	1.388	341.3	14
1972	8.5	1.280	10.9	13
1973	0.0	–	0.0	11
1974	15.7	1.529	24.1	9
1975	54.7	1.297	71.0	9
1976	120.0	1.257	150.9	8

Notes:
ΔP: mean annual values of pressure drop in the production zone.
Output: total 1966–1972; 34.6 Mton; 4.82×10^{16} J.
Enthalpy range: average $1.353\,\mathrm{MJ\,kg^{-1}}$, -9% in 1966, $+13\%$ in 1973.

spreads out initially with an apparent horizontal diffusivity of roughly about $5 \times 10^{-3}\,\mathrm{m^2\,s^{-1}}$ but contracts with an apparent diffusivity no more than $2 \times 10^{-4}\,\mathrm{m^2\,s^{-1}}$.

Ground subsidence

As a result of the discharge experiment the ground surface developed a dimple of maximum amplitude of about 200 mm, with effects centred on the major discharge area but measureably affecting a region of about 1.5 km radius: see Figures 13.9 and 13.10. The vertical velocity increased to about $150\,\mathrm{mm\,yr^{-1}}$ at the end of 1969. Following the test, the central region remained depressed and continued to fall, at a very slow rate, while the periphery began to recover, particularly in the west.

The subsidence development roughly follows that of the discharge, particularly up to 1971. The slow (negligible) recovery suggests that the recharge has been inhibited as a result of a net loss of reservoir fluid. This observation is confirmation of the notion about the deep recharge being through a permeable throttle.

Mass and gravity changes

During exploitation, as at Wairakei, there have been changes in the local gravity field. These can be interpreted by using the expression

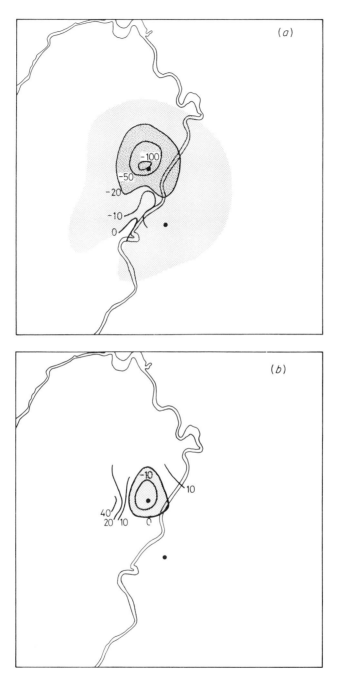

Fig. 13.9. Subsidence produced during the Ohaki experiment. Rate in mm yr^{-1} : (*a*) June 1970 to end of discharge phase, January 1971; (*b*) February 1975 to February 1976, with small renewed discharge.

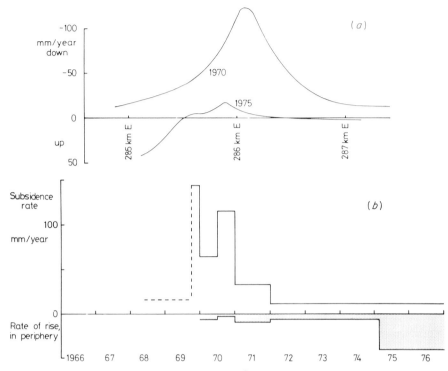

Fig. 13.10. (*a*) Subsidence profiles (line BB[1] of Fig. 13.1) as in (*a*) and (*b*) of Figure 13.9. (*b*) Rate of subsidence, amplitude as a function of time.

for the average gravity change, g', over an anomaly area, A, namely

$$g' \doteq 2\pi G \Delta M / A,$$

where M is the corresponding mass change.

If all the fluid discharged during the Ohaki experiment (or rather the interval of the gravity measurements) is simply withdrawn from the production volume, without any replacement, $\Delta M = -35$ Mton. If the area is taken as that of the entire Ohaki—Broadlands field so that $A \approx 10 \, \text{km}^2$, we have $g' = -1.5 \, \mu\text{N kg}^{-1}$. If the area is taken as that of the bulk of the production in the Ohaki area, with $A = 3 \, \text{km}^2$, we have $g' = -5 \, \mu\text{N kg}^{-1}$.

Measured values of the gravity change are shown in Fig. 13.11. In spite of a rather high "signal to noise" ratio there is a distinct pattern in the observations.

(i) Negative values are found in the Broadlands area and around the margin of the Ohaki area. The values are typically $-0.5 \, \mu\text{N kg}^{-1} \pm$ 100%. Here there has been a net mass loss. It would appear as if recharge/discharge $\approx 1/2$.

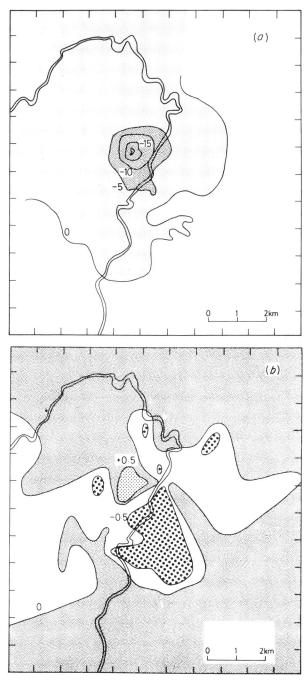

Fig. 13.11. Ohaki mass change. (*a*) Total vertical ground displacement, contour values in centimetres, from March 1968—April 1974. (*b*) Gravity changes from August—October 1967 to July 1974, corrected for vertical ground displacement, g (1974) $-g$ (1967), in μN kg^{-1} (= 0.1 milligal). Mean errors about $\pm 0.2\,\mu$N kg^{-1}. Negative values indicate a net loss of mass (from Hunt and Hicks, 1975). The extreme values are, in μN kg^{-1}: Ohaki area (NW) + 2.1; Broadlands area (SE) -0.9.

(ii) Positive values are found in the Ohaki area. The values are typically $1.0\,\mu\mathrm{N\,kg^{-1}} \pm 100\%$. Here there has been a net mass gain. The obvious explanation is that additional water has penetrated the system. A net amount of about ($5/3 \times$ discharge) would be required, equivalent to $2.6\,\mathrm{ton\,m^{-2}\,yr.^{-1}}$

(iii) Weak positive values surround the area. All but a few of these observations are within the limits of error so that it is doubtful if anything significant can be deduced.

The data strongly indicate substantial mass rearrangement in the production volume other than can be explained solely on the basis of simple withdrawal from a closed reservoir.

13.3. IDENTIFICATION OF THE RESERVOIR

Preliminary estimate of local reservoir parameters

Capacitance

As an extreme assumption consider the Ohaki discharge as taken from a closed reservoir without recharge (recharge resistance, $R = \infty$). Then the volume of fluid in the reservoir, expressed as volume of liquid water, $Q' = Ch$. The total discharge of 3.10^7 ton with a head loss equivalent to 15 bar requires

$$C = \Delta Q'/\Delta h \approx 0.2\,\mathrm{km^2}.$$

For a connected porosity of, say, 0.1, this corresponds to a reservoir cross-sectional area of about $2\,\mathrm{km^2}$. This value is similar to that of the cross-sectional area, $2.6\,\mathrm{km^2}$ within the $250°\mathrm{C}$ isotherm at $-700\,\mathrm{m}$ depth.

Clearly there has been recharge so that the above estimate of C is at best an order of magnitude.

Recharge resistance

At the opposite extreme, assume that all the discharge is taken from recharge (capacitance, $C \equiv 0$). This gives $R \approx 150\,\mathrm{m}/0.25\,\mathrm{m^3\,s^{-1}}$ $\approx 600\,\mathrm{w}$. Clearly there has been a net withdrawal of fluid from the reservoir during the discharge experiment, so that this estimate of R is at best an order of magnitude.

If the proportion of the total discharge which comes from recharge is ξ, we find:

$$R = 600/\xi \, \text{w}; \qquad C = 0.2(1 - \xi) \, \text{km}^2.$$

Thus the value $R = 600$ w is a lower bound; the value $C = 0.2 \, \text{km}^2$ is an upper bound (in this simple steady state view).

Summary

The high-level reservoir sampled during the Ohaki experiment has the following gross parameters: capacitance, $C \approx 0.1 \, \text{km}^2$; supply resistance, $R \approx 10^3$ w; thermal volume, $\approx 10 \, \text{km}^3$; natural power, $F \approx 100 \, \text{MW}$; characteristic reservoir temperature, $\theta = 300°\text{C}$ and gas content, $\lambda_0 \approx 0.02$. The Ohaki experiment was of insufficient duration to require (or identify) Φ, F.

(i) Single R, C, single-phase model

Consider a reservoir of capacitance, C, supplied from a source at pressure, P_0, through a fixed resistance, R, and with an imposed discharge at equivalent volumetric rate, $Q = Q(t)$ as described in section 11.2. Conservation of matter requires

$$\frac{dh}{dt} = -\frac{h}{\tau} + \frac{Q}{C}, \qquad \tau = RC.$$

In the steady state $h = -RQ$. Here h is the head relative to that of the steady state with $Q \equiv 0$.

A typical result is shown in Fig. 13.12(a) for: $R = 1000$ w, $C = 0.2 \, \text{km}^2$; and the Ohaki discharge $Q(t)$. The time-scale $\tau = 3.2 \, \text{yr}$. For this case the initial drawdown and recovery are both too slow.

Passable fits are found in the range $R = 500-1000$ w, $C = 0.1-0.25 \, \text{km}^2$.

(ii) Single-phase empirical $R(t)$, C models

An obvious method of improving the fit, in view of the apparent change in system time-scale with time is, for example, to take $R = R(t)$, i.e. time-dependent. The simplest version of this is to have: $0 \leqslant t \leqslant t_1$, $R = R_1$ and $t > t_1$, $R = R_2$. Choosing, for example, t_1 to correspond to the end of the main discharge phase, namely $t_1 = 6 \, \text{yr}$ we obtain results such as those shown in Fig. 13.12(b). The improvement is not startling, but there is quite a good fit with $R_1 = 800$ w, $R_2 = 1500$ w and $C = 0.15 \, \text{km}^2$.

Strictly empirical models like this merely suggest the direction in which a more physical model might be developed.

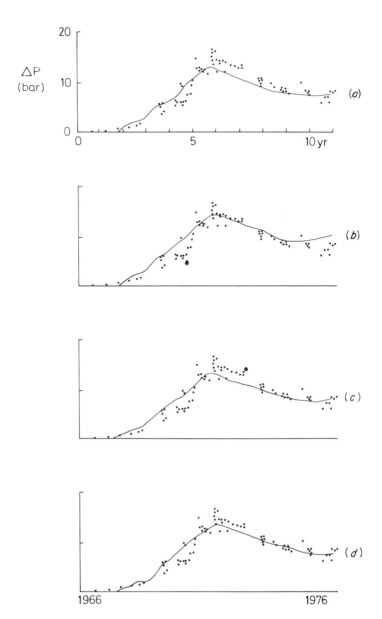

Fig. 13.12. Pressure loss, ΔP, as a function of time for Ohaki discharge for various models. (*a*) Simple RC model: $R = 1000$ w, $C = 0.2\,\mathrm{km}^2$. (*b*) RC model with R − change: $R_1 = 800$ w, $R_2 = 1500$ w, $C = 0.15\,\mathrm{km}^2$. (*c*) Simple two-phase model: $T_0 = 260°\mathrm{C}, \lambda_0 = 0.02, P_0 = 80$ bar; $R = 1000$ w, $V = 2\,\mathrm{km}^3$, $\epsilon = 0.1$, (*d*) Two-phase, $R(e)$ model: $T_0 = 280°\mathrm{C}$, $\lambda_0 = 0.02$, $P_0 = 80$ bar; $R_* = 150$ w, $V = 1.5\,\mathrm{km}^3$, $\epsilon = 0.1$. The solid line gives model values.

(iii) Simple two-phase model, constant supply resistance

With the isothermal model (end of section 11.5) quite good fits can be obtained near $R = 1000$ w, $V = 2$ km^3, $\epsilon = 0.1$ for assumed $T_0 = 260°$C, $\lambda_0 = 0.02$. Model results are shown in Fig. 13.12(c). Although the fits are good, it is noticeable (as is the case for the R, C model (ii) above) that it is not possible to obtain a very good fit to both the depressurizing and the recovery phase.

The reservoir volume $V = 2$ km^3 corresponds to a capacitance, at $e = 1$ and $P_1 = 92.2$ bar of $C_1 = 0.17$ km^2. This is a similar value to that suggested from the simple R, C model.

The overall area of the Ohaki hot spot is about 10 km^2 with productive depth at least of 2—3 km, suggesting a volume of the high-level reservoir of at least 25 km^3. Clearly, only a small portion of the reservoir has been sampled. This production volume, taking the area of the region of production as 3 km^2 requires a production zone of thickness about 0.7 km. The Ohaki production bores have the following mean values: cased depth, 615 m; uncased length, 450 m; total depth, 1055 m. This strongly suggests that the production during the discharge experiment was entirely from fluid in the immediate vicinity of the bores.

In passing, therefore, it should be remarked that even this massive test does *not* give any indication whatsoever of the long-term performance of the high-level reservoir as a whole.

(iv) Two-phase model, supply resistance $R = R(e)$

Now consider the case in which the supply resistance is no longer a (piecewise) constant, but is a permeable throttle through which all the source discharge passes. Make the assumption that the saturation in the throttle is the same as that of the high-level reservoir, so that $R = R(e)$. We use the form of the discussion above. Fig. 13.12(d) shows the model results for $R_* = 500$ w, $V = 2$ km^3, $\epsilon = 0.1$, $T_0 = 260°$C, $P_0 = 80$ bar. The slower recovery phase is now fairly well represented.

Whereas the models (i)—(iii) are essentially linear, this model is strongly non-linear because of the non-linear behaviour of the supply resistance. The choice of source conditions is then more critical, since this sets the range of saturation, e, and hence, R.

In the region for which $C \approx$ constant, the behaviour of the reservoir itself is independent of e — and in particular the initial undisturbed value. Other effects such as those of permeable throttles will, however, be a function of e.

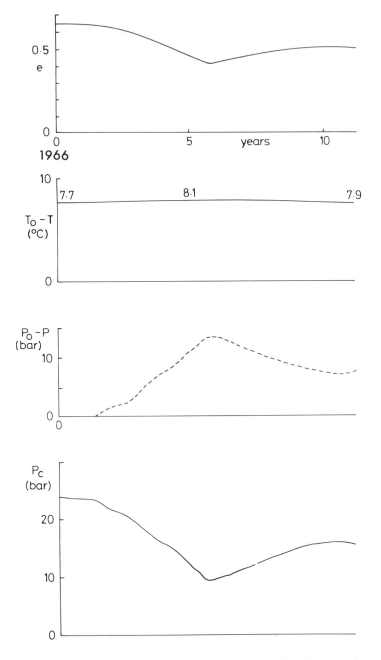

Fig. 13.13. Reservoir quantities, e, $T_0 - T$, $P_0 - P$, P_c, for two-phase $R(e)$ model: $T_0 = 280°C$, $\lambda_0 = 0.02$, $P_0 = 80$ bar, $R_* = 150$ w, $V = 1.5\,\mathrm{km}^3$, $\epsilon = 0.1$.

Acceptable fits can be found in the band: $T_0 = 260$–$280°C$, $\lambda_0 = 0.01$–0.02, $P_0 = 60$–80 bar for appropriate values in the range $R = 300$–700 w, $V = 0.8$–2.5 km^3, all with $\epsilon = 0.1$.

The variation of the reservoir quantities for one of the many possible combination of parameters is shown in Fig. 13.13. It is striking how nearly constant is the reservoir temperature, T, as will also be P_s. The behaviour is affected by the saturation variation and the changes in system pressure are virtually entirely changes of partial gas pressure in the vapour phase.

Role of gas content

In order to emphasize the powerful role of a soluble gas in the performance of a partially depressurized reservoir, Fig. 13.14 shows the pressure response for an Ohaki-like system with the same output function but for different values of λ_0. The role of the gas is dramatic and could be ignored in practice only for a gas function less than 0.005 or so.

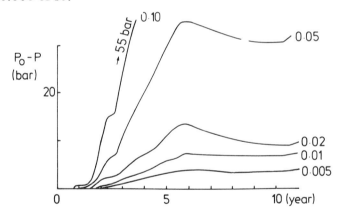

Fig. 13.14. Pressure response of Ohaki model with Ohaki experiment discharge: $T_0 = 260°C$, $V = 2$ km^3, $\epsilon = 0.1$, $P_0 = 80$ bar for $\lambda_0 = 0.02$, 0.05, 0.1, but $P_0 = 60$ bar for $\lambda_0 = 0.005, 0.01$.

Local reservoirs

The behaviour of some bores, in particular their very slow recovery when shut down, suggests that the reservoir as a whole behaves more like a collection of local reservoirs. This effect can be sufficiently extreme so that local steam "bubbles", in which $e \approx 0$, can develop around individual wells.

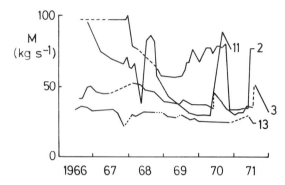

Fig. 13.15. Output of four Ohaki production wells during the discharge experiment. Note that these discharges are not at uniform WHP.

A persistent feature of the bore behaviour has been a progressive fall in output; see Fig. 13.15. This is to some degree obscured in practice, because of the numerous tests involving opening and closing bores and alterations of well-head pressure. Some of this fall has been produced by calcite deposition, for example in part in bore BR11, but undoubtedly some of the fall arises from changes induced in the field itself, as seen for bore BR13.

The response of two extreme wells are shown in Fig. 13.16:

(i) BR15. This is a non-producing well in the main production area, 2418 m deep, cased to 1800 m with bottom temperatures near 280°C. Its muted and delayed response suggests that it penetrated a pre-existing steam bubble of very low e.

(ii) BR25. This is a good well, with high apparent permeability, 1248 m deep, cased to 600 m with bottom temperatures near 300°C. The very rapid drawdown and negligible recovery suggests that it has locally drained its production region producing a very dry steam bubble.

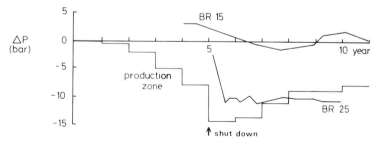

Fig. 13.16. Response of two Ohaki bores, BR15 and BR25. For comparison the annual mean pressure drop of the production zone is shown.

The role of the permeability inhomogeneties is pronounced. Wells such as BR25, penetrating a pocket of relatively high apparent permeability in a low permeability ambient are clearly preferable to those in the reverse situation and in many ways are ideal.

The Ohaki system under continuous exploitation

It is of interest to consider the response of a system like that at Ohaki, not only to a discharge pulse but to a steady discharge such as might be used under exploitation. Figure 13.17 shows the predicted

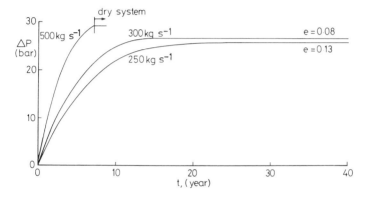

Fig. 13.17. Model (iv) behaviour under steady discharge; pressure drop as a function of time. $T_0 = 280°C$, $\lambda_0 = 0.02$, $h_0 = 1.21\,MJ\,kg^{-1}$, $P_0 = 80$ bar, $e_0 = 0.66$, $P_1 = 104$ bar.

behaviour for a particular case for model (iv) — gassy reservoir and throttle supply. The system reaches a steady state after: 20 yr at $250\,kg\,s^{-1}$ (700 MW); 17 yr at $300\,kg\,s^{-1}$ (840 MW); but at $500\,kg\,s^{-1}$ (1400 MW) it goes dry ($e = 0$) after 8 yr; and incidently model (iv) is no longer valid.

14. Vapour Systems: Larderello and Karapiti

In retrospect we see that the descriptions of the liquid system at Wairakei and gas system at Ohaki are largely confined to those of the high-level reservoir and surface zone. This is particularly the case for our third characteristic type of system, one dominated by the presence of a surface zone of saturated steam.

The properties of such a zone are most simply revealed in steaming ground where the steam zone is a mere shallow cap on a liquid system. When the steam zone is sufficiently thick its behaviour dominates as at Lardarello. The origin of such a deep steam zone is a matter of great interest and fortunately a study of the Karapiti area, adjacent to Wairakei, shows how such a zone develops.

14.1. STEAMING GROUND

Areas of steaming ground, ranging in size from patches of 10^2 m^2 to large regions of about 1 km^2, are a prominent feature of the surface zone of hydrothermal systems in which the water table is locally below but not too far below the ground surface. They appear as rather wispy irregular streaks of steam rising slowly to a few metres altitude and are particularly noticeable in cold or damp weather especially in oblique lighting. They are also recognizable from the relative absence or stunting of the ambient plants. Indeed, the areas can be mapped rather well by mapping the plant distribution either directly on the ground or from aerial photographs.

In the hotter regions the surface zone has a distinct structure.

(i) A thin uppermost zone, typically about 5 cm thick, is slightly firmer than that below. Where there is a mat of soil and roots this can easily be pulled off from the hotter layer below.

(ii) A middle zone of thermally altered ground is soft, sticky, damp and claylike in texture. The ground has the appearance of broken up clay, homogeneous impermeable lumps with numerous small cracks and joints.

302

(iii)   A bottom zone of hot water-saturated ground is as found throughout the thermal area.

Another characteristic feature of the hotter regions are patches of small fumaroles and vents. Collapse pits up to 5 m deep and of order 10 m across occur, often with small fumeroles and vents on their sides and bottoms — and usually with no pools or other direct evidence of groundwater. Derelict and inactive pits are common suggesting that activity is locally sporadic and shifts about the area.

The variation of temperature with depth is shown for a typical case of warm ground in Fig. 14.1. Temperatures reach the local boiling point at depths of a few centimetres to a few metres, below which the temperatures are nearly constant. The depth varies roughly inversely with the vapour flux. This suggests that the dominant mechanism is an upward flow of vapour for which the temperature is depressed near the surface by conductive cooling to the surface.

Steaming ground can be considered to be the result of filling in a surface pool with earth and lowering the water table. The deeper the water table the less intense the area becomes till the upward flow of the vapour is completely impeded and the heat flow becomes conductive (Fig. 14.2). It will be assumed that the vertical mass flux is ϵm, where ϵ is the porosity of the rock permeable to the vapour, and m is the evaporative flux rate. (See section 16.4 and Fig. 16.15.) The vapour pressure p_2, near the interface, is considered to be the external ambient vapour pressure p_0 (effectively zero) plus the pressure difference required to drive the vapour through the overburden of the porous rock. For a shallow overburden of thickness less than 10 cm this pressure is negligible and the heat flux is given directly by ϵm

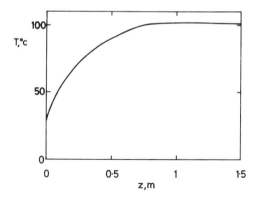

Fig. 14.1. Steaming ground temperature, T, in $^\circ$C as a function of depth, z, in metres, for strongly steaming ground.

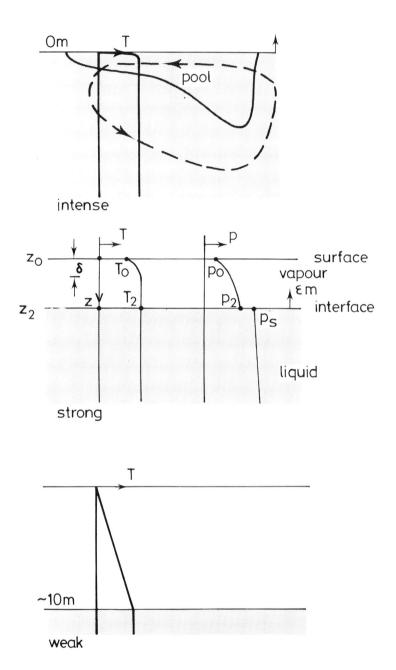

Fig. 14.2. Schematic diagrams of steaming ground for intense, strong and weak areas, together with the nomenclature used in the text.

with $p_2 \approx 0$; otherwise it is necessary to consider the details of the flow of the vapour through the overburden by means of the equations of continuity, flow and energy as below. The role of the ambient air is completely ignored.

The vapour is compressible. If q is the magnitude of the vertical flux velocity of the vapour, continuity requires:

$$\rho q = \text{constant} = \epsilon m,$$

so that Darcy's relation

$$q = \frac{k}{\mu} \frac{dp}{dz},$$

can be written, assuming the vapour behaves as a perfect gas,

$$\frac{d(p^2)}{dz} = 2\mu\epsilon m T/\beta k.$$

It is sufficiently accurate to integrate this expression with $T = $ constant $= T_2$, the interface temperature. The temperature is lowered near the surface within a zone of thickness of scale $\delta = K/\epsilon mc$, where c is the specific heat of the vapour. To evaluate δ the flux ϵm must be known; this is derived below. Hence our present estimate of δ must be justified *a posteriori*. For example, with $K = 0.8\,\mathrm{W\,m^{-1}\,K^{-1}}$, $c = 2\,\mathrm{kJ\,kg^{-1}\,K^{-1}}$; $\epsilon = 0.1$, $k = 0.1$ darcy, $Z = 10\,\mathrm{m}$; if $T_2 = 120°\mathrm{C}$, $\delta = 2.6\,\mathrm{m}$ and hence there will be a large temperature gradient near the surface — the temperature over most of the layer being close to T_2; but if $T_2 = 80°\mathrm{C}$, $\delta = 260\,\mathrm{m}$ and the temperature distribution will be nearly linear. Except in cases like the latter,

$$p^2 - p_0^2 = 2\mu\epsilon m T_2 (z - z_0)/\beta k.$$

Writing,

$$\tilde{p} \equiv p_0/p_s, \qquad \xi \equiv m/m_0, \qquad S \equiv \chi\mu\epsilon T_2 Z/\beta k p_s,$$

where $Z = z_2 - z_0$, we have

$$p_2^2 - p_0^2 = 2\xi S p_s^2$$

and hence

$$\xi = (1 + S) - [(1 + S)^2 - (1 - \tilde{p}^2)]^{1/2}.$$

The heat flux $(\chi\xi\epsilon p_s E_s)$ can now be calculated. Fig. 14.3 shows the heat flux as a function of T_2 for various Z, and Fig. 14.4 as a function of z for various T_2 for a typical case in which $\epsilon = 0.1$, $k = 0.1$ darcy, $\tilde{p} = 0$. (Other values can be accommodated by scaling the values of the flux and Z in proportion.) With this combination of values the calculations fit the Wairakei field data given by Thompson *et al.* (1961) sufficiently well. The values of ϵ, k, are rather small, since both the effective evaporative area and the permeated volume

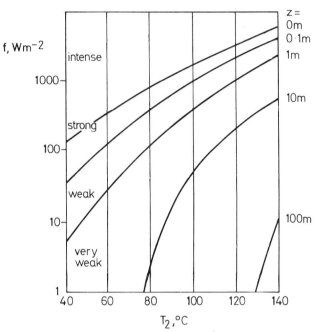

Fig. 14.3. Steaming ground heat flux, f, in $W\,m^{-2}$, as a function of water table temperature, T_2, in $°C$, for depth to water table $Z = 0$, 0.1, 1, 10, 100 m; $\epsilon = 0.1$, $k = 0.1$ darcy.

arise from the cracks and joints of the rock rather than from the pores of the rock elements themselves. It is seen that large values of S substantially reduced heat flux and that for $Z \gtrsim 100$ m, even with high T_2, the output is small. Inspection of the maps of shallow temperature surveys, such as that given by Thompson (1960) for Wairakei, shows that the steaming areas are confined to patches within the $5\,W\,m^{-2}$ contour: outside this contour the ground temperature will be established by heat conduction or by downward-moving cold groundwater. The labels in the figure, intense, strong, weak, very weak, correspond to those of field measurement of heat flow in steaming ground (Banwell, 1957; Benseman, 1959).

A number of other aspects of steaming ground should be noted.

(1) When the flux is large the ground surface will not be at ambient, rather its temperature will be determined by meteorological factors together with those discussed above.

(2) In practice, even when Z is small, the steam will tend to flow up the more permeable paths or vents, rather than uniformly as here.

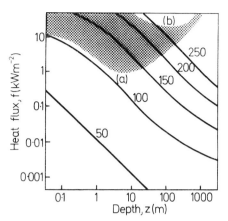

Fig. 14.4. Steaming ground output characteristic: f, heat flux $(W\,m^{-2})$; Z (metres), depth, to water-table for a variety of temperatures, $50-250°C$, at the water-table. The shaded region indicates the conditions for which phreatic eruptions may occur. Above this region steady steaming ground will be very unlikely since the ground is explosively unstable. Curve (a): SVP = lithostatic load. Curve (b): SVP = lithostatic load + finite strength (here taken at a low value of 30 bar). Data used: porosity, 0.1; permeability, 0.1 darcy. An output of 0.01 kW m^{-2} would not be noticed by a casual observer; an output of 1 kW m^{-2} is found in intense steaming ground.

(3)   Below the water table there must be a horizontal transport of water; for example, each 1 kg of water at $120°C$, say, has enough energy to produce 0.06 kg of steam at $80°C$, say; the 0.94 kg of water must be transported away. This flow may be the general circulation, or a local circulatory system. This circulation requirement will produce extreme patchiness.

(4)   At low intensities some of the steam may be condensed, for example by conductive loss to the surface.

(5)   A frequent occurrence is a steaming vertical surface, such as the steaming cliffs at Waiotapu (Lloyd, 1959). The flow of the steam through the ground is not at all dependent on gravity; it is maintained by the pressure distribution established between the water—air interface and the ground surface. Thus steaming cliffs are in no way different from steaming ground.

Steaming ground is not restricted to thermal areas, the same mechanism will occur even in normal areas. Indeed, except in arid regions, the bulk of the energy excess which enters the upper few metres of the ground from solar radiation is dissipated into the atmosphere by the evaporative mechanism of steaming ground.

14.2. THE TUSCAN HYDROTHERMAL SYSTEMS

Observation, before the time of exploitation, revealed in the Larderello area of Tuscany, Italy, areas of "soffioni" and "lagoni" (Fig. 14.5). Soffioni are dry-fumaroles (temperatures up to $185°C$ were measured), lagoni are hot pools, generally muddy. Notable also were the cold discharges of gas as "mofeta" (CO_2 emanations), and "putizzo" (H_2S emanations). The gas content is high at all depths, about 4%; in an adjacent area Mt. Amiata, 95%.

During the period of exploitation (100 years) these intense areas have nearly disappeared so that the original heat flow can only be roughly estimated. Personal observation of the remaining intense areas and inspection of old photographs give a similar impression of intensity as that of the more intense areas at Wairakei, corresponding to the zone within the $50°C$ isotherm at 1 m depth. This is a heat flux of $500\,\mathrm{W\,m^{-2}}$.

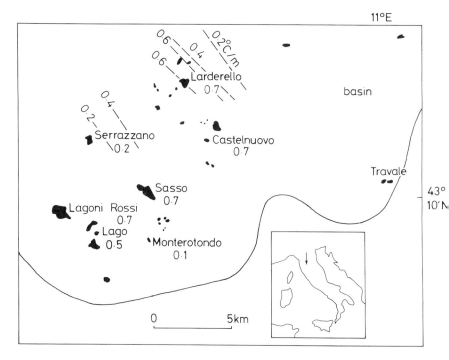

Fig. 14.5. Location of the thermal areas of the Tuscan thermal district. Area indicated in $\mathrm{km^2}$ (estimated heat flow, $400\,\mathrm{MW\,km^{-2}}$); surface temperature gradient in $°C\,m^{-1}$ (after Burgassi, 1961). Outline of ground water divide of Larderello basin (Petracco and Squarci, 1970).

Initial exploitation by shallow drilling in general revealed a hot water zone; indeed it was the boric acid in this water which provided the economic incentive for the initial development of the area and only during this century has the emphasis changed to that of power generation. Although the existence of the soffioni, especially their superheat, may have suggested an extensive steam reservoir at depth, since the soffioni occurred over only 2% of the whole Larderello zone (150—200 km^2) and also because other soffioni are known (such as Karapiti dry-fumarole at Wairakei) which do not rely on an extensive steam reservoir, the suggestion of such a reservoir would not have been justified.

Exploitation

Deeper drilling did, however, reveal (or perhaps produce) such an extensive steam zone with temperatures up to 240°C and pressures up to 40 bar at depth, the exploitation of which produced a maximum output in 1963 of about 850 kg s^{-1} from 100 km^2. Although there are now known to be other such areas in the world (e.g., The Geysers, California), Larderello has hitherto apparently been unique — other thermal areas have merely contained hot water; unfortunately this feature, which is the essence of our interest in Larderello, has received undue emphasis, resulting often in the unnecessarily exclusive prospection for such steam zones. (It is worth remembering that geothermal water of 200—250°C contains about 100 times the energy of an equal volume of steam: furthermore the bulk of the energy is stored in the rock in either case.)

During the period of intensive exploitation for power, not only have the surface manifestations disappeared, but so to a large extent has the near surface (perched) water zone. This is a deduction from observations of new bores in the exploited areas, but unfortunately no direct data are available.

A special feature of the exploitation is the gradual decline in the output of individual bores, so that in order to maintain total output new wells are needed and these cover a progressively larger area.

Recharge

The Larderello systems (Petracco and Squari, 1977) lie at the southern end of a ground water basin about 20 km East—West, 40 km North—South. The rainfall is 1.0 m yr^{-1} ± 20% with an infiltration proportion of about 0.5. Subsurface drainage is inward from the margin of this basin. The marginal water contains some tritium and has an ^{18}O composition which varies in space and time about the isotopic variation ratio of 5%, while local rain has 7%. The central steam area produces no tritium and has a uniform 0%. This suggests that recharge is largely derived from the periphery of the system and that the central steam zone has water substance of age 50 years or more — perhaps of order 10^3 yr. (The time, say, to simply refill with liquid water a volume of porosity 0.1 and depth 5 km at a discharge rate of $40 \, W \, m^{-2}$ of steam is 1.2×10^3 yr.)

If we regard a system like Lardarello merely as a deposit of heat, it is of interest to estimate its life. For a discharge of $10 \, kg \, km^{-2} \, s^{-1}$ at enthalpy $2.8 \, MJ \, kg^{-1}$, derived from rock of available thermal energy, say, $0.3 \, MJ \, kg^{-1}$, the rate of exhaustion of rock energy would be 1 m in 1 yr. Even if they were not replenished, such systems would have lifetimes of order 10^3 yr.

Temperature distribution

As it was not the practice to attempt to measure actual ground temperatures (or pressures) at Larderello (e.g., by flooding bore holes with mud after insertion of thermometer chains), it is necessary to rely on the maximum discharge temperature recorded as a function of drilled depth (Penta, 1954) and on a surface gradient traverse given by Burgassi *et al.* (1961) to obtain a rough idea of the distribution of temperature with depth. This is given in Fig. 14.6, together

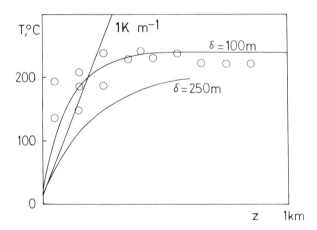

Fig. **14.6.** Temperature profiles at Larderello (after Penta, 1954). Theoretical curves for $\delta = 100$ m, 250 m.

with exponential curves corresponding to conductive cooling. The fit is rather crude, but $\delta = 100\,\text{m}$ fits the maximum temperature with depth points and $\delta = 250\,\text{m}$ fits the maximum surface gradient of about $1\,\text{K}\,\text{m}^{-1}$. The exponential form of temperature distribution immediately suggests the possibility that in the natural state there is near the ground surface a vertical transport of steam which is escaping from the steam reservoir. If this is the case, with $\delta = 100\,\text{m}$, where $\delta = K/\epsilon mc$, the corresponding heat flux is about $20\,\text{W}\,\text{m}^{-2}$. The conductive flow corresponding to $1°\text{C}\,\text{m}^{-1}$ is about 25 times normal heat flow — cf. a similar figure at Wairakei.

Steam bore characteristics

The output characteristic of a steam bore is shown in Fig. 14.7. This bore was drilled into an already heavily exploited zone and

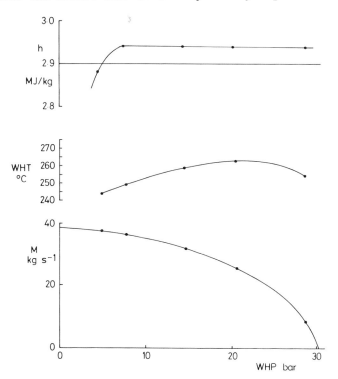

Fig. 14.7. Output characteristics as a function of well-head pressure, WHP: specific enthalpy, h, in $\text{MJ}\,\text{kg}^{-1}$; well-head temperature, WHT, in $°\text{C}$; mass discharge rate, M, in $\text{kg}\,\text{s}^{-1}$ (1963 measurement of well VC/10, from Sestini, 1970).

produces superheated steam at all well-head pressures. The specific enthalpy of the discharge is nearly independent of WHP. The well-head temperature corresponds to that of steam of the measured enthalpy and the pressure at the well-head as given in the steam tables: in other words the occurrence of a temperature maximum is simply a property of the steam.

The output characteristics are now used here to obtain an estimate of the permeability of the ground.

A simple model can be derived as follows. The approach is the same as that for bores in a fully liquid zone, although much simpler because only one phase need be considered, and is described in more detail below. As sketched in Fig. 14.8, a bore penetrates a uniform steam zone. The behaviour is controlled by the combined effects of flow in the ground and flow up the bore.

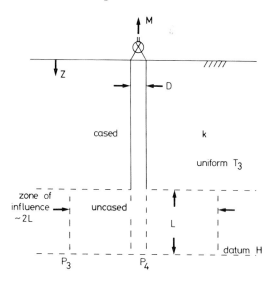

Fig. 14.8. Schema and notation for the analysis of a steam bore.

Assuming that the flow in the ground is controlled by Darcy's relation, that the expansion of the steam in the ground as it approaches the uncased section of the bore is isothermal at the ground temperature, $\tilde{T}_3$, the mass rate into the bore is:

$$M \approx \beta \pi L k (P_3^2 - P_4^2)/\mu \tilde{T}_3 \log(2L/D),$$

where

$$\beta = \beta_g = \rho_g \tilde{T}/P.$$

The flow up the bore is resisted by hydrodynamic forces and the weight of the gas column. It is readily demonstrated that the weight of the gas column is relatively negligible so that the bore pressure gradient is:

$$dP/dz = f\rho W^2/2D,$$

where f is the friction coefficient and W is the mean vertical fluid velocity. A typical value for Larderello bores is $f = 0.012$. Noting that:

$$M = \rho AW, \qquad A = \tfrac{1}{4}\pi D^2,$$

and writing

$$\rho = \beta P/\tilde{T} \approx CP,$$

then

$$dP^2/dz = fM^2/CDA^2, \qquad \text{a constant.}$$

Thus,

$$(P_4^2 - \mathrm{WHP}^2) = (fM^2/CDA^2)H.$$

The expressions for M and P_4 define the output characteristic:

$$1 - \left(\frac{\mathrm{WHP}}{P_3}\right)^2 = b_1\left(\frac{M}{M_0}\right) + b_2\left(\frac{M}{M_0}\right)^2,$$

where

$$b_1 = \frac{M_0\mu T_3 \log 2L/D}{\pi\beta LkP_3^2}, \qquad b_2 = \frac{16CfM_0^2H}{\pi^2 D^5 P_3^2}.$$

Taking $M = m_0$ when WHP $= 0$ requires $(b_1 + b_2), = 1$ an equation for M_0. Data from two bores in the Larderello area are plotted in Fig. 14.9 together with computed curves: the data from hole A fits $b_1 = 0.954$, corresponding to negligible loss in the pipe; the other fits $b_1 \approx 0.5$, the potential output of a bore in this site being reduced

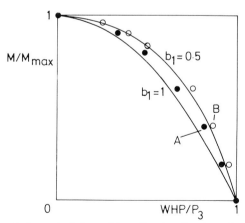

Fig. 14.9. Output characteristics of two Larderello steam bores referred to here as A, B, and theoretical curves. A: $M(\mathrm{max}) = 31\ \mathrm{kg\,s^{-1}}$; $P_1 = 32$ bar. B: $M(\mathrm{max}) = 21\ \mathrm{kg\,s^{-1}}$, $P_1 = 32$ bar.

to half because of loss in the pipe. Notice that in a particular site $b_2 \propto L^2 k^2 / D^5$. The values of permeability deduced are for bore A, 0.6 darcy and bore B, 0.2 darcy. These values are restricted to a relatively permeable zone at 400—500 m depth; in this zone the ratio of productive to sterile holes, is 15:1, elsewhere the ratio falls to 3:1 (20% of all holes in the Larderello area of 150—200 km^2 are sterile). In the upper 2 km, the average permeability for bore discharge is probably therefore no more than 0.1 darcy.

The frequency of occurrence of small values of b_2 at Larderello reveals an interesting difference in drilling practice in the Larderello steam zone and the Wairakei water zone. In the Larderello steam zone, rather large values of D, 0.3—0.4 m, give small b_2, so that the bore output is restricted solely by the ground (once D has been made sufficiently large to reduce b_2 to less than 0.2, say, larger diameter bores are unnecessary). In the wet zone at Wairakei, however, sufficient bore output has been possible with D of 0.2 m when the bulk of the loss is in the pipe and not in the ground. This difference is not a fundamental one and has arisen largely because of the lesser capacity of the drilling rigs at Wairakei.

Pressure transients

The above discussion refers to a steady exploitation of a static steam reservoir. One of the outstanding characteristics of the Larderello project however, is the absence of a steady state during the last 70 years of intensive exploitation.

(i) A change in the setting of the well-head valve results in a slow change (10—1000 h) to a new steady state. For example, Fig. 14.10 shows the growth of the shut-in pressure after closing the well-head valve on Larderello bores.

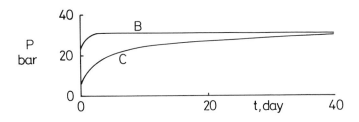

Fig. 14.10. Growth of shut-in pressure in two Larderello steam bores referred to here as B, C, after the sudden closing of the well-head valve of wells previously discharging steadily to the atmosphere (after Nencetti, 1961b).

(ii) Whereas the output of a single bore in an unexploited area remains constant over a period of 10 years, in an exploited area not only does the output of individual bores continuously decrease but so does the total output; to maintain production at Larderello a continuous drilling programme is necessary. Data from a typical Larderello bore are shown in Fig. 14.11. A rather crude average for individual outputs M is:

$$M = M_1 \exp(-0.14\, t/\text{yr}).$$

Corresponding to a fall to $0.5 M_1$ in five years.

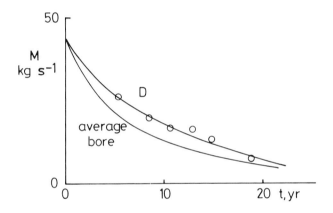

Fig. 14.11. Output as a function of time since blowing for a particular Larderello steam bore referred to here as D and the average of all Larderello bores (circa 1960).

(iii) Longer term transient behaviour is obscured by the persistent drilling over an expanding area.

While there is the possibility that these transient phenomena are related to the solution and deposition of chemicals in the porous ground, steam is not a particularly good transporter of suitable chemicals. A more obvious explanation follows the observation that at Wairakei such phenomena do not exist — water is incompressible whereas steam is compressible. The mass of a compressible fluid contained in a reservoir is a function of the pressure distribution, any change in the pressure distribution will require a mass transport to adjust the mass to its new value, a transient response will occur during the time of this flow.

Time-scale of a vapour reservoir

Consider the flow of steam at mean pressure $\bar{P}$, across a section of ground of width Z, and cross sectional area A. Then, noting that the mass flow rate $\equiv \rho_0 Q = \rho q A$, we have the flow resistance as:

$$R \sim \bar{P}/g\rho qA \approx \mu Z/\mathrm{k}g\rho A.$$

Further, noting that the total mass is:

$$C\bar{P}/g \approx \rho\epsilon V,$$

we have the capacitance,
$$C \sim \rho g\epsilon V/\bar{P},$$

The corresponding time-scale $\tau = RC \sim \epsilon\mu Z^2/k\bar{P}$, independent of A. Hence, from a measured τ, the length-scale Z can be estimated.

There is another revealing way of looking at this result. In one dimensionsal flow of a compressible fluid in a porous medium, conservation of mass requires,

$$\epsilon \frac{\partial\rho}{\partial t} + \frac{\partial\rho q}{\partial Z} = 0,$$

where ϵ is the total porosity, not just that of the joints and fractures. Hence:

$$\frac{\partial P}{\partial t} = \frac{k}{\epsilon\mu} \frac{\partial}{\partial Z}\left(P\frac{\partial P}{\partial Z}\right).$$

This is a diffusion equation in which for small changes in P we have a diffusion coefficient $K\bar{P}/\epsilon\mu$, where $\bar{P}$ is a mean pressure. Hence if fluid is removed or added to a volume of length-scale Z, the pressure will change with a time-scale τ such that,

$$\tau \approx \epsilon\mu Z^2/2k\bar{P}.$$

The volume affected by the manipulations of the well-head valve of a single well is a roughly spherical volume with Z of order L, the length of the well production zone. Take τ as the time required to effect 90% of the total pressure change: bore B gives $\tau = 12\,\mathrm{h}$, which with $\epsilon = 0.1$, $\bar{P} = 15\,\mathrm{bar}$, $Z = 100\,\mathrm{m}$ gives $k = 0.2\,\mathrm{darcy}$ in agreement with the data of Fig. 14.9; bore C gives $\tau = 600\,\mathrm{h}$ corresponding to $k = 0.003\,\mathrm{darcy}$, a very small value, similar to that of homogeneous rock.

A single bore affects only its immediate vicinity to a distance L, but with an extensive area of closely spaced bores (Larderello spacing is $400{-}500\,\mathrm{m}$), it is possible to affect the distribution of pressure throughout the steam reservoir itself. Here we notice that while pressure adjustments can be made rapidly in the more permeable zones, the ultimate response will be dominated by the movement into and out of the rock pores, for which the permeability is very small: laboratory measurements give a few millidarcy. Hence choosing the above figure of 0.003 darcy, $\tau = 25\,\text{yr}$, $\epsilon = 0.1$, we require that $Z = 2\,\text{km}$. This figure is of course exceedingly crude. Nevertheless it does support the view that beneath Larderello there is a deep steam zone.

Capacitance and resistance of a vapour reservoir

It is of interest to evaluate R and C using the order of magnitude expressions. For example, as will be shown, in the unexploited state the bulk of the reservoir was filled with saturated steam at about $235°\text{C}$, $30\,\text{bar}$ (and $\rho = 15\,\text{kg m}^{-3}$, $\mu = 1.7 \times 10^{-5}\,\text{kg m}^{-1}\,\text{s}^{-1}$). Consider first a volume of area $A = 1\,\text{km}^2$ and depth $Z = 1\,\text{km}$ of permeability $k = 0.003\,\text{darcy}$. Then $C \approx 5 \times 10^3\,\text{m}^2$, $R \approx 4 \times 10^4\,\text{w}$ and $RC \approx 7\,\text{yr}$. What is striking, in comparison with liquid reservoirs is the very small value of the capacitance, which arises simply because the vapour density is much less than that of the liquid. The resistance is large because of the low value of permeability.

Origin of the steam

There is another remarkable observation. Whereas the early wells at Larderello (and not just those penetrating only into the superficial near surface ground water zone) gave steam that was either saturated or very nearly saturated, in the years of intensive exploitation the degree of superheat has steadily risen. A well-head temperature increase of as much as $50°\text{C}$ has been reported. The maximum steam enthalpy at Larderello is at present about $2.87\,\text{MJ kg}^{-1}$, while the maximum possible enthalpy of saturated steam is $2.80\,\text{MJ kg}^{-1}$.

The presence of superheated steam is found now not only in the output of early wells but from the beginning in new wells drilled within the exploited zone: some new wells have discharge temperatures of $260°\text{C}$ at WHP of $20\,\text{bar}$, a superheat of about $50°\text{C}$. The bore field temperature maxima are shown in Fig. 14.12. The initial temperatures (presumably near the bottom of the wells) were 160–190°C and rose to more or less steady values of 200–250°C after a transition period of about 15–30 years (presumably determined by the particular rate of development of the field).

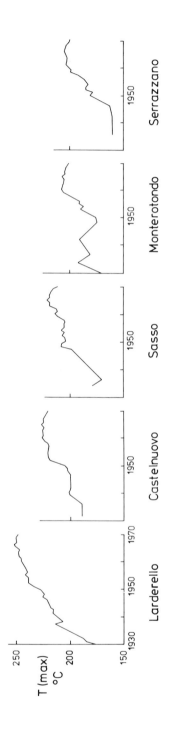

Fig. 14.12. Maximum discharge temperature as a function of time for individual Tuscan steam fields, (after Sestini, 1970).

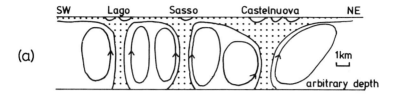

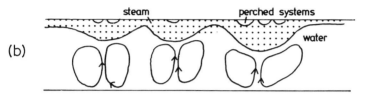

Fig. 14.13. Possible schemas for the Tuscan hydrothermal systems.

At this point we must modify the view of a static steam reservoir and consider the origin of the steam. The broad features of the various possibilities have been discussed in previous chapters; the only novel feature so far suggested is the large steam deposit or reservoir. There are two possibilities sketched in Fig. 14.13:

(a) The system is a "dry-spring". A section through the Larderello zone will be as sketched in Fig. 14.13(a); steam columns of cross-sectional area of order $0.1\,km^2$ rise from depth, arbitrarily drawn as $5\,km$, and in the upper $2\,km$ spread out to affect a surface area of order $10\,km^2$. Each of the zones is considered to have a separate identity, although this is not a necessary requirement.

(b) The system is a "wet-convector" but with evaporation occurring at the water surface as in "steaming ground": a schema similar to that of "steaming ground" but in which the water table is at great depth (or order $2\,km$). It is therefore possible to consider much higher water surface temperatures than for normal steaming ground.

There are several difficulties with view (a):

(1) It is necessary to invoke extreme possibilities; indeed on this basis it is most likely that Larderello is a "collapsed system". In fact, the maximum steam pressure found at Larderello in holes mostly to $1\,km$, but some at $2\,km$, is only 40 bar, but to avoid collapse the ground pressure must exceed 100 bar at

1 km and 200 bar at 2 km. We may suggest that this is merely a transient state, but a transient state like that of (a) is not possible; a steam column of diameter 1 km with a pressure defect of 100 bar, even with k as small as 0.01 darcy, will collapse in about three years; and Larderello was known to the Romans.

(2) Apart from this, in the steady state there would be no thermodynamic distinction between (a) and (b) provided the maximum steam enthalpy in both cases is the same. Both possibilities are compatible with this value: in (a) notice that above 2.8 MJ kg^{-1} steam can rise from any depth and under an isenthalpic expansion remain superheated; (b) is discussed below. If, on the other hand, the enthalpy was considerably in excess of this, possibility (b) would be excluded.

During the period of exploitation the enthalpy has continuously increased, while in areas newly prospected the enthalpy is found near that of saturation. This fact is not compatible with possibility (a); indeed an increase in discharge would produce a fall in enthalpy. Even if it were suggested (in spite of the facts) that a steady state has not yet been reached, an increase in enthalpy is still not possible. Any model which does not allow a phase separation will be faced with this difficulty of the enthalpy increase.

Since neither of these difficulties arise with possibility (b) and since, further, the behaviour of the enthalpy will be shown to follow naturally, this possibility is henceforth accepted. It is therefore seen that the Tuscan hydrothermal systems at depth are not necessarily fundamentally different from other wet-convectors.

It is further most interesting to note in support of this view that at Wairakei the discharge from some of the shallow bores, during the period of intense exploitation, progressively increased in enthalpy till they produce superheated steam.

Behaviour of the steam zone

A one-dimensional model of a possible steam zone is sketched in Fig. 14.14. The essence of the model is the spatial separation into two distinct phases, water and steam, at a distinct depth. The steam may be saturated or become slightly superheated as it rises towards the surface zone.

A small modification is convenient; whereas for steaming ground we take the ambient vapour pressure $p_0 = 0$ on $z = 0$, here we

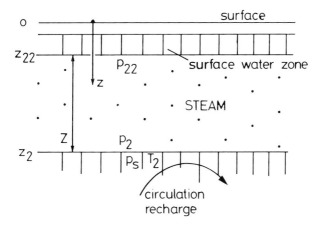

Fig. 14.14. The steam zone; one-dimensional model.

consider $p_0 = p_{22}$, the pressure on $z = z_{22}$, where p_{22} is determined, for example, either by a superimposed layer of water or by the exploitation pressure established by bore holes.

In the original state there are good indications that $\tilde{p} \equiv p_{22}/p_s$ was near unity. Thus the output may have been considerably restricted through the term $(1 - \tilde{p}^2)$. In this case, $p_2 \approx p_s$, the gas enthalpy is nearly equal to the saturation value at T_2; for example, a maximum of $2.8\,\mathrm{MJ\,kg^{-1}}$ at $235°\mathrm{C}$.

During exploitation it has been engineering practice at Larderello to set the bore WHP = 5 bar, and as has been show, loss of pressure as the fluid ascends a bore is negligible so that ground pressure p_{22} has been reduced from values at least as large as 30 bar to 5 bar. Hence, $\tilde{p}$ changes from near 1 to 0.1–0.2, ξ increases — an order of magnitude increase in heat flow is easily possible; and there is a fall in p_2/p_s. Inspection of the steam tables shows (Fig. 14.15) in the region of interest, at the same temperature, an increase in vapour enthalpy with reduction in pressure. For example, the observed increase to $2.87\,\mathrm{MJ\,kg^{-1}}$ for $T_2 = 235°\mathrm{C}$ is given by a change in p_2 from 30 bar to 20 bar.

It is important to emphasize that the main assumption is that the evaporated vapour which is in the immediate vicinity of the evaporative surface is at pressure $p_2 < p_s$, but at the same temperature as the water surface T_2. Further in a system in which $p_2 < p_s$, evaporation can proceed at temperatures higher than $235°\mathrm{C}$ without passing through a mixed zone (water and steam).

Keeping in mind the enthalpy increase that has followed exploi-

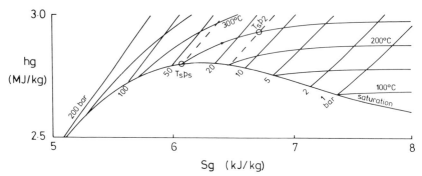

Fig. 14.15. Portion of the enthalpy—entropy (Mollier) diagram for water-substance, showing the saturation line for liquid—vapour equilibrium and the region of higher enthalpy for superheated steam only.

tation, there are two possibilities: (1) there was no near surafce resistance such as a perched water zone; (2) there was such a zone.

(1) *Unconfined zone:* In this case $Z = z_2$ is the depth of the evaporative surface below the ground surface: at the ground surface the vapour pressure is $p_0 \approx 0$. The simple model gives the pressure distribution as,

$$p^2 = 2\xi S p_s^2 z/Z,$$

where $\xi = m/m_0$ is the ratio of the mass flux to that from an open free surface; and $S = \xi \mu \epsilon T_2 Z/\beta k p_s$ is a dimensionless measure of the resistance of the overburden to flow. Figure 14.16 shows such

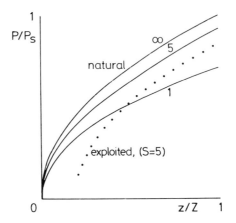

Fig. 14.16. Pressure distribution in an unconfined steam zone: in the natural state (full line) for $S = 1, 5, \infty$; and (dotted line) for $S = 5$ exploited at depth $0.2Z$ at $p = 0.125p_s$.

pressure distribution for $S = 1, 5, \infty$; notice that because of the parabolic form, much of the pressure loss occurs near the ground surface. At Larderello exploitation has been developed by setting the pressure in the rather permeable horizon at 400—500 m depth near 5 bar where with this model the natural pressure was 16 bar (assuming $p_s = 40$ bar); that is, $\tilde{p} = 0.125$ and p_0 must now be replaced by $\tilde{p} p_s$. Further, the discharge is now passing through a thinner layer; in fact if $Z = 2$ km in the natural state and we take the exploitation level as 400 m, then in the exploited state $Z = 2.6$ km and therefore S is correspondingly smaller. Taking the case $S = 5$ in the natural state, in the exploited state $S = 4$. The new pressure distribution, is also given in Fig. 14.16. We notice that p_2/p_s has changed from 0.917 in the natural state to 0.845 in the exploited state; this corresponds to an increase in heat flow of 2 times (and also an increase in enthalpy).

The increase of heat flow is more pronounced with small values of S.

(2) *Confined zone:* In this case Z is the depth of the evaporative surface below the bottom of the confining zone, i.e., $Z = z_2 - z_{22}$, and the vapour pressure p is considered to be defined by conditions in the confining zone. Figure 14.17 shows values calculated for the natural state for a layer with $S = 1$ and $\tilde{p} = 0.895$ corresponding to a heat flow reduction $\xi = 0.05$. During exploitation, the behaviour of the confining layer is unimportant because the pressure at $z = z_{22}$ is now defined by the exploitation. Also shown is the pressure distribution for an exploitation pressure $0.125 p_s$; we notice p_2/p_s has changed from 0.950 to 0.755 corresponding to an increase in heat flow of 5 times.

There is little in principle to choose between these two possibilities, and certainly there is no fundamental difference between them. Both indicate that if the explanation (b) is used, values of S cannot exceed 10, since then the changes in p_2/p_s from natural to exploited state would be too small.

Perched water zone

A popular geological idea hitherto has been that the steam was confined in the reservoir by a "cap-rock". This concept was invoked following the idea that if the steam was not confined it would all leak away. Indeed this is partially true: an unconfined system contains only two-thirds of the mass of steam of a confined system. It is, however, of interest to note that, in so far as "cap-rock" is necessary,

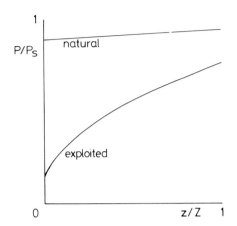

Fig. 14.17. Pressure distribution in a confined steam zone. Two cases, both for $S = 1$, with pressure p at the bottom of the confining layer: natural state, $p = 0.895p_s$; exploited, $p = 0.2p_s$.

a liquid-water-saturated layer is sufficient. If the existence of an extensive perched water zone is granted we may ask if there are any restrictions on its thickness.

The water zone is apparently in an unstable position, its lower surface will develop "tongues" of fluid which descend into the steam zone; the development of these instabilities has been studied by Saffman and Taylor (1958). Since the steam zone is everywhere above the boiling point of the descending fluid, there exists the possibility that evaporation will use up the descending fluid before it has reached $z = z_2$. In ground of small permeability and a steam zone of sufficient superheat, the perched water zone could be stable.

14.3. RESPONSE OF A STEAMING GROUND REGION TO ADJACENT EXPLOITATION

The Karapiti thermal area is a region of steaming ground within the Wairakei system. During the period of exploitation dramatic changes have occurred which are not only of great interest in themselves, through the information they give about the mechanism of steaming ground, but also through the implications for long-term prediction of behaviour of field and plant during exploitation. The presentation here is based on a detailed study by Allis (1978).

Activity is largely in a $0.3 \, \text{km}^2$ area generally higher than the surrounding region, about 3 km south of the production borefield.

Prior to the 1950s, when development began, the activity was in small areas of mudpools and moderate steaming ground and one large feature, the Karapiti "blowhole", with a total output of 40 MW (first described by Hochstetter, 1864 and Grange, 1937). Changes were occurring naturally. The Karapiti blowhole temperature was 101°C in 1927, 114°C in 1950. The earliest detailed observations and measurements are from 1946 aerial photographs. There are several large apparently extinct craters. These changes were possibly dominated by the integrated effects of climatic variation, particularly rainfall, together with progressive alteration of the ground. These changes, however, were minor compared to the dramatic effects after exploitation.

Exploitation at Wairakei has produced a transition to a new hydrological state in an interval of about 20 years. System pressures have been drawn down more or less monotonically. During this time the Karapiti area has produced a strong heat pulse. The data shown in Fig. 14.18 summarize the change. The power of the surface system increased by an order of magnitude from the undisturbed value of 40 MW to 400 MW followed by a gradual decline, to 200 MW, in 1978. The surface activity follows that of the system pressure change but with a lag of about two years. During the peak activity output was dominated by intense fumarolic activity.

During this interval the details of the area have changed considerably. Schematic maps of the area are shown in Fig. 14.19 and a very sketchy diary of events is given in Table 14.1. Detailed changes were already noticeable in 1960 and have thereafter been pronounced. These effects are greatly increased steaming ground activity and the appearance of a number of new fumaroles often initiated with minor eruptions.

Some measurements of individual features are shown in Fig. 14.20:

(a) Subsurface run-off has decreased dramatically as shown by the temperature of the main stream draining the area. Clearly, at least the upper part of the surface zone has been partially dried out.

(b) The discharge temperature of the Karapiti blowhole was high during the heat pulse, suggesting higher temperatures in the fumerole's source volume (possibly at greater depth). The fluctuations, however, are probably not very different from those in the undisturbed state.

(c) The behaviour of feature E (mudpool and fumarole) is again probably a local effect of the interaction of the fumarole and superimposed pool. When the fumarole is hot, with a small mudpool, direct discharge of part of the fumarole supply is possible. A slight

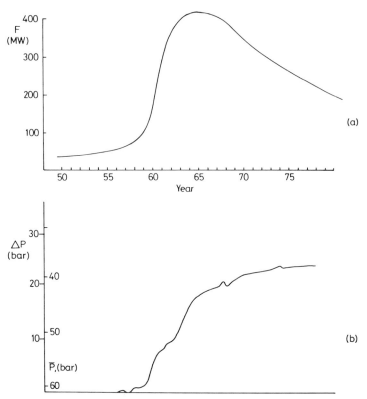

Fig. 14.18. Gross response of Karapiti thermal area to adjacent exploitation (a) Total heat flow F, from the area in MW. (b) For comparison, the loss of system pressure, ΔP in bar, obtained from the mean aquifer pressure at 274 m below sea-level in the production area. Actual pressures, $\bar{P}$ are indicated (After Allis, 1978).

Table 14.1. *Diary of events in Karapiti thermal area.* (The items A-G referred to are shown in Fig. 14.19)

1952 and earlier:	natural state, 40 MW
1950–1957:	minor normal changes
1954:	small eruption and new fumarole 100 m SE of Karapiti blowhole
1955:	(Wairakei Geyser Valley mass flow much reduced)
1957:	increased intensity
1958:	
E:	unknown in natural state: active mudpool, slow increase
1960:	
K:	vent expanded to 1.3 m², 120°C, 35 MW
K:	max 38 MW
E:	separate mudpool and fumarole

(*Continued facing page*)

July		blowout in discharging drillhole on Poihipi road, 1 km SW Karapiti blowhole
1961:		
	E: April	eruption, in two coalescing mudpools with fumaroles, 5 m and 10 m deep, area 120 m^2, 13—21 MW
	C: Sept.	previous small deep crater and fumarole, erupted to form mudpool and fumarole, 34 MW. (Ceased on eruption of B, Dec. 1967)
1963:		(Mass rate of withdrawal maximum at Wairakei, 2.3 ton s^{-1})
1964:		
	D: March	erupted, 5 coalescing craters, 1250 m^2, 99 MW initially. Peak total power 420 MW
1966:		
	F: Jan.	2 new fumaroles in 4 m diameter craters, gradually coalescing to a pit 600 m^2 × 10 m deep with 2 fumaroles. Initial power 50 MW
1967:		
	B:	blew out, Dec. 1967: 116 MW for first few months
	K:	simultaneous rapid drop (in 2 months) in output: 102°C, 14 MW,
	D:	decreased to 20 MW
	C:	ceased
1968:		
	D:	up to 67 MW
		Total power 380 MW: 140 MW, steaming ground; 240 MW, fumaroles and mudpools
1969—1975:		continued heightened activity
1976:		
	G: April	initial small, deep, very active crater
1978:		
	D:	20 MW
	G: June	eruption to form crater 600 m^2 × 10 m deep. 15 MW continuous output
1979:		
	B:	6 MW: hill, 50 mN (gradually) increased activity
	K:	107°C, 7 MW
	F:	single vent, 1 m^2, in crater wall, 40 MW most powerful feature. Total power 220 MW: 100 MW, steaming ground; 120 MW, fumaroles and mudpools

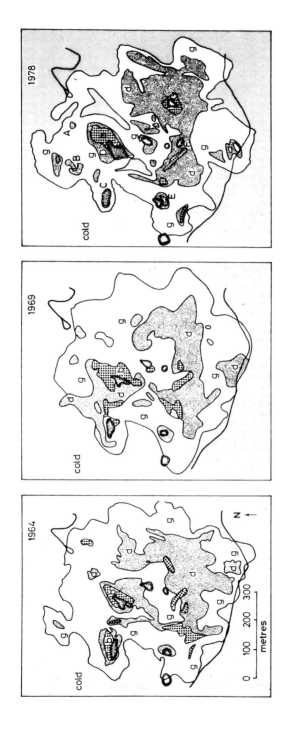

Fig. 14.19. Simplified heat flow maps of the Karapiti thermal area for 1964, 1969, 1978 (from Allis, 1978). Shading indicates the intensity grade of the steaming ground (after Dawson and Dickinson, 1970); b = bare ground, 5000 W m^{-2} (97°C at 7 cm depth); d = algae, moss cover, 500 W m^{-2} (80–90°C at 7–15 cm depth); g = stunted leptosperms — small pines cover, 500 W m^{-2} (25–80°C at 15 cm depth). Various features are labelled A–G.

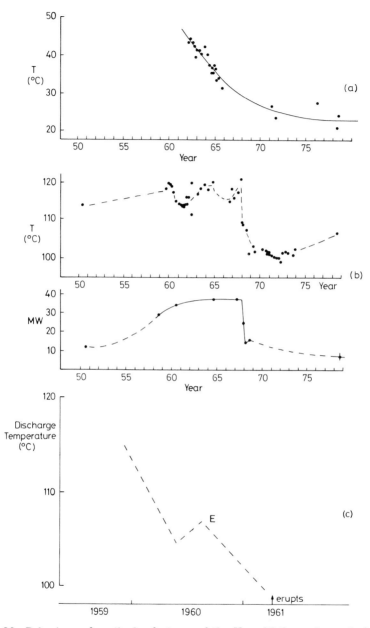

Fig. 14.20. Behaviour of particular features of the Karapiti thermal area during adjacent exploitation. (*a*) Main surface drainage, temperature of Waipouwerawera stream. (*b*) Discharge temperature and power of Karapiti blowhole. (*c*) Temperature of a new large fumarole, feature E.

fall-off in supply, possibly with extra rainfall allows the fumarole to be swamped by the pool. Once temperatures beneath the pool build up again a small hydrothermal eruption may be possible. This increases the pool size, and if it is large enough a new equilibrium is reached.

The total output in 1978 of 220 MW is made up of about: 100 MW, steaming ground; 120 MW, 11 major fumaroles, 2 mudpools; 2 MW, a minor hot water seepage to the Waipouwerawera stream (from temperature rise from $17-28°C$). Fluctuations in output and its general decline are dominated by variations in the fumaroles.

Mechanism

Pressure changes in the deep parts of the Wairakei system are quickly transmitted throughout the volume with a time-scale less than 100 days. This suggests that the sole direct effect of the drawdown of Wairakei system pressures on the Karapiti region is through a loss of pressure at depth and corresponding flow of deep liquid water, more or less horizontally towards the production discharge zone. This further implies that the Karapiti behaviour is a consequent local effect. Consider, therefore, the model sketched in Fig. 14.21.

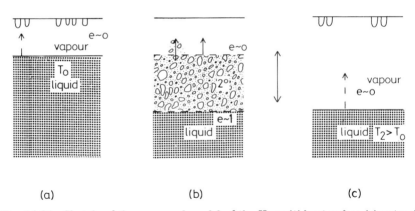

(a)   (b)   (c)

Fig. 14.21. Sketch of the proposed model of the Karapiti heat pulse: (a) natural state; (b) intermediate state; (c) final state.

(a) In the natural state, a vapour zone lies above a shallow liquid zone. In the vapour zone, the liquid fraction $e \approx 0$, except near the surface where the localized effects of rainwater infiltration are met. The interface between the vapour and the liquid is relatively sharp. The liquid at and below the interface is at temperatures below the BPD.

(b) In the intermediate state, an extensive, two-phase zone separates a near surface vapour zone from a deep liquid zone. In the two-phase zone progressive flashing occurs in the ground until the fluid is boiled off.

(c) In the new equilibrium (not yet reached at Karapiti), a similar situation exists except that the liquid—vapour interface is at greater depth.

Possible temperature profiles are shown in Fig. 14.22. Purely for the purpose of discussion, let us assume that the liquid zone is at the

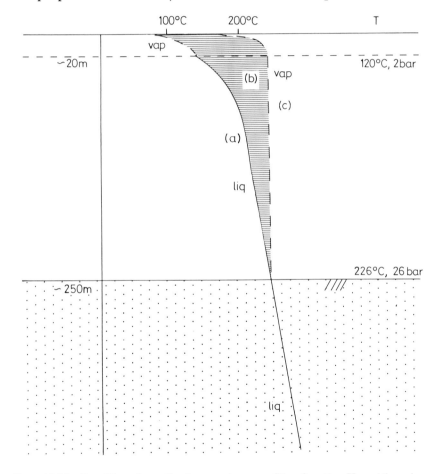

Fig. 14.22. Possible schematic temperature profiles for the Karapiti region: (a) natural state; (b) intermediate state; (c) final state; liq = liquid; profile presumes BPD; vap = vapour, profile produced by flow of steam. Shaded region indicates zone in which rock is heated up.

BPD and that the region below the boiling zone retains its temperature. The curves are drawn for an assumed initial and final interface temperatures of $120°C$ and $226°C$ corresponding to SVP of 2 bar and 26 bar.

Two processes control the rate of the transition.

(i) The rock above the base of the boiling zone has its temperature raised. The time-scale of this process can be estimated from

$$\tau_* \sim \rho_* c_* h \Delta\theta / \bar{F},$$

where h is the depth of the zone, $\Delta\theta$ the mean increase in temperature; $\bar{F}$ is the mean power passing through the system. for $h \approx 250\,\text{m}$,

$$\Delta\theta \approx 50°C, \qquad \bar{F} \sim 200\,\text{MW} \quad \text{we have} \quad \tau \sim 1\,\text{yr}.$$

(ii) The depressurized zone will tend to boil off from the top down, rather like the starting phase of opening a bore. But because of the compressibility of steam this will have a time-scale of order

$$\tau \sim \epsilon\mu h^2 / k\dot{\bar{P}}$$

where $\bar{P}$ is the mean pressure. For $\epsilon = 0.1$, $\mu = 4 \times 10^{-5}\,\text{N m}^{-1}\,\text{s}^{-1}$, $h = 250\,\text{m}$, $k = 0.1$ darcy, $\bar{P} = 15$ bar we have $\tau \approx 0.05\,\text{yr}$. The transition is therefore controlled by the thermal buffering of the rock.

If the pressure change ΔP were produced instantaneously the power output would in a time τ_* rise to a quasi-equilibrium until discharge came from the deeper levels of the two-phase zone and the increased resistance of a thicker vapour zone began to impede the output.

Depending on the permeability of the ground and the pressure change, the final output may be higher or lower than the natural value.

As the transition develops the flow rate may reach values such that shallow eruptions occur. In the description of steaming ground this was found to be when the heat flux reached $1-10\,\text{kW m}^{-2}$. The Karapiti area has clearly passed into this state.

Part IV

DISCHARGE FROM HYDROTHERMAL SYSTEMS

The surface zone of a hydrothermal system in which the interface between the deep system and the hydrosphere and atmosphere is established is complex because of the variety of possible modes of discharge. The discharge systems are dominated by the thermodynamic properties of water-substance, in particular through the presence of two phases, liquid water or steam, whether by evaporation or partial boiling.

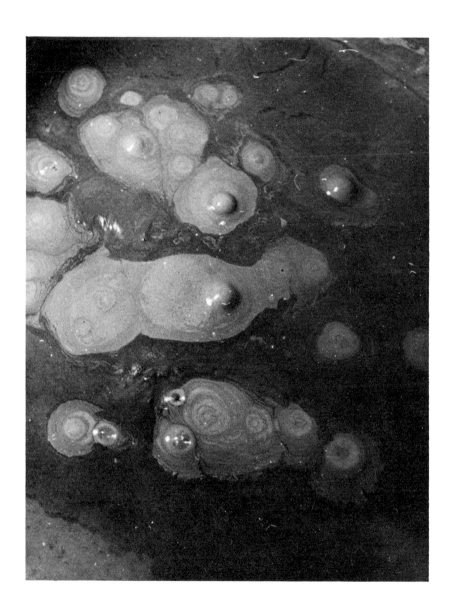

PLATE. Laboratory model of a surface discharge system, a boiling mud pool. Oblique view looking down onto the surface of a 30 cm deep layer of potter's clay heated from below. The container cross-section is a circle of diameter 45 cm. The situation simulated here is for moderately wet mud in which the formation of steam bubbles, which rise from the bottom of the wet layer and grow as they rise, is the main transfer mechanism.

15. General Features of Hot Springs

The nature of hot springs and the surface zone in which they are embedded is revealed largely from superficial observations at the surface. These observations strongly suggest, however, that the variety of hot springs arise in a shallow surface zone in which direct interaction with the atmosphere can occur.

15.1. THE SURFACE ZONE: GROSS FEATURES

The discharge system brings fluid which has been heated to the surface where it is discharged. Near the surface, extending to depths of order 100 m is the surface zone, the region where those processes occur which control or greatly modify the character of the discharge. Below this zone there will generally be an extensive body of water at a fairly uniform temperature, a temperature which changes only slowly with depth. This is the water of the deep part of the discharge system cooling a little by lateral heat loss as it rises from depth.

The surface zone will show considerable variation of permeability from place to place since the upper layers of the earth are less affected by the unifying effect of compaction. This variation will be increased by the enhanced effect at the surface of faulting produced by local ground movements. As the water rises through the surface zone it will tend to flow along the more permeable paths. A further influence which will tend to localize the flow to certain restricted paths will be the effect of topography. Clearly, for example, a spring-type discharge cannot occur unless the topography is such that the water-table can cut the ground surface. Even though the less intense pool-type discharges, such as non-discharging pools, fumaroles, or steaming ground, occur when the water-table is below the ground surface the further the water-table is depressed below the surface, the less intense these discharges tend to become. The net effect of permeability variations and the effects of the topography on the local water-table will be a strong tendency to produce a flow pattern rather like that in a sequence of increasingly branching pipe-like structures.

Each branch of this dendritic structure will contain relatively rapidly moving hot water, rising toward the surface. The surrounding rock will tend to be heated by the presence of the hot water in the adjacent dendrites. In addition, cold ground water in the surrounding rock will tend to move, due to the difference in hydrostatic head between the hot water in the dendrite and the surrounding cold water, some of which will enter the dendritic structure and cool the margins of the rising hot fluid. This cold ground water may have entered the system as meteoric water by percolating down to the water-table from the surface quite recently, but the hot water rising from depth may have entered the hydrothermal system a long time ago.

Figure 15.1 is a diagram illustrating these features. It is a vertical section of the surface zone of a thermal area. On the left, where the

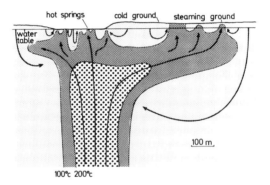

Fig. 15.1. Surface zone of a wet spring system.

water-table cuts the ground surface, there are hot springs; but on the right, where the water-table is depressed below the surface, there is a region of steaming ground. The rising hot fluid has moved the iso-therms near to the surface; the difference in hydrostatic head be-tween the cold surrounding ground water and the hotter water of the discharge system will tend to elevate the water-table in the discharge area.

An actual section, near Wairakei Geyser Valley is shown in Fig. 15.2. This was obtained from early borefield measurements and presumably gives a good representation of undisturbed ground conditions. There is extreme spatial variation and boiling is occurring at a few spots.

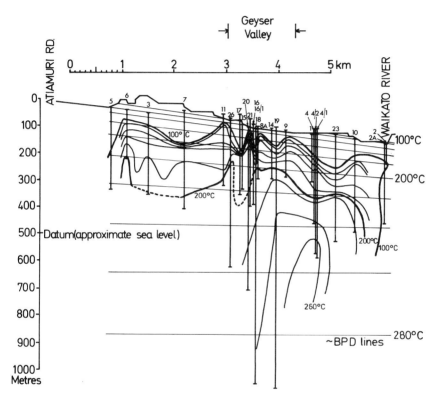

Fig. 15.2. Surface zone temperatures measured near and nearly parallel to the Wairakei Geyser Valley (drawn through bores 5 and 23, plane at 064° True, approximating the D-line, adapted from Banwell (1957)).

15.2. THE DOWNWARD MOVEMENT OF COLD GROUND WATER

A characteristic feature of the surface distribution of heat flux in a thermal area is the relatively sharp boundary between the hot areas of large heat flux and the surrounding cold areas of practically zero heat flux. This sharp boundary between hot and cold fluid is also sometimes found at depth on the margins of the discharge system. The effect at the surface is that of a group of isolated hot patches surrounded by cold ground. In fact, shallow temperature surveys reveal very little more than what is immediately obvious by visual inspection of the surface. This does not mean, however, that hot water is only available immediately below the hot patches. On the contrary, hot water can often be found below the surface zone between such hot patches.

The cause of this effect is the downward movement of cold ground water in the immediate surroundings of the rising hot water of the surface zone. The magnitude of this effect may be calculated with the following simplified model. Ground water moves downward, at volumetric flux rate q, due to the pressure difference between the column of cold ground water and the hot water of the discharge system, to a depth h where it flows away horizontally. (The effect of this horizontal flow on the temperature of the ground water is ignored in this simple approximation.) Below this level let the upward heat flux be f. The solution of the energy equation which satisfies $\theta = 0$ at $z = 0$ and $K(d\theta/dz) = f$ at $z = h$ is:

$$\theta = \frac{f\delta}{K\exp(h/\delta)} \left[\exp(z/\delta) - 1\right], \qquad \delta = \kappa/q.$$

The surface heat flux $f_0 = K(d\theta/dz)$ evaluated at $z = 0$ is:

$$f_0 = f\exp(-h/\delta).$$

The surface heat flux is hence reduced by a factor $\exp(-h/\delta)$ below the value that would be found if there was no descending ground water. It is also worth noticing that the temperature at the bottom of the current, obtained with $z = h$, has been reduced approximately by the factor δ/h. This factor can be estimated in the following way. The flow is produced by the pressure difference between the hot and cold columns and at depth z; this pressure difference is $(\rho_0 - \rho)gz$ where ρ_0 is the density of the cold ground water and ρ is the density of the hot water in the discharge system. Hence by Darcy's relation:

$$q = k(\rho_0 - \rho)g/\mu$$

For example, when the hot water temperature is $100°C$, $\rho_0 - \rho = 0.04\,\mathrm{ton\,m^{-3}}$, so that using $\mu = 0.01$ poise, a value appropriate to the cold column, and $k = 0.1$ darcy, the average value in the surface zone at Wairakei, we obtain $q = 4 \times 10^{-6}\,\mathrm{m\,s^{-1}}$ and hence with $K = 1.2\,\mathrm{W\,m^{-1}\,k^{-1}}$, $\delta = 6.7\,\mathrm{m}$. Even with such a small value of h as $20\,\mathrm{m}$, $h/\delta = 3$ and the surface heat flux is practically zero.

15.3. GENERAL COMMENTS ON THE CONSTITUENTS OF HOT SPRINGS

Early notions about geothermal systems were of necessity dominated by ideas arising from studies of hot springs. Because of the unusual nature of the chemical constituents of these springs it was assumed that the fluids and their constituents arose at great depth and travelled directly with little change to the surface. This idea is quite wrong.

A great amount of data is available on the chemical constituents of geothermal fluids. Unfortunately, the interpretation of this data is difficult and often misleading, not only because of the multitude of constituents and the variety of their interactions in the wide range of thermodynamic conditions within a hydrothermal system, but because of a variety of possible sources of the constituents. The constituents may be produced solely from within the system itself, that is by leaching of the local rocks; or brought into the system from the exterior either in the surface-derived recharge—discharge system; or from depth being carried to the vicinity of the hydrothermal system within the magma and rock of intrusive bodies or from associated volatiles. When one considers the variety of essentially *ad hoc* schemes and models that have subsequently been shown to be wrong one is left sceptical about the usefulness of studies of this type. As I have elsewhere been led to remark: "How much do we learn about how to cook an egg by studying the chemical constituents of the water in which it is boiled?" This is an extreme view. We are still faced with the problem of describing the chemical data. In view of the rapidity of chemical reactions perhaps it is best to try to specify the thermodynamic conditions by physical studies and then consider how those conditions will affect the chemistry, rather than the other way around.

Among the variety of pools two main types are recognized: near neutral or alkaline pools often with various chlorides as the principal constituent; and acid pools which may contain chlorides but are often characterized by hydrogen sulphide and sulphates. The acid pools have sulphate concentration enhanced in the ratio 20—200, whereas the chloride may be low by a factor of 100. The chloride is no great problem, either the direct supply of deep chloride is small or it is heavily diluted with surface water. The sulphate, on the other hand, requires a distinct mechanism for its concentration. Clearly a spring which immediately discharges the water from depth by overflow cannot accumulate enough sulphate even if all the H_2S were oxidized. Equally, if there were no discharge by overflow, chloride could accumulate in the pool.

The nature of the waters from hot springs can be characterized in a variety of ways. The most elaborate and complete is by means of the chemical constituents. Otherwise, some gross or average measurement is required. The electrical measurements of pH and electrical conductivity are useful for this purpose. A number of such measurements are collected in Fig. 15.3. It is immediately clear that there is

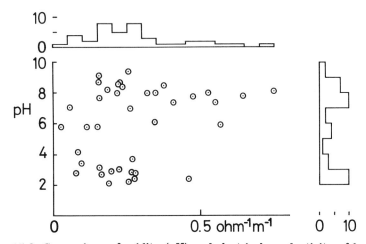

Fig. 15.3. Comparison of acidity (pH) and electrical conductivity of hot spring waters of the Rotorua–Taupo district (Data from Thompson, 1957). Also shown are histograms of pH and electrical conductivity.

negligible relationship between these two quantities. This is not surprising since the electrical conductivity is largely a function of the total ionic mobility of the ions in solution while the pH is a function only of the hydrogen ion concentration. If however we consider the distribution of these quantities separately there is a pattern in the data. The frequency distribution for electrical conductivity is unimodal with most frequent values of $0.15–0.3$ ohm^{-1} m^{-1} and skewed towards low values. On the other hand, the frequency distribution for pH is strongly bi-modal with modes near $2–3$ and $7–8$. Here two distinct processes must be at work. There are two fairly distinct types of water: (i) neutral or sub-alkaline; (ii) acid. This distinction has been known and recognized ever since hot springs have been studied. It is worth noting that in most areas both types of water can be found. Therefore the processes which in the same area produce these different waters must be local processes related to the conditions of individual springs.

The higher chloride springs are generally sub-alkaline, are hotter and have the greater discharge. No cool springs have high chloride content.

Chemical Identification of Hot Springs

The following is based on Wilson (1966). In Table 15.1 there is a compilation of data on the constituents of the waters of some discharges in the Wairakei district. Certain constituents are more or less ubiquitous: these are the ones entered in the following tables. Certain constituents occur merely as traces or as isolated occurrences: polythionic acids, White Island and Ketatahi; As, Waiotapu; Hg, Ngawha; Ge, Te Aroha; Au, White Island. Much of the material is being reprocessed from material deposited with the earlier volcanics which underlie the hot springs; sulphur is a notable example.

The best chemical indications of high temperatures at depth are given by the Na/K ratio which for example varies at Wairakei from 9 to 35 — low in hot areas, high in cool areas. These constituents of the hot water are leached from the country rock in a manner such that at higher temperatures there is an increased tendency for K to replace Na in feldspars and also K together with silica and alumina forms the adularia feldspar.

Wairakei Geyser Valley was characterized by boiling springs and geysers of clear chloride water. The Waiora area a few kilometres south is characterized by muddy acid sulphate springs with one geyser of mixed chloride—sulphate water. Also there are some dry fumaroles of which the largest was Karapiti blowhole.

If we compare the mass concentrations in Champagne Cauldron with those in the West bores the following broad pattern emerges.

(i)   There is a group of "inert" constituents whose proportions are merely changed by apparent dilution: Cl, Na, F, B, As, Br, Li, SO_4 with ratios of 0.79, 0.82, 0.79, 0.81, 0.79, 0.88, 0.87, 0.86. These constituents presumably remain in the water phase.

(ii)   A group of constituents are enriched: K, Rb, Mg with ratios of 1.43, 1.78, 2.1. This suggests that the water which feeds Champagne Cauldron is derived from a hot part of the hydrothermal system.

(iii)   A group of constituents which are depleted: SiO_2, Cs, Ca, NH_4 with ratios 0.56, 0.68, 0.25, 0.20. No single explanation presents itself. Presumably the SiO_2 and Ca are progressively lost by deposition at lower temperatures — the silica data shows a strong correlation with temperature. The solubility of silica at $250°C$ is 490 PPM (Kennedy, 1950), comparable with that of the deep bore water. Silica

Table 15.1. *Liquid water constituents, Wairakei district, New Zealand.*

	Geyser Valley					Waiora			Bores	
	Dragons mouth	Haematite geyser	Champagne cauldron	Eagles nest	Devil's inkpot	Heavenly twins	Devil's eyeglass	Shallow (No. 9)	East	West
U_0, (kg s^{-1})	0–10	0.75	120	0–7	~0	7	–	2.7	30	64
Cl$^-$ (PPM)	1840	1420	1770	1400	1270	110	660	1170	2140	2240
Na$^+$ (PPM)	1140	950	1070	920	840	90	410	1080	1280	1310
Cl/F	161	141	164	148	140	~10^3	~10^3	265	162	165
Cl/B	24.2	24.4	24.5	24.3	24.7	19.1	20.7	22.3	24.2	24.0
Cl/As	1020	1000	1010	920	1030	1100	1400	900	960	1000
Cl/Br	–	780	1000	820	800	80	440	890	870	900
Cl/NH$_4$	1250	3800	1200	1720	3450	15	84	31	4100	4800
Cl/SO$_4$	185	100	188	150	130	0.7	6.9	147	178	172
Cl/I	7800	9000	–	7600	–	200	2000	–	16500	13500
SiO$_2$ (PPM)	–	–	294	–	254	–	318	350	402	521
Na/K	22.9	22.4	17.8	29.2	23.0	10.8	15.5	24.8	12.7	9.8
Na/Li	28.6	26.7	29.8	29.6	27.2	36.5	32.4	43.2	29.1	27.0
Na/Rb	2300	–	3600	2500	–	4000	2000	–	1850	1600
Na/Cs	3600	–	2500	2400	–	20000	20000	–	3400	2900
Na/Ca	70	78	56	79	94	–	20	78	117	180
Na/Mg	–	460	1400	320	460	–	102	370	520	520

Note: B as boric acid; As as arsenious acid. The values for bores East and West are averages.

is close to saturation. The NH_4 ratios strongly suggest that depletion has occurred in steaming ground. It should be noted that NH_4 is exceptionally high in the Waiora valley and shallow bores there. I have no adequate explanation for the Cs ratio, it would appear to have been deposited near the steaming ground interface since the amount in the perched Waiora springs is extremely small.

To some degree the constituents must characterize the source rocks from which they are drawn. This can be the earlier deposited volcanics or currently active intrusive bodies. In view of the difficulties of describing the distributions of the main constituents of hot springs alone, deductions about the source rocks from hot spring minor constituents are uncertain. For example, the small As and Cs content of water at Rotorua may suggest that these are older hydrothermal areas now depleted. Deductions of this type are not only *ad hoc* but are in the argument category of a very small tail wagging a big dog.

It is of interest to compare the data from the Wairakei area with those of springs from other areas. Some relevant data are collected in Tables 15.2 and 15.3.

At Waiotapu the constituents of the most voluminous discharge from Champagne Pool are not very similar to the deep bore water; the bore water is closer in constituents to the rather feeble Postmistress Spring. At Rotokaua the mixed chloride—sulphate waters are rather acid probably owing to the dissociation of bisulphate at low temperatures. Warm springs, whether normal springs unrelated to volcanism or those in nearly extinct hydrothermal systems, have low silica and generally high Na/K (note however the rather low value for Hanmer).

Simple chemical code

It is convenient, if for no other reason than putting the information in some sort of order, to have a numerical labelling system for things characterized by a mass of rather undigested data. Various schemes occur to mind but such schemes are especially useful when devised for a particular area and a particular type of investigation. Just to give the idea, consider the following. A code is constructed by adding together values for the existence of a particular property. The boundary values are chosen to give a division appropriate to the property.

Table 15.2. *Liquid water constituents, Rotokaua and Waiotapu.*

	Rotokaua							Waiotapu					
	9	6	163–168	Lake	82	Parariki (1)	Parariki (4)	Champagne pool	Ngakoro	Postmistress	Lady Knox geyser	Waikite spring	Bores
U_0 (kg s^{-1})	~0.05	7	~4	20	5	2.5	~0.05						
Cl$^-$ (PPM)	1520	1350	180	390	500	420	710	1880	1780	670	820	100	~1100
Na$^+$ (PPM)	970	990	280	280	390	350	530	1150	1100	480	440	180	~700
Cl/F	9.3	9.6	16.8	10.5	8.8	11.8	12	226	176	—	84	24	100
Cl/B								21	22	29	24	24	30
Cl/As								720	480	1260	—	920	~1000
Cl/Br								620	670	770	—	6000	~700
Cl/NH$_4$	405	430	470	216	200	179	240	70	320	290	59	45	~800
Cl/SO$_4$	10	7	6.5	2.7	1.1	1.3	2.6	51	82	26	—	3	30
SiO$_2$ (PPM)	355	340	280	195	330	330	304						450
Na/K	16	16	18	15	14	14	15	12	54	36	38	33	~20
Na/Li	38	38	36	42	48	47	45	37	57	33	22	26	~45
Na/Ca	81	154	77	31	45	28	39	67	59	94	—	62	~500
Na/Mg	150	90	100	100	40	35	87	470	620	740	—	74	~350

Table 15.3. *Constituents of waters of hot discharges of New Zealand* (PPM *and ratios*).

	Cl	Na	SiO_2	Cl/SO_4	Cl/NH_4	Na/K
Tokaanu	3090	1820	300	100	1000	18
Taupo Spa	1600	1010	120	1.7	400	24
Wairakei bores	2240	1310	520	172	4800	10
Rotokaua	1420	990	400	2	430	16
Ohaki	1070	900	305	54	110	19
Ngatimariki	430	410	250	0.5	220	42
Orakei Korako	430	320	430	0.75	400	19
Waiotapu	1880	1150	450	44	75	12
Waikite	110	180	—	3	24	33
Waimangu	960	600	740	3.5	70	26
Whaka	570	460	440	13	800	13
Rotorua	750	600	225	16	400	33
Kuirau	320	350	320	11	1100	19
Tikitere	320	600	250	12	35	40
Waitangi	360	300	220	20	500	32
Kawerau	1100	690	330	210	90	11
Te Aroha	570	3120	120	1.6	86	67
Ngawha	920	690	110	5	0.2	20
Banks peninsula	280	230	—	19	2500	34
Morere	15500	3000	25	1600	5000	130
Tarawera	710	2160	42	23	150	94
Hanmer	500	360	49	18	58	16
Maruia	140	870	65	11	29	38

Let us give the values:

20: Cl > 1000 PPM (or Na > 600 PPM);
10: Na/K < 20;
4: SiO_2 > 150 PPM ;
2: Cl/SO_4 < 40
1: Cl/NH_4 < 100

This gives possible codes of 0, meteoric water, to 37. In effect this is an octal system and has the advantage over a mere ordered list in that a code can be uniquely decoded. Some codes which may be verified from the above data are listed in Table 15.4.

What is most striking about this data is that in any thermal district more or less all possible codes are obtained. This is strong evidence that springs are dominated by local shallow processes.

General comments on the water composition

The following is after Grange (1937). Although in this book little explicit attention is paid to the details of the composition of the discharge, it is useful to keep in mind the chemical nature of the waters:

Table 15.4. *Hot spring chemical codes.*

0	Meteroic water
1	
2	
3	Te Aroha; Maruia
4	
5	
6	Waiotapu, Postmistress; Ngatimariki; Rotorua; Waitaugi.
7	Waikite; Tikitere
10	
11	
12	
13	
14	
15	
16	Rotokaua springs, except 6, 9; Kuirau
17	Waiora
20	Morere
21	
22	Tarawera*; Taupo Spa
23	Hanmer
24	Wairakei geyser valley; Waiotapu, Ngakoro
25	Wairakei shallow bores
26	Waiotapu bores 1, 3, 5; Orakei Kbrako; Whaka
27	Waimangu
30	
31	
32	Tokaanu
33	Ngawha*
34	Wairakei deep bores
35	Waiotapu, Champagne Pool; Kawerau
36	Rotokaua springs: 6, 9; Waitapu bores 4, 6, 7
37	Ohaki*

Note: * = borderline case.

(1) Total dissolved matter is typically in the range 100–20000 PPM.

(2) The main anions, in typical order of abundance, are chloride, sulphate, bicarbonate. Springs with dominant chloride, sulphate or mixed chloride and sulphate are the commonest — much of the subsequent discussion refers to these. Bicarbonate waters seems to occur mainly where local rocks contain carbonates. Notable occurrences of dominantly bicarbonate springs are rare in New Zealand (alkaline spring, steaming cliffs, Rotomahana; one at Mokai; one on Te Huia stream, Atiamuri Road). Nevertheless, bicarbonate is an abundant constituent of many chloride waters. The acidity of the

waters is very crudely related to the presence of sulphate as indicated in Fig. 15.4. In effect, these pools contain weak sulphuric acid. Most chloride pools are neutral or mildly alkaline, although a few are acidic — notable examples are at Waiotapu, Inferno and Frying-pan flat at Waimangu and on White Island. It is noteworthy that the acidity is not necessarily uniform throughout a pool. For example a large pool Roto-a-Tamaheke, Whaka, has a pH of 8.7 where it is bubbling vigorously, but in a shallow turbid region near the margin the pH is 3.3 (Grange, 1937). This is perhaps the clearest evidence for the origin of acidity through oxidation by atmospheric oxygen.

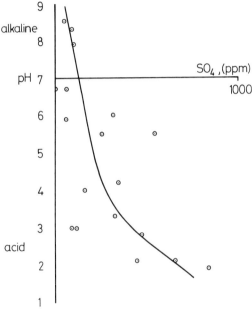

Fig. 15.4. Acidity of hot springs; pH related to the sulphate content, in PPM, for some springs of the Taupo district (Data from Grange, 1937, p. 107).

(3) Silica is highest in chloride waters, reaching 750 PPM. It can be transported in solution or colloidal form.

(4) The cations seem to be related to the nature of the water and its effect on the ambient rocks. The most abundant cations of crustal rocks (Na, K, Ca, Mg, Fe, Al) are present but in rather different proportions from those in the rocks. Typically the main cations are Na and K in amount up to 1000 PPM and 100 PPM in hot chloride waters. The others Ca, Fe, Al, Mg are generally below 10 PPM. Somewhat more abundant Fe and Al are found in sulphate waters

occasionally of order 100 PPM. Where limestones occur in the ambient rocks Ca and Mg are abundant in the water reaching typically 200 PPM and 100 PPM.

(5) Dissolved matter released as gas varies widely in amount. Some springs produce no gas at all and even in those with abundant gas the contribution to the mass discharge is usually minute. It is worth emphasizing that, even for dry fumaroles, by far the dominant constituent is steam, the contribution from gases dissolved in the deep water being small.

Of the dissolved gases CO_2 is dominant being typically 80—100% of the exsolved gas. When CO_2 is not dominant or relatively low the other main gas is N_2 (reaching 88 vol.% in Hinemoa's Bath, Mokoia Island, 38°C). Oxygen is depleted relative to nitrogen from its atmospheric value owing to ready incorporation in silicates but can reach 10 vol.%. In quiet acid pools much of the nitrogen and oxygen may have come directly into the pool from the local atmosphere, the oxygen being depleted by oxidation reactions in the pool. Hydrogen is carried (other than in the water) mainly as CH_4 and H_2 and, perhaps surprisingly owing to its ready detection because of its potent smell, H_2S is rarely present in more than 1 vol.%. But H_2S can be as high as 15 vol.% as at Tikitere.

(6) Where sulphate-rich pools have locally small or no discharge by overflow, salt incrustations develop, commonly found as whitish patches a few centimetres thick on the margins of hot ponds. This material is composed of sulphates of Ca, Al, Na, Fe, K, Mg, in usual order of abundance, Ca being highest, the sulphate typically more than 70%. These salts can be obtained by the action of weak sulphuric acid on rhyolites, or where the Ca is high on basalt (as at Waimangu).

Sinter

One of the commonest deposits produced by hot springs is sinter. It is a nearly ubiquitous deposit of alkaline springs and to a small extent or not at all from acid springs. Normally it occurs in overflow channels or around the mouths of fumaroles and geysers. Its composition is that of common opal, a hydrated silica, usually of a creamy grey colour but it can be pink if ample Fe is present (example 2% Fe_2O_3 at Papakura geyser, Whaka) or black if Mn is present (example, 4% MnO at Black geyser, Wairakei). An analysis (Grange, 1937) of the sinter from Papakura geyser, Whaka, gives in mass parts per thousand: SiO_2, 863; total H_2O, 95; Al_2O_3, 18; CaO, 6; Na_2O, 5; K_2O, 3; Fe_2O_3, 3; total S, 2; TiO_2, 1; BaO, 1; traces of MnO, P_2O_5, ZrO_2 but none of SrO, Cr_2O_3, CO_2.

An estimate of the time required to produce a given deposit is straightforward in principle. For example, consider a spring discharging $1\,kg\,s^{-1}$ of water which loses $100\,PPM$ of SiO_2 in flowing over on a small terrace of present volume $100\,m^3$. The time required is then about 80 years.

Some very large sinter terraces are known. Perhaps the most famous were the Pink and White Terraces (destroyed by a volcanic eruption in 1886) of Lake Rotomahana. They rose from the lake as a broad terrace of about 20 steps, extending over a distance of $240\,m$, to the upper terrace and main basin at a height of $24\,m$ above the lake. From various descriptions it is estimated that the deposit had an area of $2 \times 10^4\,m^2$ and a mean thickness of $4\,m$, a volume of $8 \times 10^4\,m^3$. No discharge measurements are available, but from the descriptions as an intermittent boiling spring of area $350\,m^2$ and depth $2\,m$ or more, in which the water level rose, with central eruptions to heights of 3—5 m until overflow occurred, followed by a fall of water level lasting several hours, one would guess that the discharge averaged less the $50\,kg\,s^{-1}$. This figure corresponds to filling the upper part of the basin of volume $750\,m^3$ in one hour and a duty cycle of 1:5. With the above figures, this corresponds to an age of at least $1280\,yr$. These figures are very crude but do suggest an age of order 10^3 yr.

Hydrothermal alteration of rocks

A striking feature of a thermal district is the effect of the hot saline waters and steam on rocks to produce extreme metasomatic alteration. Again this subject is not strictly within the framework of this book but a comment in passing is necessary. The alkalis and alkaline earths are progressively removed and some silica is added, first producing opalized feldspars in a hard rock of unaltered texture, and finally a kaolin clay. Many of these clays are brightly coloured. Deposits altered by alkaline waters are less common and produce montmorillonites.

16. Hydrology of the Surface Zone

The vital agent in a hydrothermal system, water-substance, originates at the surface and passes through the system before returning
to the surface. If attention is concentrated on the role of the water-
substance, the system is seen to be driven and controlled effectively
by the supply, flow and discharge of the water. In particular we
concentrate on how the water enters and how it leaves the system.

This chapter focuses attention on the water by describing the
system and its parts by means of a variety of equivalent circuits
based on data from natural systems in the Taupo district. The validity
of the ideas is limited by the small amount of data. It is most unfor-
tunate that very little attention has been given to comprehensive
hydrological studies of natural spring systems. What follows should
therefore be treated as a sort of hydrological cartoon.

Most of the clues to the nature of the hydrology come from
observations on the temporal behaviour of individual springs, groups
of springs, and the surface discharge of an entire thermal area. Considerable insight into the mechanics comes from regarding a spring,
or a group of springs, and the immediate surroundings as a reservoir
supplied by the overall ground flow, referred to here as seepage.

16.1. ASPECTS OF THE SURFACE ZONE

Seepage

The importance of ground seepage is revealed in some measurements made in the relatively undisturbed Wairakei Geyser Valley in
August, 1954 (data from Banwell, 1957) at two weirs between which
active discharges occurred. The data and schematic arrangement are
shown in Fig. 16.1. Let us assume an input, q_0, the total driving the
visible hot springs and flow into the stream through the stream bed;
and a seepage from the stream through the stream bed into the
ambient ground. Take the seepage temperature, θ_1, to be a mean
stream water temperature of $25°C$ and the input temperature, θ_0, to

be $100°C$. Conservation of mass and energy then gives $q_0 = 98.5$ $kg\,s^{-1}$ and $q_1 = 40.7\,kg\,s^{-1}$ (a value of $\theta_0 = 80°C$ requires 134.3 $kg\,s^{-1}$ and $76.5\,kg\,s^{-1}$ respectively, and so on). Seepage is pronounced and makes an important contribution to the mass budget.

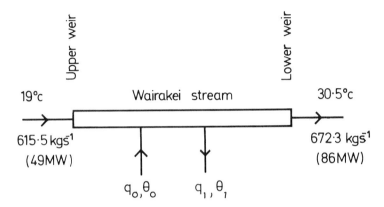

Fig. 16.1. Mass and heat balance between two weirs, 650 m apart, in the Wairakei stream, for August 1954. (Data from Banwell, 1957)

Role of water table level

An interesting set of measurements, have been made at the Taupo Spa thermal area beside the Waikato river whose level is artificially controlled. These are reproduced in Fig. 16.2. Springs well above the river level are little affected, but those close to it are strongly influenced. For example, both the discharge and the temperature of the Waipikirangi geyser rise and fall with river level. Similarly the important role of seepage is revealed by the corresponding rise and fall of the level of Tiger Pool, a non-discharging pool — but with the temperature being now out of phase with the fluid level. Presumably, two effects are operating: the change of differences of head and changes in the seepage resistance.

Effect of a major disturbance

Further useful clues to the system behaviour are possible following some marked disturbance. A nice example of this is provided by detailed observations of a spring in the Wairakei Geyser Valley in 1952–1954. This is shown in Fig. 16.3.

The spring, number 174, is about 50 m from the stream at a relative altitude of 9 m. During that time, bore output was raised rather

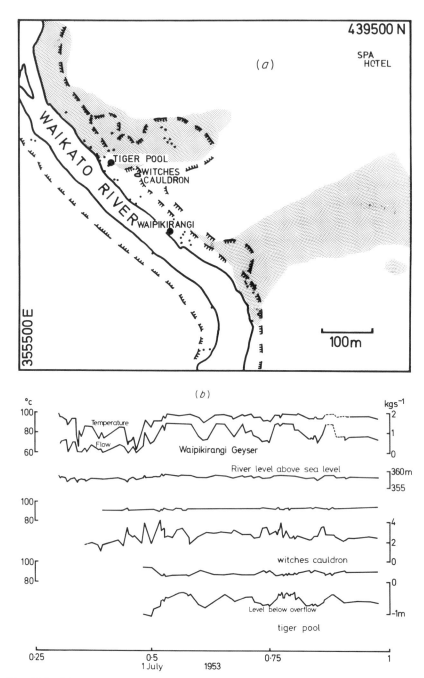

Fig. 16.2. Taupo Spa thermal area: (a) locality; (b) measurements of selected springs during 1953: temperatures, in °C; discharges, in kg s^{-1}; levels, in metres. (After Thompson, 1957)

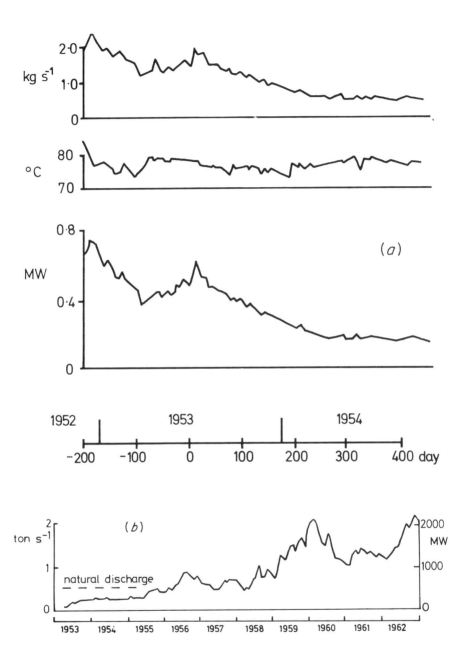

Fig. 16.3. (a) Discharge of a small spring, number 174, Wairakei Geyser Valley: mass discharge, in kg s⁻¹; temperature, in °C; and power, in MW (data from Thompson 1957). (b) For comparison, Wairakei bore discharge rate (data from Grindley, 1965)

sharply from a low value of 0.2 to about 0.7 of the natural discharge of the whole Wairakei system and was held there until mid-1955. Within an interval of about 9 months, what appears to have been a new equilibrium level was achieved. This suggests a time-scale of 3—4 months.

This was not an isolated occurrence. Data for some other springs in Wairakei Geyser Valley are shown in Fig. 16.4. Whereas springs at low levels were apparently unaffected by this disturbance (example, spring 18, a vigorously boiling superheated pool 5 m from and 3 m above the stream), springs at high levels were profoundly altered. Spring 29 behaved as follows: till November 1952, periodic; then steady; October 1953, output decreasing; April 1954, output ceased with a rapid drop of water level to 0.5 m below the overflow level and has continued fall since then. Spring 113 (opal pool) behaved similarly: till November 1952, a steady discharge of about $1 \, \text{kg} \, \text{s}^{-1}$; followed by a continuous decrease of output; January 1954, output ceased with a rapid drop of water level. Some changes occur naturally, for example Spring 55 switched from steady to periodic in February 1953.

Hot spring fluctuations

Individual hot springs and other discharges are variable in their properties. For example, Fig. 16.5 shows for a selection of springs of good discharge in Rotorua the temperature range plotted against the maximum temperature. All the springs show temperature fluctuations, typically 5—25°C. There is a very rough correlation of range and maximum temperature suggesting that the cooler springs are more influenced by local ground water conditions. This notion is confirmed by the observation that after rain the springs warm up, but in dry weather spring temperatures fall.

Rainfall and thermal area variability

The overall hydrology of thermal areas is rather poorly understood, this is particularly so for the problem of how the surface systems are fed from depth. Some observations have been made and those for the Waiora valley, of catchment area $2.3 \, \text{km}^2$, shown in Fig. 16.6, are particularly interesting. Various features will be noted in turn.

(1) The outstanding observation is the variability. In this data fluctuations occur on times from less than 1 day to more than a year.

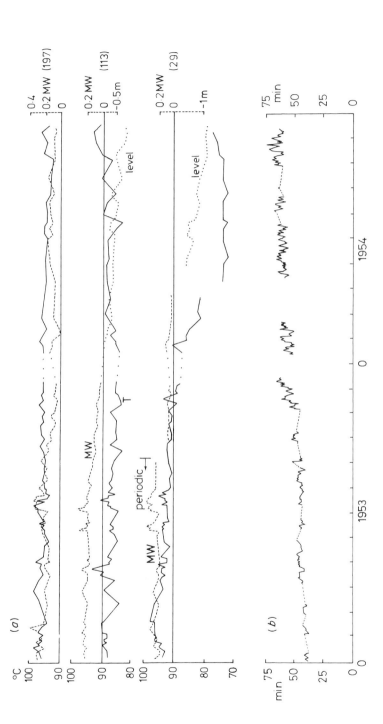

Fig. 16.4. Discharge from springs in the Wairakei Geyser Valley during 1953–1954 (data from Thompson, 1957). (a) Springs 29, 113, 197: full lines, temperature, in °C; dashed lines, heat discharge rate, in MW; dotted lines, where discharge drops to zero, the level of the water below the lip. (b) Bridal Veil geyser period, in minutes.

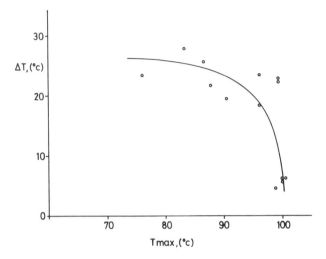

Fig. 16.5. Temperature range of selected hot springs at Rotorua, New Zealand, ΔT = maximum temperature — minimum temperature, against maximum temperature, T_{max}. (Data collected 1927, 1928 and given by Grange (1937), p. 114)

(2) The longer term fluctuations clearly have a strong annual component — typically: temperature, $\pm 2°C$; discharge ± 7 kg s^{-1}.

(3) The annual temperature variation is the most regular, following closely the monthly mean air temperature but with a lag of 0—1 month, and an amplitude reduced in the ratio about 0.3. This strongly suggests that these waters come largely from shallow depths, averaging about 2 m, and that temperatures of the waters below these shallow depths have remained constant.

(4) The annual discharge variation is marked, but not as regular as that of the temperature. The discharge is roughly out of phase with the temperature but this is rather incidental since the discharge is clearly dominated by the rainfall pattern, but with a lag of about 3 months. This lag indicated the response of the regional subsurface reservoir which maintains the Waiora valley systems.

(5) Short-term fluctuations over intervals of a few days are pronounced and have amplitudes comparable with those of the annual fluctuations.

(6) Temperatures respond immediately to rainfall, dropping as soon as rain starts and recovering quickly in a day or two after it ceases. This is presumably a mere cooling and subsequent reheating of the uppermost ground, a further clue that much of the stream discharge is of very shallow origin.

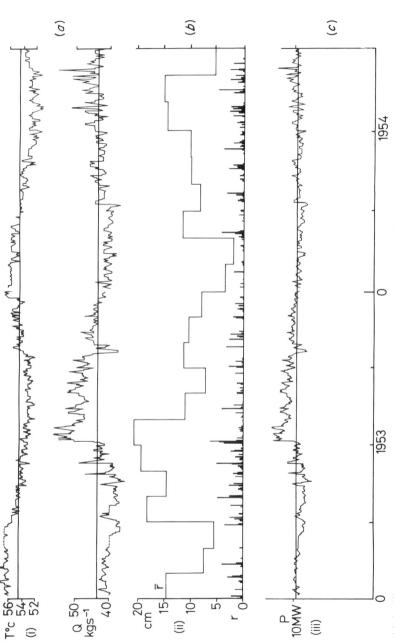

Fig. 16.6. Waiora valley stream flume measurements for January 1953—September 54. (*a*) Temperature *T*, in °C; mass flow *Q*, in kg s⁻¹; (*b*) daily rainfall *r*, in cm and monthly total, *r̄* in cm; (*c*) power, *P* = *QcT*, in MW relative to 0°C. (Data from Thompson, 1957) The mean values are: 54°C; 43 kg s⁻¹; 11 cm rain per month; 9.7 MW.

(7) The response of the discharge to rain is delayed. There is a sharp increase after about 2 days, with a more gradual fall lasting about 7 days. This lag indicates the collective response of the local reservoirs feeding individual discharge systems. A few of the discharge spikes are negative, especially towards the end of the relatively dry summer. I have no explanation for this.

16.2 MODEL

Mass input

For the purpose of the following discussion consider the reservoir shown schematically in Fig. 16.7. The reservoir has capacitance, C, thermal volume, Φ, and a net leakage resistance, R. Fluid is supplied to the reservoir at a given equivalent volumetric rate, q_s, from some other reservoir system, and if the reservoir is open to the surface, by surface water input at a rate rC. Fluid leaves the reservoir at a rate, d, to supply some other reservoir system or by means of an artificial discharge and at h/R, as natural leakage to the external surface system, where h is the pressure head.

The surface water input to a particular reservoir may come from an adjacent stream or lake or directly from rainfall. The streams and lakes can be modelled as reservoirs (see for example Donaldson, 1974). In any event the ultimate source is rainfall.

The rainfall input is handled as follows. Let the mean net rainfall be u in units $m\,s^{-1}$. For example $u = 4 \times 10^{-8}\ m\,s^{-1}$ corresponds to

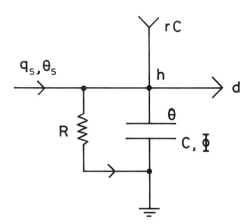

Fig. 16.7. Reservoir equivalent circuit element.

$1.26 \, \text{m yr}^{-1}$. If this falls onto an area, A, of ground with porosity ϵ, the volumetric rate is uA and the increment to the water table level rate of rise, $r = u/\epsilon$. Conservation of matter and energy require:

$$\frac{dh}{dt} = r + [q_s - (d + h/R)]/C,$$

$$\frac{d\theta}{dt} = [q_s\theta_s - (d + h/R)\theta]/\Phi$$

In a steady state:

$$h = R(rC + q_s - d), \qquad \theta = q_s\theta_s/(d + h/R).$$

The pressure level, h, is controlled by the net input rate. If, for example, only the discharge, d, is changed then $\Delta h = -R\Delta d$. This is the familiar steady drawdown. If for example only the rainfall input is changed then $\Delta h = RC\Delta r$. Thus levels will fluctuate in proportion to the rainfall input.

Representation of a spring

As shown schematically in Fig. 16.8, a flowing spring can be represented quantitatively by its supply resistance, R_s; the capacitance, C, of the reservoir affected by the spring; seepage resistance, R; overflow resistance R_0; and an overflow valve, V. The valve is closed for

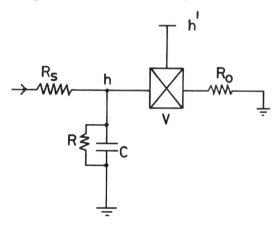

Fig. 16.8. Equivalent circuit of a single spring showing: supply resistance, R_s; the local leaky reservoir C, R; the overflow "valve", V, with set overflow level, h' and overflow discharge resistance, R_0. If $h < h'$, the overflow valve is closed, there is no overflow, merely leakage at rate h/R; otherwise h is held at h' and, in addition to the leakage h'/R, there is overflow at rate h'/R_0.

$h < h'$, open for $h \geqslant h'$ where h' is a predetermined level usually set by the lip of the spring. Furthermore, since during overflow $(h - h')$ is very small, usually less than 1 cm, in quantitative models it is sufficiently accurate to set $h \equiv h'$ during overflow.

It is also worth noting that when overflow ceases, then the capacitance of a spring is important. If the components are small, the corresponding time-scale will be small and the fall of level within the mouth of the spring will be rapid. For example in a small spring with seepage of $1\,\mathrm{kg\,s^{-1}}$ at 10 m above a stream, so that the internal resistance is 1000 w, with a well sintered mouth of capacitance $10\,\mathrm{m^2}$, the level in the vent will fall to its new level in about 8 hours.

If a spring has a continuous overflow its own internal capacitance plays no role in the dynamics of the spring; the spring being full, the capacitance is effectively switched out of the circuit. Yet it is apparent that most springs have a fairly lively response to the rainfall input: after rain the discharge returns to its base level and does not hang on the higher rate. This implies that the spring draws also on an adjacent local reservoir which may drive local seepage or a local group of springs. If this local group has a time-scale of about 5 days then an individual overflowing spring within the group will respond to rainfall as found. Thus a minimal model for a single spring for an interval of a few years, requires three elements representing the particular hot spring area, the local group of springs, and the spring itself.

Cascade model for a wholly wet system

Consider $(n + 1)$ leaky reservoirs connected in series as indicated in Fig. 16.9. Then conservation of matter and energy require, taking the fluid density to be constant:

$$q_j = (h_j - h_{j+1})/S_j,$$

$$Q_j = q_{j-1} - q_j,$$

$$\mathrm{d}h_j/\mathrm{d}t = [Q_j - d_j - h'_j/R_j]/C_j + r,$$

$$\mathrm{d}\theta_j/\mathrm{d}t = [q_{j-1}\theta_{j-1} - (q_j + d_j + h'_j/R_j)\theta_j]/\Phi_j$$

where r is the equivalent rainfall mass flux taken directly into the reservoirs, C_j and Φ_j are the flow and thermal capacitance. Notice that in the steady state, $\mathrm{d}/\mathrm{d}t = 0$, the quantities do not depend on the capacitances. Further note that the temperatures θ_j are means over the inputs and outputs of the element — the model does not require mixing within the reservoirs. If leakage is to level b_j take

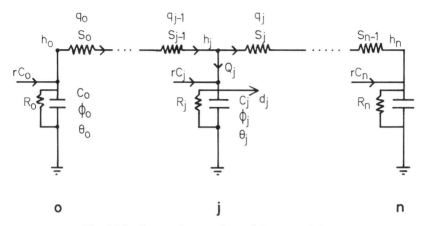

Fig. 16.9. Reservoir cascade model nomenclature.

$h_j' = h_j - b_j$. Calculations use the equivalent temperature θ, such that the specific enthalpy $h = c_0\theta$.

Where a number of reservoirs are in cascade, not only is there drawdown in that reservoir but in all the others, although of lesser amount. In a crude but effective manner this simulates the bowl-shaped depression of the water table, pronounced near discharging systems but extending into the ambient water table.

The outstanding feature of hydrothermal systems emphasized by the equivalent circuits used here is that the system operates under conditions determined by the input of meteoric water. We are dealing with flow-driven systems — in electrical parlance we have current drive rather than voltage drive. The flow inputs are determined by the rainfall and the areas of the relevant catchment. These inputs then determine the pressures in the system.

Note that the representation of a particular feature can often be given without referring to all the elements of the system. A spring system sees the recharge system simply as a source of constant discharge rate. The large reservoirs have such long time-scales that the surface discharge through a group of springs can be represented by a single resistance.

16.3. MECHANISM OF THE SURFACE ZONE

The surface zone has two prominent features: it has variable output and it is a zone through which not only discharge but also recharge occurs. An outline in very schematic form, representing the system and its parts by means of a number of equivalent circuits, allows a description of the relation of the surface zone to the system at depth, the role of fluid recirculation and the effects of steaming ground. The gross behaviour of hot springs is then considered.

The presentation is for a hypothetical system, but one similar to a system based on measured data from Wairakei in the natural state. The purpose is to show the interaction of the various possible mechanisms by referring to a particular type of system.

The system as a whole

The surface zone will be identified as a superficial feature of the discharge system imbedded in the top of the high-level reservoir. The main reservoir properties will be identified by measurements which include a field drawdown experiment, as described for example at Wairakei (Chapter 12). To proceed, it is necessary to see the system in its hydrological setting.

Identification of the model system is made with certain data.

(i)   The field discharges rates of energy and matter. At Wairakei these are: power, $F = 1000$ MW; total equivalent volumetric discharge rate of $0.9 \, \mathrm{m^3 \, s^{-1}}$, approximately equally as steam, spring overflow, and seepage.

(ii)   The reservoir gross parameters. At Wairakei these are: capacitance: $C_* = 1.86 \, \mathrm{km^2}$ net resistance, $R = 160$ w. Here we will take $C_* = 2 \, \mathrm{km^2}$, $R = 150$ w.

(iii) Net meteoric input rate, r. Here I use the value $r = u/\epsilon = 10^{-7} \, \mathrm{m \, s^{-1}}$ based on an annual rainfall of $1.2 \, \mathrm{m \, yr^{-1}}$, a guess of 50% runoff and surface porosity, $\epsilon = 0.2$. Unfortunately the net runoff has not been measured.

(iv) Hydraulic levels. At Wairakei, no extensive data are available for the natural system. Fortunately, however, there is a compilation early in the exploited life and this is shown in Fig. 16.10. A number of features are of interest:

(a) There is a pronounced and roughly eastward pressure gradient.

(b) Maximum pressures, relative to the more or less fixed Waikato river level, are about 100 m head.

(c) In the Wairakei Geyser Valley, where the bulk of the overflow occurred, the data suggest that the level was about 60 m head. Inspection of the topographic contours however indicates an elevation of about 30 m. Quite arbitrarily, in the model below I use 50 m head as the near overflow level.

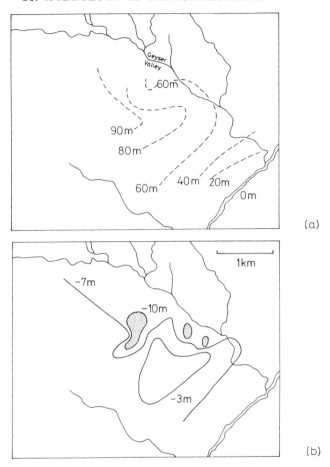

Fig. 16.10. Piezometric level at Wairakei, meter head relative to Waikato river level. (a) mid-1955; (b) rate of loss in metre head per year from mid-1955 to mid 1958. (Adapted from Studt, 1958)

The overall behaviour of the system is represented by the equivalent circuit of Fig. 16.11:

(1) R_s, C_s represent the cold water reservoir from which the recharge is drawn. For an assumed catchment area of $100\,\text{km}^2$ and surface porosity 0.2, the capacitance, $C_s = 20\,\text{km}^2$ which with meteoric input rate $r = 10^{-7}\,\text{m s}^{-1}$, has an input of $2\,\text{m}^3\,\text{s}^{-1}$. If, as measured, the deep recharge to the surface is $0.6\,\text{m}^3\,\text{s}^{-1}$, the leakage of $1.4\,\text{m}^3\,\text{s}^{-1}$ with an estimated head of $100\,\text{m}$ requires $R_s \approx 70\,\text{w}$. For an estimated system pump pressure of $1000\,\text{m}$ the hot reservoir supply resistance is $S_0 \approx 1700\,\text{w}$. The cold reservoir time-scale $R_s C_s$

is about 45 years. Thus the deep recharge will be nearly constant over intervals of order 10 years and will be controlled by climatological processes of time-scale 10^2 years and more.

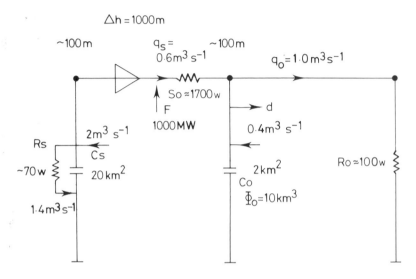

Fig. 16.11. Equivalent circuit, gross system.

(2) R_0, C_0 represent the mushroom of hot water. The net values of resistance and capacitance are determined by withdrawal tests of the reservoir. Since the supply resistance $S_0 \gg R_0$, the mushroom is largely decoupled hydraulically from the supply reservoir and $R_0 \approx R$, $C_0 \approx C_*$, the net observed values. The hydrological time-scale, $R_0 C_0$, as directly measured, is of order 10 yr. The thermal time-scale, Φ_0/q_0, is about 300 yr (where the thermal volume of 10 km^3 is estimated for a mushroom head of depth 1 km, deep porosity 0.1, capacitance 2 km^2). The hydraulic time-scale is that which has been measured. The thermal time-scale is compatible with the very small fall in deep bore temperatures of order $1°\text{C/year}$.

(3) The surface zone. In this representation, in which details are ignored, the surface zone is represented solely by the flow $1.0 \text{ m}^3 \text{ s}^{-1}$ through the discharge resistance, R_0. Discharge through the surface zone would cease, in this model, for an artificial discharge of rate d_* when $h_0 = 0$, that is when $d_* = q_s + r C_0$, which here is $1.0 \text{ m}^3 \text{ s}^{-1}$. For $d > d_*$, cold water would then flow in the reverse direction from the surface zone into the mushroom and lower its mean temperature with time-scale Φ_0/d.

Operation of the surface zone

The view of the surface zone presented so far has been very schematic since its behaviour has in effect been lumped together with that of the high-level reservoir — the mushroom. This is appropriate and adequate when we are attempting to identify the overall parameters from the results of field discharge experiments. If we wish to look in more detail into the operation of the surface zone itself, it is necessary to recognize the major processes controlling its behaviour. Here I describe the processes found in a wet system like that at Wairakei. These are: access of surface water; recirculation; role of steaming ground; spring overflow; seepage.

Identification of the model of the surface zone is made with certain data:

(i) Equivalent discharge rates of about $0.3 \, \text{m}^3 \, \text{s}^{-1}$ each of steam, spring overflow, and seepage. (It is fortuitous that these values are about equal). There is considerable uncertainty in these figures.

(ii) The natural variability of the zone as correlated with variation of rainfall indicates time-scales of order 100 days.

(iii) The response of the surface zone to exploitation is shown in a fall in spring output which effectively ceases after about 5 years, although steaming ground continues little changed.

The surface zone can now be represented by a circuit like that shown in Fig. 16.12. The surface discharge, till now represented as a simple leakage (as shown in Fig. 16.11 through resistance R_0), is considered as occurring through a number of elements.

(1) C_0. The deeper parts of the mushroom not directly affected by the surface zone is represented by a simple capacitance C_0 supplied with meteoric input at rate rC_0 and deep recharge of given and fixed rate q_s. Any artificial discharge is considered to be taken from this reservoir.

(2) S_1, S_2, R_1, C_1, R_2, C_2 and the associated components represent the surface zone itself.

(3) R_1, C_1. Laboratory model experiments and the pipe model of Wairakei indicate that fluid is recirculated within the higher levels of the system. This fluid flows through the reservoir shown as R_1, C_1. For convenience we write $C_0 = \xi C_*$, $C_1 = (1 - \xi)C_*$. The value of ξ is not known and data are shown for $\xi = 0.5$. At Wairakei the predominant discharge mode is that of steaming ground. I envisage that this arises in two ways: firstly as "direct" steam production from the

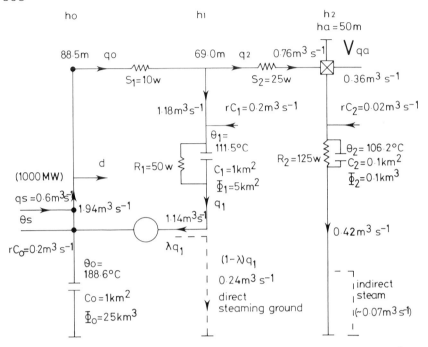

Fig. 16.12. Equivalent circuit, surface zone. Element values for $r = 10^{-7}$ m s^{-1}, $\lambda = 0.83$.

fluid flowing in the recirculation system; secondly as "indirect" steam production from the fluid flowing out of the system as seepage. Thus for the more important direct steam, if the flow through R_1 is $q_1 = h_1/R_1$, let a fraction $(1 - \lambda)$ be discharged as steam and λ remain to be recirculated.

(4) V. Spring overflow is represented by the valve, V.

(5) R_2, C_2. These represent the springs and associated seepage of the system. Some of the seepage, of rate h_2/R_2, is envisaged as producing "indirect" steam. Of course, steam will arise from the overflow, too, but this will not be counted as steaming ground. The indicated value of $C_2 \approx 0.1$ km^2 is chosen to give a hydrological time scale of order 100 days.

Conservation of matter and energy gives the following system of equations.

Deep mushroom

$$dh_0/dt = r + [q_s + \lambda h_1/R_1 - (q_0 + d)]/C_0$$
$$d\theta_0/dt = [q_s\theta_s + \lambda h_1\theta_1/R_1 - (q_0 + d)\theta_0]/\Phi_0$$

Recirculation system

$$dh_1/dt = r + [q_1 - h_1/R_1]/C_1$$

$$d\theta_1/dt = [q_0\theta_0 - q_2\theta_1 - \{\lambda\theta_1 - (1-\lambda)E/C\}h_1/R_1]/\Phi_1$$

where $E = E_* + c_s(\theta_1 - \theta_*)$, and for $\theta_* = 100°C$, say, $E_* = 2680$ kJ kg^{-1}, $c_s \approx 2$ kJ kg^{-1} K^{-1}.

Spring system

If $q_2 > h_a/R_2$, then $h_2 = h_a$ and $q_a = q_2 - h_a/R_2$. Otherwise

$$dh_2/dt = r + [q_2 - h_2/R_2]/C_2 \quad \text{and} \quad q_a = 0,$$

and in both cases,

$$d\theta_2/dt = [q_2\theta_1 - (q_a + h_2/R_2)\theta_2]/\Phi_2.$$

There is the further constraint that the maximum amount of direct steam is such that the temperature of the recirculated fluid is greater than zero. Thus there is a minimum value of λ such that

$$\lambda_{min} \leqslant \lambda \leqslant 1 \quad \text{where} \quad \lambda_{min} = 1 - c\theta_1/E.$$

Surface zone behaviour, natural state

The behaviour of the steady state surface zone, for the parameters shown in Fig. 16.12, is indicated by the data of Fig. 16.13 where various quantities are shown as functions of the total reservoir capacitance C_* for a given power input of 1000 MW and $\lambda = \lambda_{min}$ (0.83 at $C_* = 2$ km^2). The system temperatures θ_0, θ_1, decrease as C_* increases, because at larger C_* there is more cold meteoric water passing through the system. The total discharge, $q_s + r(C_* + C_2)$ is a linear function of C_*. This is made up of a constant seepage rate over the range of interest (with persistent overflow); a nearly constant direct steam discharge rate; and an overflow rate nearly proportional to C_*. If we use these characteristics to identify the Wairakei surface zone we note: (i) the total steam rate is 0.31 m^3 s^{-1} which includes indirect steam from seepage at the largest possible rate of 0.07 m^3 s^{-1}; (ii) overflow rate 0.36 m^3 s^{-1}, somewhat larger than that measured; (iii) seepage rate 0.4 m^3 s^{-1}, also larger than that measured, but probably underestimated by measurement; (iv) the surface zone temperature of 111°C is similar to that indicated from the Karapiti fumaroles — temperatures above the surface boiling point are indicated by the high proportion of naturally occurring boiling springs.

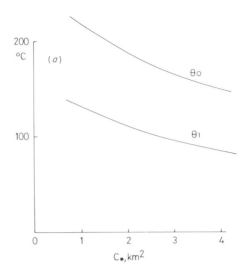

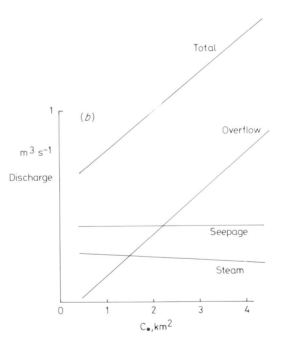

Fig. 16.13. Characteristics of natural, steady state, surface zone as a function of C_*. (a) Reservoir temperatures, in $^\circ$C; (b) Discharge rates, in m^3 s^{-1}.

A value of $C_* = 1.5\,\mathrm{km}^2$ provides a somewhat better fit to the quoted measurements.

It is worth noting that manipulating the model with various values of λ suggests that at Wairakei $\lambda \approx \lambda_{\min}$. In other words, the direct discharge of steam in steaming ground is at about its maximum possible rate. The steam discharge rate is controlled by the availability of energy and not by the resistance of the surface zone to the flow of steam.

Note also that since $rC_2 \ll h_2/R_2$, the seepage is effectively independent of C_2. While overflow is occurring, the predominant effect of the spring reservoir is through its thermal capacitance with time-scale $\Phi_2/(\text{seepage rate}) \sim 10\,\mathrm{yr}$. Thus, fluctuations of temperature in a spring system must arise through the operation of individual springs or small groups of springs with appropriately small thermal volumes.

Surface zone behaviour, exploited state

Minor exploitation began at Wairakei about 1950 but the main production phase started about 1953.0. There is evidence of continued variability in the spring system but not until 1954 was there a factor clearly related to exploitation. The artificial equivalent discharge rate $d \approx 0.13\ (\mathrm{t}/(\mathrm{year}) - 1953.0)\,\mathrm{m}^3\,\mathrm{s}^{-1}$ for the years 1953– 1963. For this discharge rate, the system of Fig. 16.12, with $C_* = 2\,\mathrm{km}^2$ responds as shown in Fig. 16.14. After 5 years (by 1958.0) overflow has ceased. The direct steam discharge rate has hardly

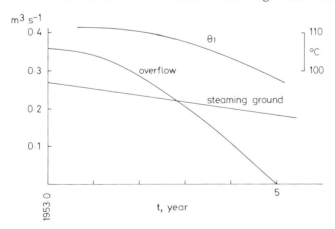

Fig. 16.14. Response of the surface zone to exploitation. Discharge rates, in $\mathrm{m}^3\,\mathrm{s}^{-1}$, and temperature, in $^\circ\mathrm{C}$, as functions of exploitation time, t, in years.

changed; the surface zone temperature, θ_1, has fallen from $111.5°C$ to $98.9°C$. Unfortunately, detailed observations throughout this interval were not made and an important characterization of the system has been lost. Some observations, referred to above, were made in 1953—1954. Already by mid-1954 the overflow rate is reduced. At that time in the field the higher level and marginal springs were becoming erratic or had ceased.

16.4. MECHANISM OF A HOT SPRING

Much of the fluid discharged at the surface is derived directly from the surface zone either as steam or seepage. Some of the fluid however is temporarily retained in relatively small reservoirs associated with pools, flowing springs, fumaroles and geysers. As a prelude to the detailed consideration of the behaviour of these individual discharge systems, in the following chapter, it is possible at the level of detail of this chapter to make some general statements that apply to all hot springs.

Consider springs connected directly to the surface zone or indirectly, as perched systems, supplied in part of wholly with steam from the surface zone. The matter is considered to be discharged as liquid water, steam produced by evaporation, water, and steam produced by boiling.

Evaporation

Pool-type discharges are maintained by evaporation at a water—air boundary located at the surface of a pool, in a void at depth, or in the pores and joints of the rock. It is assumed that the rate of evaporation, m, per unit area of interface is determined by the temperature of the water and the conditions in the air in the immediate vicinity of the interface. In the steady state, the resistance of the country to fluid entering the pool will not affect the discharge. Banwell (1957) has obtained experimental data for m for open surfaces up to $20\,\mathrm{m}^2$, shown in Fig. 16.15 fitting the empirical formula:

$$m = \chi(\Delta p/p)(1 + \eta u), \qquad \Delta p = p_s - p_2,$$

where p is the atmospheric pressure, p_s is the saturated vapour pressure at the interface temperature T_2, p_2 is the water vapour pressure in the adjacent atmosphere just above the interface, u is the air velocity (in $\mathrm{m\,s}^{-1}$) and $\eta \approx 0.5\,\mathrm{s\,m}^{-1}$. Near the boiling point, bubbles

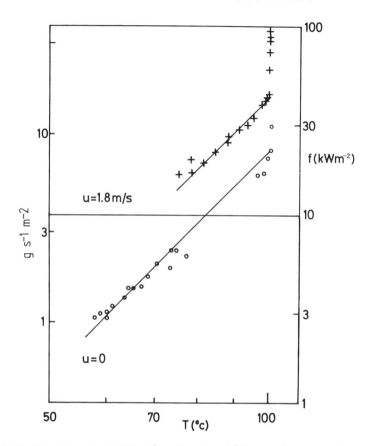

Fig. 16.15. Heat loss, f, in $kW\,m^{-2}$, referred to $0°C$, as a function of temperature, T in $°C$ from a water—air interface by water vapour transport — evaporation
at the free surface except near $100°C$ where boiling occurs. Data at ambient-
temperature $10°C$ from Banwell (1957); $\circ$, an outside pool at Rotorua of area
$20\,m^2$ in still air; $+$, similar laboratory data for an area $38.5\,cm^2$ with air speed
of $1.8\,m\,s^{-1}$. Note that the data is plotted with log—log scales but with the f scale
expanded by a factor of 4. Lines of unit slope, as drawn, then correspond to
$f/T^4 = $ constant.

of steam will arise in the body of the water producing a transfer of
many times m, which accounts only for the evaporative loss directly
from the interface. Here variations in p and the effect of the air
velocity will be ignored so that a minimum value of m is:

$$m = m_0\xi; \qquad m_0 = \chi p_s, \qquad \xi = (1 - p_2/p_s),$$

$$\chi \approx 8 \times 10^{-3} \ kg\,m^{-2}\ s^{-1}.$$

(The corresponding enthalpy loss is 2.68×10^3 kJ kg^{-1} at $100°$C, giving $\chi E = 21.5$ kW m^{-2}.) This relation implies a discontinuity of magnitude $(p_s - p_2)$, in the vapour pressure at the interface. This discontinuity will occur within a few mean free paths of the interface, a distance very much smaller than any other length-scale in the problem.

Operation of a hot spring

Consider the gross constraints on the operation of a hot spring by means of the schematic arrangement sketched in Fig. 16.16. A pool of surface area A, open to the atmosphere, has the following inputs and outputs measured by their mass rates:

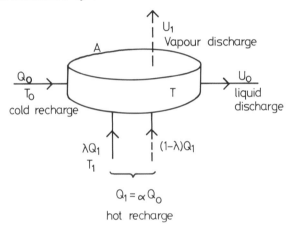

Fig. 16.16. Schematic arrangement for a hot spring.

(i) Q_0, cold recharge water, directly from atmospheric precipitation or seepage below the water table;

(ii) U_0, hot discharge water, at the pool water temperature, T, as seepage below the water table or as overflow either into a drainage channel or as liquid water thrown out of the pool during fumarole or geyser action.

(iii) U_1, water vapour discharge of enthalpy, E, produced either at sub-boiling temperature by evaporation or as a component of the flashing or boiling discharge at or above the surface boiling point.

(iv) Q_1, a deep sub-surface recharge of hot fluid of two components: (a) hot liquid water of temperature T_1 and rate λQ_1, obtained by seepage or subsurface boiling; (b) water vapour of rate $(1 - \lambda)Q_1$, obtained by subsurface evaporation at a deep water table below a perched surface zone or from subsurface boiling. We consider $0 < \lambda < 1$ to allow for the range of hot recharge from entirely water vapour to entirely liquid water. For convenience write $Q_1 = \alpha Q_0$, where $0 < \alpha < \infty$.

It is convenient to consider Q_0, Q_1, λ, A, as if they were given. The vapour discharge is obtained by energy constraints and the liquid discharge is chosen to satisfy mass conservation, so that:

$$U_0 = (1 + \alpha)Q_0 - U_1. \tag{1}$$

The conservation of energy (quantities relative to liquid water at $0°C$) requires:

$$U_0 cT + U_1 E = \alpha Q_0 V + Q_0 cT_0,$$

where

$$V = \lambda cT_1 + (1 - \lambda)E, \tag{2}$$

and E is the vapour enthalpy relative to $0°C$ and c is the specific heat of liquid water. The determination of U_1 requires the treatment of two separate cases depending on the relation of the pool temperature to the local boiling point, T^*.

(a) $T < T^*$ (to be verified *a posteriori*). This is simple evaporation with:

$$U_1 \approx \chi A (T/T^*)^4, \qquad \chi = 0.008 \, \mathrm{kg \, m^{-2} \, s^{-1}}. \tag{3}$$

Substitution of (1) and (3) in (2) produces an equation for T. It should be noted that this is of fifth order, but because of the dominance of the fourth order term $U_1 E$, it is readily solved by numerical iteration. A key dimensionless parameter is

$$\xi = (\alpha Q_0 V / \chi AE)^{1/4}$$

such that to a first approximation $\xi \gtrsim 1$ indicates boiling and for $\xi < 1$, $T \approx \xi T^*$.

If the solution obtained has $T > T^*$ then simple evaporation is incapable of transferring the heat input. The more vigorous mechanism of flashing or boiling must occur and the vapour discharge is determined as in case (b).

(b) $T = T^*$. This is boiling or flashing at the ambient boiling point. Hence, substituting for U_0 from equation (1), equation (2) with $T = T^*$ is a relation for U_1. The vapour discharge can have two extreme limits: (i) an upper limit in which the entire mass discharge occurs as vapour, namely $U_1 = (1 + \alpha)Q_0$ and $U_0 = 0$; (ii) a lower limit, not determined by equation (3) but by the physical constraint that $U_1 \geqslant \chi A$, the vapour discharge by evaporation at the boiling point. If however we deal with case (a) first this difficulty will not arise.

A hypothetical example is illustrated in Fig. 16.17. for a given pool in which only α is varied. For α small, the temperature of the

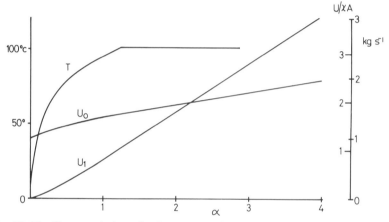

Fig. 16.17. Characteristics of a hypothetical hot pool with parameters: surface area, $A = 100\,\text{m}^2$; $Q_0 = 1\,\text{kg s}^{-1}$; $\lambda = 0.2$; $T_1 = 100°C$ and ambient temperature $10°C$ for various values of $\alpha = Q_1/Q_0$. Liquid water discharge, U_0 and water vapour discharge, U_1, in kg s^{-1}, and pool temperature, T, in $°C$.

pool varies rapidly with α, the evaporative discharge is very small and the liquid discharge is only slightly increased. Above a critical value of α, the pool is boiling. The evaporative or flashed vapour discharge increases more rapidly than the liquid discharge and at a particular value of α, here 2.2, the bulk of the discharge is water vapour. As $\alpha \to \infty$ the cold recharge, Q_0, becomes insignificant and the total discharge becomes the same as the deep recharge, so that in this case $U_0/U_1 \to \lambda/(1 - \lambda)$, here 0.25. Indeed, above the boiling point,

$$U_0 = \alpha\lambda Q_0 + \text{constant}, \qquad U_1 = \alpha(1 - \lambda)Q_0 - \text{constant}.$$

This model assumes thorough mixing within the pool. In nature this is not always the case. In shallow pools vapour can readily "blow through" the water layer so that much of the deep recharge is discharged directly to the atmosphere. When blow through occurs over a region wide compared to the depth of the pool the effect is reminiscent of water boiling in a frying pan. Even in deep pools fumarole activity may be confined to a region of the pool above the recharge inlet.

If we choose units of discharge rate, χA, and temperature, T^*, then writing $u_0 = U_0/\chi A$, etc., and $\theta = T/T^*$, we have in dimensionless form:

$$u_0 = (1 + \alpha)q_0 - u_1,$$

$$u_1 = \theta^4,$$

$$u_0\theta + \beta u_1 = q_0 v,$$

where

$$v = \theta_0 + \alpha[\lambda\theta_1 + \beta(1 - \lambda)]$$

and

$$\beta = E/cT^*.$$

Thus, the problem is defined by the four dimensionless ratios: q_0, λ, α, β. Since β is effectively fixed (only water-substance is used), then only the parameters q_0, λ, α can be independently varied — these three parameters characterize a hot spring. The behaviour of a hot spring as a function of these parameters is illustrated in Fig. 16.18. Notice that for λ small the springs are dominated by vapour discharge except at small values of α, whereas for λ large the vapour discharge is always relatively small. The sub-boiling temperature variation has about the same form in all cases, but boiling is reached at very different values of α:12, 0.3, 100, 2.6 for the examples shown.

It is of interest to mention the case $\alpha = 0$, an ordinary pool or pond of water without hot deep subsurface recharge. If the pool had a conducting impermeable lid so that $U_1 = 0$, then $U_0 = Q_0$ and the pool temperature, T, would be that of the average ambient atmosphere, T_0. On removing the lid so that evaporation is possible, $U_1 > 0$ and the liquid discharge falls to $U_0 = Q_0 - U_1$. The temperature of the pool will be determined by a balance between the loss of heat in the vapour and the heat input from the recharge and directly from the atmosphere into the pool, so that $T < T_0$. The details need not concern us, since for all the cases of interest here $\alpha > 0$ and the pool temperature considerably exceeds that of the ambient atmosphere.

Nevertheless, it is of interest in passing to note the sort of results obtained from this model for an ordinary pool with $\alpha = 0$. For example: a pool of area $125\,m^2$, so that $\lambda A = 1$, at mean ambient temperature $T_0 = 20°C$, has at $Q_0 = 0.1, 0.3, 1, 3, 10\,kg\,s^{-1}$, a pool

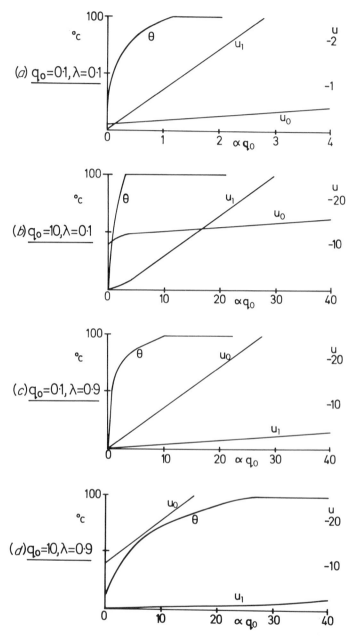

Fig. 16.18. Hot spring output characteristics. Temperature, θ; discharge as liquid, u_0 and vapour, u_1 as a function of α for values of q_0, λ. (a) 0.1, 0.1; (b) 10, 0.1; (c) 0.1, 0.9; (d) 10, 0.9.

temperature $T = 15.9$, 17.9, 19.2, 19.7, $19.9°$C. The overflow is $U_0 \approx Q_0$ in all cases, since the evaporative discharge is at most $1.6\,\mathrm{g\,s}^{-1}$. These estimates of the pool temperature will be extreme since this model assumes the ambient air to be dry and ignores any effects of heating by the atmosphere.

Solute concentration in hot pools and springs

The operation of hot springs is determined by the supply of water-substance and thermal energy. If a solute is introduced into a hot spring system it will not affect the operation of the spring unless the concentration is sufficiently high to modify the recharge or discharge rates through changing the fluid density or the rate of evaporation, or the energy requirements of chemical reactions modify the energy balance. Here I consider weak solutes in a given spring with given flow rates.

There are a variety of possible situations, such as chemically non-reacting solutes, reacting solutes, solutes which are carried solely in the liquid phase or the vapour phase, solutes which are carried in both phases, and so on. The situation of immediate interest here is for SO_4^{-2}, H_2S and Cl^-.

(i) Non-reacting solute, transported solely in the liquid phase (example Cl^-). If z_1 is the concentration in the deep recharge liquid phase and z that in the pool, conservation of solute gives:

$$z = \lambda Q_1 z_1 / U_0 .$$

This is simple dilution.

(ii) Two reacting solutes (example SO_4^{-2}, H_2S). Consider two components X, Y of concentration x_1, y_1 in the deep recharge, and x, y in the pool. Component X is transported solely in the liquid phase but component Y jointly in both phases, a ratio ξ in the vapour to that in the liquid water. Component Y is partly oxidized to produce X within the pool, the proportion, p, being determined by conditions in the pool, unit mass of Y producing σ of X. We consider the conservation of each component separately.

(a) Y. The total input of Y to the pool is:

$$Q' = [\lambda + \xi(1 - \lambda)] Q_1 y_1 ,$$

but of this an amount pQ' is converted to component X, so that the net supply is $(1 - p)Q'$. The discharge of Y is $(U_0 + \xi U_1)y_0$. Conservation of Y requires

$$y = (1 - p)Q' / (U_0 + \xi U_1) .$$

(b) X. Similarly conservation of X requires:

$$x = (\lambda Q_1 x_1 + \sigma p Q')/U_0$$

where $\sigma p Q$ is the rate of production of X from Y.
In addition to the quantities specifying the operation of the pool
we now have σ, ξ, p. The reaction $H_2S + 2O_2 \rightarrow H_2SO_4$ for $1\,kg\,H_2S$
gives $2.88\,kg$ of H_2SO_4, so that $\sigma = 2.88$. The quantity ξ has been
estimated from bore discharges to be in the range $1 - 2$. The quantity
p, with $O < p < 1$, can presumably have values anywhere within its
range — from near unity in a quiet pool with ample time for continuous aeration, to near zero in a vigorous pool with steam blow-
through or eruptive discharge.
Note that the above relations are linear and homogeneous in the
concentrations. Any appropriate concentration units may then be
used.
The effects for a small pool are illustrated in Fig. 16.19. In this case
both the chloride and sulphate concentrations increase nearly linearly
with recharge proportion, α. Although the chloride is strongly
diluted, the sulphate is concentrated, by about 25 times at the boiling

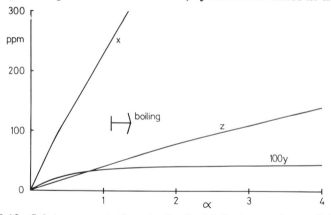

Fig. 16.19. Solute concentrations in the liquid discharge of a small hot pool as
a function of $\alpha:x, y, z$ in PPM for SO_4^{-2}, H_2S, Cl^- with recharge concentrations
x, y, z of 10, 50, 1000 PPM. $A = 1\,m^2$, $Q_0 = 0.01\,kg\,s^{-1}$, $\lambda = 0.05$, $\xi = 2$, $p =$
0.99, $T_1 = 100°C$. Note that for $\alpha > 1.1$, the pool is boiling.

point. The hydrogen sulphide level rises with α well below the boiling
point. Pools of this type are rather common with discharge concentrations x, y, z, in PPM being typically: 10—600, 0—2, 5—20. An
example is "Dragons mouth" pool at Ketatahi (described in Wilson,
1960) of area about $1\,m^2$ with x, y, z in PPM of 57, 5, ~ 0.01 and
$\theta = 76°C$ and crudely estimated $q_0 = 2.5$, $\lambda = 0.05$, $\alpha = 0.2$ and

$Q_0 = 0.02\,\mathrm{kg\,s^{-1}}$. (The values obtained in this manner must be treated with a high degree of scepticism. Detailed field measurements relevant to a total mass, energy and constituent balance of hot pools hardly exist, in spite of the vast amount of miscellaneous and random data reported).

At the other extreme are discharges like "Champagne Cauldron" (circa 1950), Wairakei Geyser Valley. This pool of surface area $260\,\mathrm{m^2}$, discharged $120\,\mathrm{kg\,s^{-1}}$ of boiling water. Temperature soundings give $108^\circ\mathrm{C}$ at 8 m depth. Water at this temperature produces $2\,\mathrm{kg\,s^{-1}}$ of steam (taking $Q_0 = 0$, $\lambda = 1$). The flow ratio $U_0/U_1 \approx 60$ is so large that negligible accumulation of solutes can occur. Yet we find that the overflow concentration of chloride as $z = 1770\,\mathrm{PPM}$, compared with the deep bore water of $1320\,\mathrm{PPM}$. Clearly a further mechanism to that of an isolated pool is required. Near surface water can accumulate solutes such as chloride in its passage beneath steaming ground. The temperature, θ_1, of the deep water feeding the steaming ground can be estimated from the measured water temperature, θ, and the proportion of steam, ξ, from the concentration ratio, $1 - \xi \approx 1320/1770$ and $\theta_1 = (1 - \xi)\theta + \xi E/c \approx 243^\circ\mathrm{C}$. This temperature is comparable with the deep temperatures and suggests that nearly all the water supply to "Champagne Cauldron" was undiluted deep water which had been depleted in enthalpy solely in supplying steam to steaming ground.

17. Particular Discharge Mechanisms

The behaviour of the fluid discharged at the surface in a thermal area is influenced by two factors:

(a) The discharge may proceed by direct flow of the fluid of the discharge system to the surface without change of phase. If the discharge system is dry, the discharge will be as strongly superheated fumaroles. Such systems are found, for example, in New Zealand on White Island and on the cone of volcano Ngauruhoe, and except for the similarity near the surface of strongly superheated steaming ground and the steaming ground of a wet system, there is no diversity of discharge modes as is possible with a wet system. If the discharge system is wet, the discharge is a continuously flowing spring, in no way different from a cold spring. These (wet and dry) passive spring-type discharges are dominated by the flow rather than by the presence of the surface, which is merely the level at which the discharge occurs, and their features are those of the discharge system.

(b) If the fluid in the surface zone is water, the discharge may be affected by the phase change of water to vapour or steam. This may occur by flashing of water to steam within the body of a volume of water hotter than the surface boiling point, or by the evaporation of vapour at a water—air surface not necessarily at the boiling point. Both of these processes occur independently of the level of the water table, whereas, for a spring, the water table must be at the surface. Evaporation will always occur; discharges in which flashing is dominant are called flashing-type, where it is negligible pool-type.

The interaction of flow and phase change leads to a sequence of increasingly intense discharges:

(1) *Warm ground*: weak steaming ground marginal to the intense areas and running out to cold ground of zero gradient and more distant normal ground.
(2) *Steaming ground.*
(3) *Dry fumaroles.*
(4) *Surface pools* without overflow.

(5) *Springs*: wet or dry (slightly superheated fumaroles) with continuous discharge.

(6) *Geysers*: intermittent wet fumaroles and mud volcanoes.

(7) *Wet fumaroles*: bores and mud pools, with continuous discharge of wet steam.

(8) *Phreatic explosions* and hydrothermal eruptions.

Numbers 1—4 are pool-type, numbers 6—8 flashing-type discharges; but springs, though dominated by the flow, can be strongly affected by both evaporation and flashing so that they are a combination of pool-type and flashing-type discharges.

There are a number of incidental features in common:

(i)   The discharge brings hot fluid from depth thereby raising the local isotherms towards the surface.

(ii)   Since the discharge fluid is hotter than that of the surroundings the local water level will be higher, except near an intense fumarole where the large discharge will produce a local fall in water level (i.e., local draw-down).

(iii)   A steady pool-type discharge needs a continuous circulation of fluid in the ground in its vicinity to remove the partially cooled surface fluid.

The Maoris of New Zealand distinguished three types of hot springs (Hochstetter, 1864): (i) Puia, a fumarole or geyser; (ii) Ngawha, a non-erupting, possibly boiling spring; (iii) Waiariki, a hot pool suitable for bathing.

This chapter is concerned with the mechanisms of particular types of discharge. Some of this is, however, presented elsewhere: hot pools and steaming ground have already been described; wet fumaroles are presented only schematically in this chapter since they are regarded here as natural examples of bores in a liquid zone and these are treated in detail in the following chapter. (Bores in a steam zone have been described in connection with Lardarello.)

17.1. NORMAL GROUND

Surface zone of atmospheric influence

Since the earth is such a poor conductor of heat the effects on ground temperatures of variations of local atmospheric temperatures are confined to a thin outer skin. This zone has a thickness, δ, given by $\delta^2 \sim 10\tau\kappa$, where τ is the period of the temperature variation and κ is the thermal diffusivity of the ground. In the theory given by Kelvin (see Carslaw and Jaeger, 1959, section 2.12) the temperature, $T = T(z, t)$, is a function of depth, z, and time, t, for a variation of surface amplitude of zero phase at $t = 0$ and circular frequency $\omega = 2\pi/\tau$:

$$T = T_0 \exp(-y)\cos(\omega t - y),$$

with

$$y = (\omega/2\kappa)^{1/2} z.$$

Some data from near Taupo, New Zealand, illustrating these relationships is shown in Figs. 17.1 and 17.2. The monthly mean temperature is shown in Fig. 17.1. The annual variation is dominant

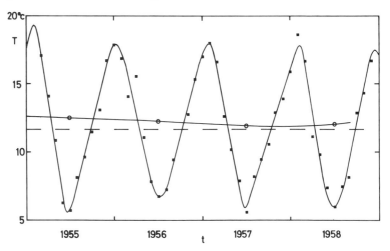

Fig. 17.1. Temperature, in °C, at Taupo, New Zealand 1955–1958. Filled squares, monthly means; open circles, annual means; dashed line, 30 year mean.

with amplitude 5.35 K and second, third and fourth harmonics of amplitude of 0.7 K, 0.6 K, 0.2 K. The monthly values over five summer days are shown in Fig. 17.2, as are values at 10 cm and 20 cm depth. The daily variation is dominant. The surface temperatures are not very sinusoidal, but already at 20 cm depth the high frequencies are being filtered out. The measured amplitude of the daily and annual variation as a function of depth is shown in Fig. 17.3. The form of the relationship corresponds to $\kappa = (3.1 \pm 0.3) \times 10^{-7}$ m^2 s^{-1} (a light pumiceous soil). An amplitude of 0.01 K is reached at: daily wave, 0.48 m; annual wave, 7.8 m. This temperature increment for a normal geothermal gradient of 25 K km^{-1} represents the temperature change across 0.4 m.

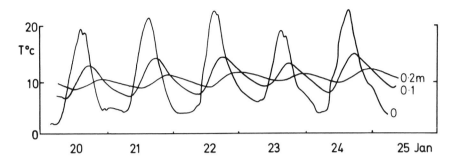

Fig. 17.2. Temperature, in °C, in shade behind laboratory at Wairakei, New Zealand at depths of 0, 0.1, 0.2 m. Continuous recording, 20—25 Jan 1959.

Thus, in stable ground unaffected by water or water vapour transport, the temperature below 10 m or so is determined by the interior heat flux and the surface temperature averaged over several years. This zone acts as a low-pass thermal buffer between the exterior atmosphere (and ocean) and the interior of the earth.

Microclimate

Within an intense geothermal area there is a detectable microclimate. The details will be complex, being dependent on the ambient weather as well as the extra input of heat and water vapour and the consequent effects on the vegetation. Solely to put into perspective the power level required for a detectable microclimate, let us assume calm conditions, a uniform surface, turbulent convection and no radiation effect. Then using the free convection relation for the sublayer thickness, δ, the heat flux in the air, $f = K\theta/\delta$ where θ is the excess temperature at the surface. At Wairakei, for example the mean heat flux is about 0.5 W m^{-2}. This requires $\theta = 0.35$ K. Perhaps it is fortuitous that this crude estimate is similar to that measured by Robertson and Dawson (1964), but it does indicate the order of magnitude of the effect.

17.2. STEAM PLUMES AND CLOUDS

Many thermal areas are immediately noticed at a distance owing to the presence of a "steam cloud" rising above the area. The formation of these plumes and clouds will be described together with rough but useful methods of estimating the discharge which produces them.

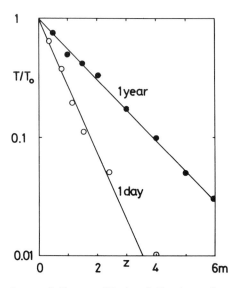

Fig. 17.3. Temperature variation amplitude relative to surface amplitude, T_0, as a function of depth, z, in metres, for periods of 1 day and 1 year, at Wairakei, New Zealand.

The key point is that, while the cloud is delineated by the presence of water droplets, they contribute little to the mass of the cloud except at low temperatures. Various estimates of the mass contribution of the droplets give $0.2-1.0 \, \mathrm{g \, m^{-3}}$ with typical values near $0.5 \, \mathrm{g \, m^{-3}}$. It is relevant to note that the average liquid water content of cumulus clouds is usually less then $1 \, \mathrm{g \, m^{-3}}$ (Warner and Squires, 1958). As a rough rule of thumb excess water of amount $0.01 \, \mathrm{g \, m^{-3}}$ is sufficient to form a distinct visible trail or patch of cloud composed of droplets of radius of order 0.5 micron and greater (Appleman, 1953).

Some distance from the source of the "steam", owing to the vigorous entrainment of ambient air into the rising cloud, the temperature within the cloud approaches that of the ambient air. Within the droplet region the air is saturated.

Direct discharge estimate

The volumetric flow rate can be estimated from the area of cross-section and the velocity, directly in the field, by measuring angular displacements or timing the passage across known land marks, or from a sequence of photographs (Wilson, 1939). Noting that water vapour is considerably less dense than dry air at the same temperature and pressure — density at $100°C$ and 1 atmosphere pressure is for dry air, $0.946 \, \text{kg m}^{-3}$ and for dry steam $0.598 \, \text{kg m}^{-3}$ — we can crudely estimate the buoyancy of the steam cloud solely from its water vapour content. Thus the mass flow of water vapour is $Q = \pi \rho' b^2 w$, where b is the local radius of the steam column, w its local mean vertical velocity and ρ' is the density contribution of the water vapour, approximately $\rho' \approx 0.6(T/100)^4 \, \text{kg m}^{-3}$ at temperature T in $°C$. In a steady state, Q will be independent of height in the steam column.

For example, at Ketatahi springs in 1937, Wilson refers to two main steam columns of diameters $50 \, \text{m}$ and $100 \, \text{m}$, a total area of $10^4 \, \text{m}^2$. A velocity of $2.5 \, \text{m s}^{-1}$ and a vapour density of $1 \, \text{g m}^{-3}$ (at a temperature of $20°C$) corresponds to a mass flow of water vapour of $25 \, \text{kg s}^{-1}$. On White Island in 1938, with the main discharge from four columns of diameters 20, 30, 60, $120 \, \text{m}$ of total area $2.10^4 \, \text{m}^2$ and velocity $3 \, \text{m s}^{-1}$, at vapour density $8 \, \text{g m}^{-3}$ (at a temperature of $30°C$) we have $500 \, \text{kg s}^{-1}$.

Plume height

The quantities which specify the steam column, provided the Reynold's number, wb/ν is large are: vapour mass flow rate, Q; acceleration of gravity, g; density of the ambient air, ρ; vapour density, ρ'; height, z. By analogy with experimental studies and dimensional analysis of buoyant plumes (Turner, 1973, section 6.1), we have:

mean plume radius: $b = b^* z$;

mean vertical velocity: $w = w_* (gQ/\rho z)^{1/3}$;

mean density deficit: $\rho' = \rho'_* Q z^{-2} (gQ/\rho z)^{-1/3}$.

Here b^* is a constant, but the quantities w_*, ρ'_* are functions of the dimensionless radial coordinate $\xi = r/z$, where r is the horizontal distance from the vertical axis of the plume, varying as $\exp(-\beta\xi^2)$ with $\beta \approx 80$ (Rouse, Yih and Humphreys, 1952). The quantities b^*, w^* averaged across the plumes are determined experimentally to be: $b_* \approx 0.12$, corresponding to a half-angle of spread of about $7°$; $w_* \approx 2.6$; $\rho'_* \approx 8.7$. These expressions only apply well away from the source, but they indicate clearly that reasonably accurate estimates of Q from observations of the steam cloud require local measurements.

Plume measurements

Laboratory measurements of visible plume height are shown in Fig. 17.4(a). The steam was produced in an insulated electrically heated container and the height estimated using strong axial illumination. Discharges below the boiling point, produced from surface evaporation, reach a much smaller visible height than those produced

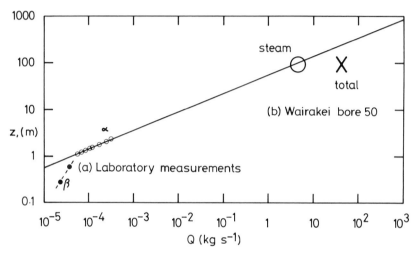

Fig. 17.4. Steam plume visible height, z, in metres, as a function of discharge rate, Q, in $\mathrm{kg\,s^{-1}}$. (a) Laboratory measurements at ambient temperature $22°\mathrm{C}$, orifice area $4\,\mathrm{cm^2}$. The filled circles are for runs below the boiling point. The solid line has slope $2/5$ and corresponds to $Q/z^{2/5} = 4.3 \times 10^{-4}\,\mathrm{kg\,s^{-1}\,m^{-2/5}}$. ($b$) Data for Wairakei bore 50 (1961) shown: steam, $5\,\mathrm{kg\,s^{-1}}$; total discharge $43\,\mathrm{kg\,s^{-1}}$; estimated plume height, $95\,\mathrm{m}$.

from boiling. In these measurements, for plumes of visible height more than a metre, we have $Q/z^{2/5} = k$, a constant, corresponding to the plume model in which the visible plume terminates at a fixed ρ'. The measured value of $k = 4.3 \times 10^{-4}\,\mathrm{kg\,s^{-1}\,m^{-2/5}}$. Presumably this value will be affected by the relative humidity of the ambient air, but by a small amount when the humidity is low.

If we now compare the laboratory data with measurements of larger discharges, as shown in Fig. 17.4(b) for Wairakei bore 50, we find, in spite of an extrapolation of order 10^5, rather good agreement. This is provided only the steam fraction is considered. Unfortunately, in practice, the proportions of water and steam will not necessarily be known. Nevertheless, a suitable estimate can usually be made so that the total output can be estimated to within perhaps a factor of 2 solely from a measurement of the visible plume height.

As an example, the steam cloud above White Island, New Zealand during April 1951 reached an altitude above the collapsed crater floor of about 600 m. This corresponds to a total discharge of 0.4 ton s^{-1}. This is comparable with estimates made in 1939 of 0.45 and 0.255 ton s^{-1} (Wilson, 1939, 1959).

The half-angle of spread of the laboratory plumes near the source was measured as 6°. It is rather noticeable, however, that the visible plume above a certain height has a rather more columnar shape.

Isolated steam clouds or steam thermals

After an isolated eruption a steam cloud is seen rising above the discharge area. We can consider this isolated packet of air, vapour, and water droplets as a type of thermal. Ordinary thermals have been studied in nature and in the laboratory and theoretical models constructed using similar ideas to those in the study of continuous plumes. It is found (Turner, 1973) that the overall shape is slightly oblate with volume $V = nb^3$, where b is the extreme horizontal radius and $n \approx 3$ and that the expansion of the thermal follows $b = \alpha z$, with $\alpha \approx 0.25$. Assuming that these results apply to the steam thermal then the mass of water vapour is $M = n\rho' \alpha^3 z^3$, where ρ' is the vapour density. Following the discussion of plumes, we assume that the visible steam thermal will disappear when it has been sufficiently diluted for ρ' to fall below the ambient saturated vapour density.

As an illustration consider the ambient air temperature to be 10°C, for which $\rho' \approx 0.6 \times 10^{-4}$ kg m^{-3}, and if the cloud disappears at $z = 100$ m, then $M = 2.8$ kg. A cloud which rises to 1 km needs $M = 2.8$ ton.

It is of interest to note that at the upper end of the discharge range from thermal areas the discharges are comparable to those in small cumulus clouds. These have vapour transports from 10^2 kg s^{-1} for a small cumulus to 10^5 kg s^{-1} for a large "congestus" cumulus. Hurricane cumuli can have transports exceeding 10^6 kg s^{-1} (see, for example, Malkus, 1960).

17.3. DRY FUMAROLES

The discharge of a dry fumarole can be described as if it were a hot spring, apart from the complication of the hydrodynamic effects of expansion of the compressible vapour through the constriction of the vent or orifice. A schematic arrangement is shown in Fig. 17.5 and the system can be described with the following equations.

$$p_s = p_s(T_1) \approx p_*(T_1/T_*)^4 \tag{1}$$

$$Q = \chi A_1 (p_s - p_1)/p_* \tag{2}$$

$$\rho_1 = \xi p_1 /\tilde{T}_1 \tag{3}$$

$$\rho_0 = \rho_1 (p_0/p_1)^{1/\gamma} \tag{4}$$

$$q = Q/\rho_0 A_0 \tag{5}$$

$$p_1 = p_0/[1 - \tfrac{1}{2}\rho_1 q^2 /np_1]^n \tag{6}$$

$$T_0 = T_1 - \tfrac{1}{2}q^2 /c_p \tag{7}$$

$$p_0 = p_* \tag{8}$$

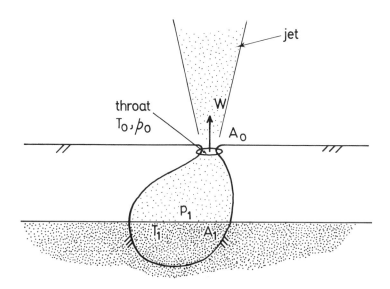

Fig. 17.5. Schematic diagram of a dry fumarole.

(Note: temperature in °C, expect for absolute temperature, $\tilde{T}_1 = (T_1 + 273.16)$ K. $\xi = 2.2 \times 10^{-3}$ K s^2 m^{-2}, $c_p = 2.06$ kJ kg^{-1} K^{-1}, $\gamma = 1.324$ for steam.)

For the moment we presume that the discharge is produced by evaporation, equations (1) and (2) where $p_* = 1$ bar and $T_* = 100°$C, at density, ρ_1, given by the pressure, p_1, and absolute temperature, T_1, immediately above the free water surface, equation (3). If the vapour expands adiabatically so that $p/\rho^\gamma = $ constant, where $\gamma = c_p/c_v$, through the throat or orifice with density ρ_0 at pressure p_0, equation (4), the velocity at the throat is given by equation (5).

Along a stream line the Bernoulli equation is

$$\tfrac{1}{2}q^2 + \int dp/\rho = \text{constant}.$$

For an adiabatic expansion, evaluating the integral from $p = p_1$, where $q = 0$,

$$\int dp/\rho = n(p_1/\rho_1)(p/p_1)^{1/n},$$

$$n = \gamma/(\gamma - 1), \qquad 0 < n < 1.$$

Thus, if p_0, q are the pressure and velocity at the throat and the velocity deep in the chamber is nearly zero, we have equation (6). Immediately downstream of the orifice within the emerging jet, the pressure will be close to atmospheric so we take $p_0 = p_*$.

For an adiabatic process with a perfect gas, equation (7) is simply a statement that the enthalpy is conserved. Notice that if $q = 0$ we have $p_0 = p_1$ and $T_0 = T_1$, but otherwise both p_0, T_0 are changed by amounts depending on the kinetic energy of the discharge.

Note that in this model the output $Q = 0$ for $T = T_*$. At first sight this may seem extreme, but then the system is saturated water vapour in a large chamber in which there is a small hole across which the pressure difference is zero. There will be a small transfer at the orifice largely because of turbulent mixing into the ambient atmosphere of amount proportional to A_0. This emphasizes the essential geometrical feature of a fumarole system of this type, namely that $A_0/A_1 \ll 1$. Notice that, for $(T_1 - T_*)/T_* \ll 1$, q is small; from equation (6) $p_1 \approx p_0$ so that $\rho_0 \approx \rho_1$ and hence

$$p_1 = p_0 + \tfrac{1}{2}\rho_0 q^2,$$

which is the Bernoulli equation for an incompressible fluid. For most dry fumaroles the role of the dynamic pressure is negligible, that is $\tfrac{1}{2}\rho_0 q^2/p_0 \ll 1$. For example this ratio is 4×10^{-3} for Karapiti.

The dynamical effects are shown for a particular case in Fig. 17.6. Over the range $T_1 = 100-150°C$ the Mach number at the throat W/c (where $c = (\gamma p_0/\rho_0)^{1/2}$ is the speed of sound in the steam, about 480 m s^{-1} here), ranges up to 0.6. The dynamical reduction of dis-

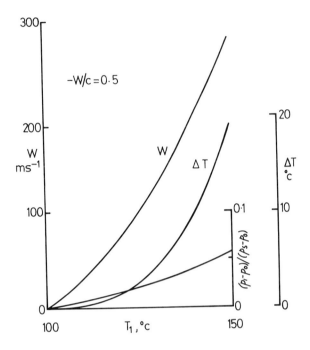

Fig. 17.6. Hydrodynamic effects in a dry fumarole. Various quantities as a function of chamber temperature, T_1, for $A_1 = 1000 \, \text{m}^2$, $A_0 = 0.2 \, \text{m}^2$. W, discharge velocity in m s^{-1} — Mach number of 0.5 is indicated; reservoir pressure p_1, in bar; pressure ratio; expansion temperature drop $\Delta T = (T_1 - T_0)$, in °C.

charge, measured by $(p_1 - p_0)/(p_s - p_0)$, which is unity in the absence of dynamical effects reaches a mere 5.7% even at 150°C with a discharge of $30 \, \text{kg s}^{-1}$. Thus taking $p_1 = p_0$ is a good approximation for natural dry fumaroles. The effect of the adiabatic expansion has a more marked effect on the outlet temperature, so that here $(T_1 - T_0)$ reaches 19°C at 150°C. This indicates that for the most intense dry fumaroles some care is need in estimating T_1 from measurements of T_0, but below about 120°C any errors in taking $T_1 = T_0$ are minor.

The discharge of one of the best known fumaroles, Karapiti has been measured (by confining the outlet in a conical cap) to be $4.4 \, \text{kg s}^{-1}$, about 12 MW, at 114°C (L. Ledger: referred to in Wilson, 1955, p. 29). The constituents of the discharge were in vol.%: steam, 95.5; carbon dioxide, 4.1; hydrogen sulphide; 0.35; ammonia, 0.04.

To produce this flow by evaporation at $T_1 = 114°C$ with $p_s = 1.7$ bar, $p_1 = 1$ bar, requires an area $A_1 \approx 800\,\text{m}^2$. It is worth noting that this area is small compared with the area of the larger hot surface pools of the district. The throat area can be crudely estimated at $A_0 = 0.2\,\text{m}^2$. Inserting these quantities in the discharge equations we find: $W = 38\,\text{m s}^{-1}$, $p_1 = 1.004\,\text{bar}$, $T_0 = 113.64°C$. Thus the chamber pressure differs from that in the ambient atmosphere by 0.004 bar and the temperature at the throat from that of the evaporative surface by $0.36°C$. These differences are negligible, so that the evaporation proceeds independently of the contraction.

The above discussion has been largely about the conditions under which the vapour leaves the chamber. These conditions will apply whatever the origin of the vapour. For simplicity this has been taken as simple evaporation. It could also arise from boiling at depth, within the porous ground but not from an open surface since discharges of this type are dry and slightly superheated.

An important feature not yet mentioned is the method of supply of matter and energy to the chamber. Water supplied at temperature T_2 can produce a mass proportion of vapour, ξ, and water, $(1 - \xi)$, where

$$\xi = c(T_2 - T_1)/(E - cT_1).$$

If Karapiti were supplied with water at say $120°C$ then 1 kg of such water could produce 0.01 kg of vapour, leaving 0.99 kg of water at $114°C$ to leak away into the surrounding country. This leakage is then a major part of the local convective system required to maintain the fumarole.

17.4. MUD POOLS

In mud pools, mud volcanoes, and hydrothermal eruptions the rock takes an active part in the system. Mud is the viscous water-saturated, decomposed rock. In mud pools the steam rises slowly as bubbles rather than by the uniform permeation of ordinary steaming ground — the mud itself is almost impermeable. The flow about the bubbles will be Stokesian and the mud transported by this flow will tend to circulate in the pool, as hot water and mud comes from depth evaporation produces the steam which accumulates in bubbles.

The mechanism of the mud pool can be viewed from the role of the steam bubbles, as sketched in Fig. 17.7. Let us assume for the moment that owing to the stirring action of the bubbles, the mud is

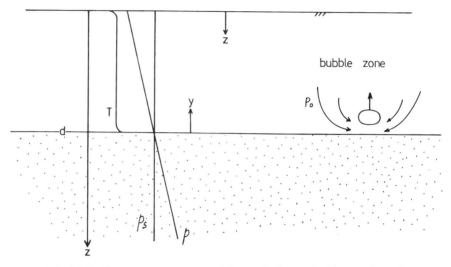

Fig. 17.7. Schematic arrangement for analysis of a boiling mud pool.

well mixed and that the temperature, T, within the bubble zone is nearly uniform. Then there will be a depth, d, at which the SVP of water at temperature T equals the hydrostatic pressure, $p = \rho_0 gz$, of the mud column. Below this depth bubbles cannot form; rather, if they formed, they would collapse. Thus, the material in the mud pool has an upper bubble zone and a lower bubble-free mud zone, separated by an interface. Bubbles will be able to form throughout the bubble zone but the bulk of the heat and mass transport arises from those bubbles which originate near the interface and grow the most.

Consider the generation of a single bubble from a localized region of the interface which has recently produced a bubble but is now quiescent.

(i)   The local fluid will have been cooled owing to the removal of latent heat. A period of time must elapse while this portion of the interface warms up with heat supplied from the slow motion in the bubble free zone.

(ii) Ultimately, when the region is again hot enough, a small bubble will nucleate. There are several possible mechanisms whereby the bubble will temporarily be stationary within the fluid matrix. For example until $(p_s - p) \gtrsim 2S/\Delta$, where S is the surface tension and the bubble radius $a \approx \Delta$, the scale of the matrix.

(iii) The bubble continues to grow trapped in the fluidized matrix but the period of gestation will end when the buoyancy of the bubble is sufficient to overcome the constraints of the matrix. Now the bubble is an independent entity and it begins to move through the matrix and accelerate upwards. As it rises more steam is evaporated from the hot water in the ambient mud into the bubble.

(iv) The bubble motion will produce a corresponding motion in the ambient mud, in particular following lift off, mud from above will be drawn into the interfacial region.

Theoretical sketch

Provided the viscosity of the mud is sufficiently high, the bubble motion will be determined by a balance between the buoyancy and the viscous forces. Treat the bubble as a sphere of radius a. The density contrast $\Delta\rho = \rho_0 - \rho_s$, between the ambient fluid of density, ρ_0, and the steam of density, ρ_s, within the bubble is determined by $\rho_0 \gg \rho_s$. Thus, the vertical velocity of the bubble is:

$$W = \frac{dy}{dt} \approx 2a^2 g/9\nu,$$

where ν is the kinematic viscosity of the mud and y is the distance the bubble has moved above the interface. Note that the coefficient 2/9 arises not only from the bubble shape but also from the detailed nature of the forces at the bubble surface, so that the coefficient is uncertain by a factor of two or so.

The rate of increase of the mass of steam within the bubble, dM/dt, is determined by the rate at which new steam evaporates into the bubble from the bubble interface so that:

$$\frac{dM}{dt} = \frac{4}{3}\pi \frac{d}{dt}(a^3 \rho_s) = 4\pi a^2 \chi.$$

For the moment taking both ρ_s, χ as constants then:

$$\frac{da}{dt} = \chi/\rho_s \equiv \beta, \text{ say.}$$

Hence

$$a = a_0 + \beta t,$$

where a_0 is the radius of the bubble at $t = 0$, the moment of lift off. The equation of motion can be integrated to give

$$y = \frac{2g\beta^2}{27\nu}(t^3 + 3\zeta t^2 + \zeta^2 t) \text{ with } \zeta = a_0/\beta$$

Except for small values of t this expression is dominated by the growth term βt since $\zeta \ll 1$. Thus throughout the bulk of the motion

$$a \approx \beta t; \qquad y \approx 2ga^3/27\nu\beta.$$

Inserting typical values $\chi = 8 \times 10^{-4} \text{ kg m}^{-2} \text{ s}^{-1}$; $\rho_s = 1 \text{ kg m}^{-3}$, we have $\beta \approx 0.6 \text{ cm s}^{-1}$. Thus for example with $\nu = 10 \text{ cm}^2 \text{ s}^{-1}$:

$$1\,\text{s}: \quad a = 0.6 \text{ cm}, \quad y = 0.03 \text{ m};$$
$$3\,\text{s}: \quad a = 1.8 \text{ cm}, \quad y = 0.8 \text{ m};$$
$$10\,\text{s}: \quad a = 6 \text{ cm}, \quad y = 30 \text{ m}.$$

Clearly values of y of order 10 m will require a full treatment of the compressibility of the steam within the bubble but the quantities are of the order of magnitude of those found in the field.

The power flux in the bubble zone is $F = nME$, where n is the rate of production of bubbles per unit area, M is the mass of a bubble, and E is the specific enthalpy of the steam in the bubbles. Notice that since M increases with height so does F. This emphasizes the important role of the stirring action of the bubbles in recharging the upper portion of the bubble zone with wetter, hotter mud.

The energy carried by a single bubble is:

$$e = \frac{4}{3}\pi a^3 E\rho_s = 18\pi E\beta\nu y/g.$$

This suggests that this simple model will break down if ν is too large. This is to be expected, since it is implicit in the model that the mud is completely fluidized, which implies that ν is less than some critical value.

17.5. SUPERHEATED SPRINGS

Intermediate in intensity between ordinary hot springs and geysers and fumaroles are springs in which the water in a surface zone is hotter than the local boiling point. Such springs have been found in Yellowstone (Allen and Day, 1935) and in New Zealand (Grange, 1937). There are 10 located in various geothermal areas of the Taupo district: 4 at Tokaanu; 2 at Wairakei — notably the now derelict Champagne Pool; 1 each at Rotokaua; Orakei Korako; Waiotapu; and the Cauldron at Whakarewarewa.

A possible explanation is as follows. The key to their behaviour is the temperature as a function of depth, shown for two examples in Fig. 17.8. There is a near surface zone of temperatures above the BPD — for spring α of depth 0.36 m; for spring β of depth 1.75 m. Below this superheated liquid zone, although increasing with depth, temperatures are below the BPD. Convective motions within the pool produce an upwelling of deeper hotter fluid which is sufficiently hot to boil in a surface zone as sketched in Fig. 17.9. The upwelling is not, however, sufficiently vigorous to permit eruptive activity. The regions of upwelling are visible as the most vigorously boiling portions of the pool, frequently of sufficient strength to dome up the surface a few centimetres. Superheated fluid spreads out horizontally from the upwellings, so that often the entire surface region of the pool is superheated.

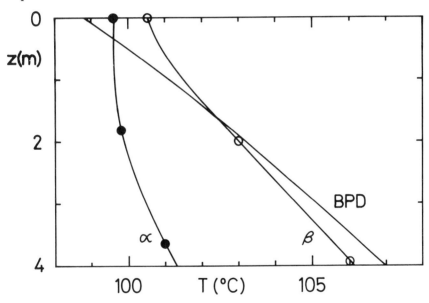

Fig. 17.8. Superheated pool temperature profiles: depth, z, in metres, against temperature, T, in °C, compared with the local boiling point with depth, BPD. Both springs at Tokaanu. (Data from Grange, 1937: α, item 7; β, item 8, Koro-koro spring)

17.6. PERIODIC NON-ERUPTIVE SPRINGS

Some springs which discharge liquid water have a pronounced periodicity of output. This is particularly apparent for marginal springs and is sometimes observed in an area where water levels have

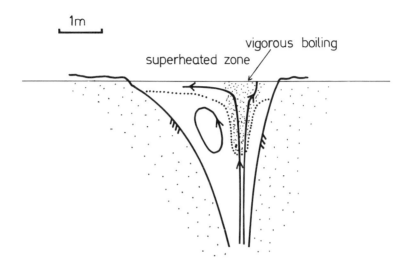

Fig. 17.9. Schematic arrangement of a superheated spring. Similar to Korokoro spring, Tokaanu − β of Fig. 17.8 − vertical section along biggest diameter, walls rather hypothetical.

been reduced for springs which previously had steady flow. These springs are distinguished from eruptive springs like geysers and wet fumaroles by not involving flashing in the discharge vent.

A possible arrangement is indicated in Fig. 17.10 which shows a reservoir and vent supplied with hot water and from which seepage

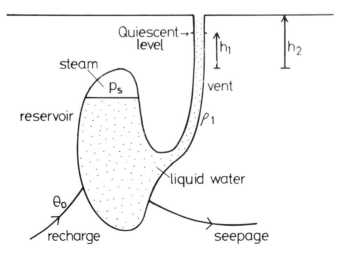

Fig. 17.10. Periodic spring, schematic arrangement of reservoir and vent.

occurs. The vent needs to draw from the lower part of the reservoir deep enough for the BPD to be reached within it. Consider the sequence of events following a minimum in the discharge.

(1) The reservoir is as full as possible and relatively cold. Deep hot water at temperature θ_0 passes into, through and out of the reservoir. The temperature of the reservoir rises.

(2) The BPD is reached in the upper part of the reservoir and boiling commences. The fluid in the chamber is similar to that in a superheated spring except that it is enclosed. This is possible if the vapour pressure, p_s, at temperature, θ_0, exceeds $\rho_1 gh_1$, where ρ_1 is the density of fluid in the vent and h_1 the excess fluid height in the vent above the top of the chamber. Then the pressure difference $(p_s - p_1 gh_1)$ will drive fluid up the vent. If p_s is sufficiently large, somewhat greater than $\rho_1 gh_2$, overflow occurs. A volume of steam collects in the upper part of the reservoir and the water level in the reservoir falls.

(3) Hot fluid continues to enter the reservoir, but the upper fluid in the reservoir is somewhat cooled in proportion to the steam produced. Ultimately, if the recharge rate is less than the total discharge together with the rate of steam production, the reservoir temperature will fall below the BPD and the steam volume will collapse. The overflow will rapidly fall and possibly cease before returning to phase 1.

17.7. WET FUMAROLES

Among the most vigorous springs are some which more or less continuously discharge water-substance into the air well above the level of the enclosing pool or vent. In a geothermal area there are innumerable small vents of dimension of order a centimetre or so discharging steam, but larger fumaroles are relatively uncommon, for example in the Rotorua-Taupo district there are only ten: Taheke (2), Rotomahana, Waiotapu (3), Te Kopia, Karapiti, Waihi (2). They are numerous on White Island and at Ketatahi. Most are found in areas of poor ground water supply. The vent diameter at the surface is generally in the range 0.3—3 m. Two types can be recognized: (i) dry fumaroles which discharge slightly superheated steam only — the commonest; (ii) wet fumaroles which discharge a saturated mixture of liquid water and steam — except for their steady output, these are otherwise similar to geysers.

Most fumaroles have no direct connection with volcanoes. The majority occur in geothermal areas within which hydrothermal systems operate. Their behaviour is controlled by very local conditions in a surface zone of hot water and rock. Some fumaroles are intimately associated with volcanoes since they are embedded within the volcanic structure, but even here the majority are driven by parasitic hydrothermal systems. Fumaroles of these two types may exist in a particular area for of order 10^3 yr, as in Solfatara crater, north of Naples, the gross features of which have changed little since Roman times. A further type, transient or rootless fumaroles, occur on new lava and ash flows as short-lived superficial features which obtain their water from seepage from surface streams into the lava and subsurface waters heated by the lava, as in the Valley of Ten Thousand Smokes (Allen and Zies, 1923).

It is remarkable that fumaroles are uncommon or absent on some volcanic structures. For example the Hawaiian shield volcanoes, though highly permeable with water saturated bases, have only minor fumarolic activity when the volcano is active, but otherwise they are absent.

The visibility of a dry fumarolic or steaming ground discharge can be enhanced dramatically by releasing a small amount of smoke from a flame or engine. Condensation occurs onto the nuclei in the smoke.

Consider a fumarole as sketched in Fig. 17.11, made up of a more or less open vent embedded below the water-table in a region in which at least over part of its depth the ground temperature, T_1,

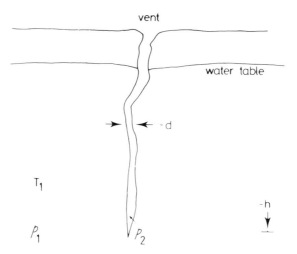

Fig. 17.11. Schematic diagram of a wet fumarole.

exceeds the surface boiling point. Suppose there is no discharge. Then at any given level the pressure in the vent fluid is $p \approx p_1$, the ambient ground pressure. Suppose the vent fluid is disturbed. If p is reduced sufficiently, local boiling will commence and if there is an overflow p will continue to fall. Also the pressure difference $(p_1 - p)$ will drive water from the country into the vent. Isotherms will rise and boiling will accelerate. Flashing will spread in the vent until a balance is established between the rate of water entering from the ground and the discharged from the vent.

This process of "starting up" a fumarole can be illustrated nicely in a diagram as sketched in Fig. 17.12, which shows the relationship of the mass flow rate as a function of pressure at the bottom of the vent for the two parts of the system.

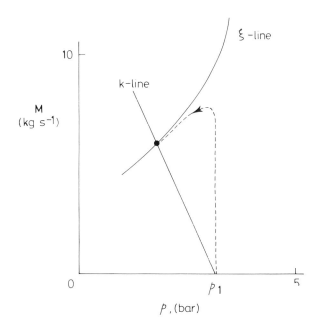

Fig. 17.12. Output characteristic diagram of a wet fumarole.

(i) The vent alone. For a given discharge rate in principle we can integrate the hydrodynamic flow equations down from the surface to obtain p. This curve is labelled "ξ-line" since on it the steam fraction is a constant (if the inlet fluid is all of the same temperature).

(ii) The ground supply alone. For a given ground inflow rate in principle integration of flow equations (namely Darcy's relation) from infinity into the vent wall gives p. This linear relation is called a "k-line" since along it the permeability is taken as a constant. The slope of the k-line,

$$\mathrm{d}m/\mathrm{d}p \sim \nu \log(h/d)/kh,$$

is inversely proportional to permeability.

On the diagram, the track of the start up is shown. Initially the system behaves as if it has an infinite supply of water and the track is close to a line with $k = \infty$. The system comes into equilibrium at the point where the k-line and ξ-line intersect. If there is no such point, such as when the permeability is very small, a steady discharge is not possible.

Natural wet fumaroles are much less powerful than a typical bore. For illustration, model data appropriate to a large natural fumarole is shown in Fig. 17.13. Notice that for small ground resistance, R, the output is determined by the hydrodynamics of the vent, but for a sufficiently large vent and moderate ground resistance, the output is determined by the ground resistance.

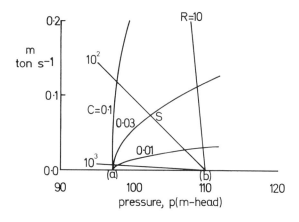

Fig. 17.13. Mass output characteristic for a discharging fumarole: m is mass output rate in ton s^{-1}; p is pressure at base of vent in metres head of hot water; depth of vent, 100 m; deep water temperature, 150°C. Curves (a) Vent characteristics, with zero output at point (a) for vent cross-section areas, C, in m^2, equal to 0.1, 0.03 and 0.01. Curves (b) Reservoir characteristics for various resistances, R, in $w = \mathrm{s\,m^{-2}}$, equal to 10, 10^2, 10^3. The point S would be the steady operating point for $R = 10^2$, $C = 0.03$. Natural systems, apart from a few famous ones, rarely have outputs exceeding 0.01 ton s^{-1}.

17.8. GEYSERS

The intermittent wet fumaroles known as geysers are certainly the most distinctive and intriguing phenomena found in a thermal area. The best known geyser regions are Yellowstone National Park, Iceland and North Island of New Zealand, but they occur at many other places over a range of altitudes from sea level to 5 km. Geysers vary greatly in character both in space and time and although an individual geyser may be nearly periodic for long intervals it may also have long periods of quiescence. The eruptive discharge may total a few grams thrown a few centimetres upward or reach of order 10^3 ton thrown 500 m as at Waimangu, New Zealand (circa 1899—1904). Individual geysers have life-times which range from 1 to 10^3 yr, but geyser activity in that region continues for much longer. The few extinct geysers that have been explored show an interconnected collection of small chambers and fissures in some cases extending to depths of 100—200 m. The key to the understanding of the operation of geysers was given by Bunsen as a result of his studies of Geysir, Iceland (see for example: Thorkelsson, 1928; Sherzer, 1933; Barth, 1950).

Mechanism

Here a geyser is regarded simply as a wet fumarole which is intermittent because of relatively small ground permeability and whose behaviour is modified by the existence of reservoirs and ground leakage.

Consider first the system of Fig. 17.14 in which a void made of a vent and perhaps a reservoir penetrates ground water at temperatures above the surface boiling point. Let us follow the sequence of events after the moment the discharge ceases by plotting in Fig. 17.15 the discharge, M, against the pressure, P_4, at the bottom of the void.

(1) At the cessation of the discharge the vent may be nearly completely empty; P_4 will fall to near atmospheric (point A). Water percolates into the void, and the void begins to fill. If there is a large reservoir, the filling time may be considerable; nevertheless, the water level continues to rise at approximately an exponentially decreasing rate. P_4 also rises.

(2) The temperatures in the vent and adjacent ground at the cessation of the discharge are shown. This is similar to that for a fumarole. Initially the refilling is with colder water, which has been

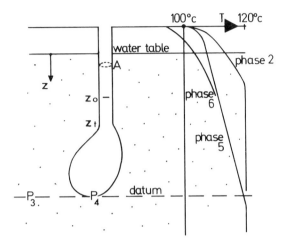

Fig. 17.14. Diagram of a simple geyser; temperature distributions corresponding to phases referred to in Fig. 17.15.

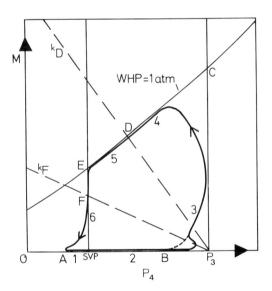

Fig. 17.15. Geyser cycle; mass output characteristics shown by mass output rate, M, as a function of pressure.

cooled during the flashing discharge (surface cold water may also enter) so that a discharge is not immediately possible. However, as refilling proceeds, deep water enters the void and the temperature rises, as shown for the phase (2) temperature distribution. With some geysers the ground water pressure, P_3, is sufficient to raise the water level to the top of the vent and produce an overflow; in this case, temperatures will continue to rise in the vent as more deep water enters the void and consequently P_4 will fall a little. If no water over-flow is possible, temperatures can continue to rise by conduction from the adjacent hot ground and by free convection in the vent — a slow process. The system is now near point (B). In both cases the temperature will rise until at some level z_0 the temperature reaches the BPD.

(3) Flashing commences at z_0 and fluid begins to well out of the vent, pressures fall and flashing rapidly spreads to levels below z_0 and to all levels above; the discharge for a short time rises rapidly, perhaps to its maximum possible value at point (C).

(4) If the k-line is like that of k_D and lies between (E) and (C) the discharge will quickly move to the steady discharge point (D). In this case geyser action is not possible and the sequence of events (2) and (3) is like that of blowing a bore. But if the k-line is like that of k_F, a steady discharge is not generally possible. In this case, as flashing extends to deeper levels, P_4 falls and the discharge point moves along the ξ-line CDE corresponding to WHP = 1 bar.

(5) If a reservoir exists at and below $z = z_t$, where the throat depth $z_t < z^*$, the flashing level z^* enters the reservoir, for example, at (D). The discharge, now maintained from the reservoir, falls slowly from (D) to (E). The temperature distribution in the vent will be like that of a low-temperature bore. If there is no reservoir, z^* will quickly increase until the point (E) is reached. If the vent length exceeds z^*, water will remain in the vent.

(6) Above z^* flashing is occurring in the ground. If the permeability above z^* is very small, but is sufficiently large near or below z^*, no upper colder water can fall into the void and a weak discharge is possible at (F). Otherwise, colder water will begin to enter the vent and the discharge has ceased.

Some geysers are especially regular and periodic in their behaviour. The rate of discharge for one of this type, Bridal Veil geyser (Spring,

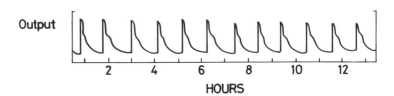

Output

```
 2     4     6     8    10    12
```
HOURS

Fig. 17.16. Bridal Veil geyser output, in arbitrary units, as a function of time, during 29 July 1954. (Data from Thompson, 1957.)

199, Wairakei Geyser Valley), measured with a weir box is shown in Fig. 17.16. At the time of measurement the period was 64 minutes.

Equivalent circuit

The gross temporal behaviour of a geyser can be represented by an equivalent circuit. This is fine, provided it is kept in mind that models of this type ignore one important aspect of actual systems. In the equivalent circuit, a single reservoir of capacitance, C, fed through a resistance, R, has a time constant, CR; but this assumes that the mere recharging of the reservoir is all that occurs. As we have just discussed, the recharge fluid may be relatively cool and need not only fill the reservoir but may be further heated. The equivalent circuit will be satisfactory provided the recharge fluids are hot or the reheating times are relatively short.

Wainui geyser

The nature of the reservoir and also ground leakage has been revealed by Benseman's (1965) intriguing and unique physical manipulations of a geyser *in situ*: Wainui Geyser at Orakei Korako, New Zealand. His ideas are simply understood by means of an equivalent circuit, Fig. 17.17. Wainui geyser is situated on the bank of the Waikato River (Fig. 17.18) and can be considered as using part of the general seepage of ground water to the river. The seepage is the flow produced by the difference of head between the ground water and the river through the ground resistance $(R_3 + R_0)$; the volume flow/sec $(Q_1 + Q_2)$ which passes through the geyser system is tapped off from the general seepage and enters the geyser system through a ground resistance R_4. The feed system thus consists of a (large) ground water capacitance, C_0, and resistances, R_0, R_3, R_4.

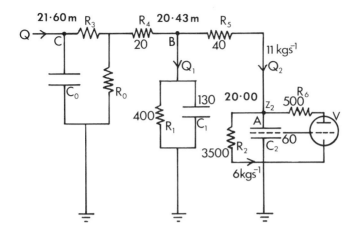

Fig. 17.17. Equivalent circuit of Wainui geyser.

The geyser system itself consists of a (large) reservoir, C_1, a (small) reservoir, C_2 — the vent — and their leakage and supply resistances; together with a triggering device, drawn as a valve V, which opens when the temperature is high enough for flashing to commence. Note that for simplicity all leakage has been taken to a common level.

The behaviour of this circuit can be determined by standard methods and the result compared with field observations. At the ground surface all that can be observed is the flow from the geyser and the level z_2 of the water in the vent, corresponding to the head at point (A) of Fig. 17.17. The vent of the Wainui geyser becomes very wide near the surface, its capacitance = (volume)/(unit change in level) = area of cross-section, is large and so the vent does not empty each cycle. This is convenient, since the mean level z_2 can be adjusted by damming up the vent overflow thereby slightly decreasing R_2 and increasing R_6 and the subsequent transient behaviour followed. The values based on Benseman's figures are shown in Fig. 17.17.

The various time constants are:

(1) Fall in reservoir head during discharge. R_1 and R_5 are in parallel and the time constant is $C_1 R_1 R_5 /(R_1 + R_5) = 4700$ s, there will be little fall in head during one cycle $\Omega = 900$ s, but changes in mean operating level will require a transient time of order 4700 s.

(2) Fall in vent level during discharge. A normal eruption lasts 100 s and discharges 4000 kg, equivalent to a vent resistance of

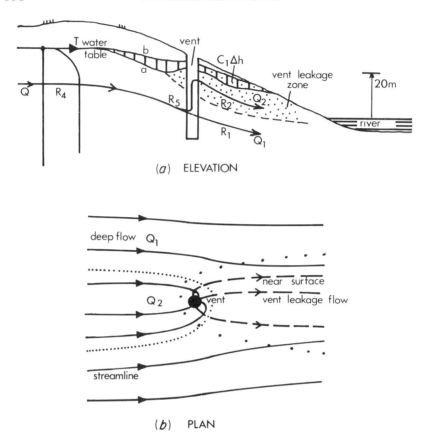

(a)   ELEVATION

(b)   PLAN

Fig. 17.18. Wainui geyser. (a) Section showing the arrangement of the vent, ground water and flushing path. (b) Horizontal projection of ground water flow at depth and near the water table.

$R_6 = 500$ w, hence the time constant $R_6 C_2 = 3 \times 10^4$ s and a very small fall in level will occur in one cycle — in fact 4000 litre/60 m^2 = 7 cm.

(3) Refilling time constant $C_2 R_5 = 2400$ s, much larger than the period, so refilling occurs close to the maximum possible rate. It is this feature of large time constants which has allowed Benseman's perturbation calculation to work so well, and permits his assumption that $(Q_1 + Q_2)$ is approximately constant throughout the cycle and independent of small changes in operating level z_2.

Benseman's chief empirical result is that the geyser period is related, *not* to the mean flow/sec Q' through the geyser vent, but to the flow/sec Q_2 into the system (R_2, C_2, V) by $\Omega Q_2 = $ constant. At first sight we would expect that the net effect of each cycle is merely to flush from the vent a volume equal to that errupted, so that $\Omega Q'$ should be constant. But $\Omega Q'$ is not constant. However, ΩQ_2 does represent a flushed volume: 10 000 litres for Wainui geyser, 2.5 times the erupted volume. The excess volume of 6000 litres must be discharged through the leakage resistance R_2. An arrangement which makes this possible is shown in Fig. 17.18. Ground water seepage is moving toward the river, but a part of it is discharged in an intermittent spring — the geyser; near the spring the isotherms will be moved toward the surface and the water level will be higher than if no spring was present. Water will enter the vent of the spring at depth and some of it will leak out again in the upper portion — this is the key point.

If the discharge level is raised, Δz_2, the adjacent water level, will also rise from position (a) to position (b). The extra volume of ground water is $C_1 \Delta z_2$ where C_1 is the capacitance of the reservoir.

Hence, after an eruption not only will the vent fluid be considerably cooled by flashing but so will a large body of adjacent ground, of total volume ΩQ_2. Only when this body has been flushed with deep hot water is the temperature again high enough for an eruption.

Description of the activity of a complex geyser

Detailed descriptions of geysers are rare but that of the "Great Wairakei" geyser is particularly interesting. It was located at the Western end of Wairakei Geyser Valley, 18 m from the south bank of the stream and almost at the same level (Fig. 17.19). Prior to about 1930 it erupted about every 10 minutes. Sometime during 1931 the period increased to about 10 hours and since May 1954 it has been inactive, presumably owing to the effects of the continuous discharge of exploration bores since 1951. The description here is for early 1951, and mid 1952 and is based on the study by McCree (1957).

The geyser basin or mouth has triangular form of area $13\,\mathrm{m}^2$, is 3 m deep and while quiescent is full of boiling water to 1 m below the rim. Features of a particular eruption and a summary of data from several eruptions are shown in Fig. 17.20 and should be referred to during the following description of the particular eruption sequence.

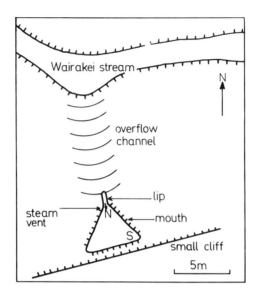

Fig. 17.19. Schematic map of layout and location of the Great Wairakei geyser (spring 59, now extinct).

(0) Prelude (0 hr). A full eruption ends after about 8 minutes with a dry steam phase. The basin is empty. Steam enters through a vent in the northern corner blowing nearly horizontally over the rocks in the floor of the basin. The sequence of events timed from the end of the preceding eruption is as follows:

0.5 hr : steam visibly wet;
1 hr : water visible in vent;
4 hr : water covers the basin floor;
10 hr : first overflow, premonitory phase established.

(1) Premonitory phase (10 hr). The basin is full. Jets of water rise 1—2 m in two locations, N and S. The S jets are continuous and pulsate with a period of 1 s except that they are muted or cease during and for 10 s after the N jets erupt. The N jets erupt at intervals of 80 ± 20 s and play for 25 ± 10 s.

(2) False play (11—15 hr). This starts as an N eruption but in a few seconds the overflow gradually increases and jets spurt to heights of 10 m or more. The jet continues for about 50 sec. The basin water level falls during this time (it does not empty) in a large event by 1.2 m, corresponding to a loss from the basin alone of 16 ton. This is a discharge rate of 0.3 ton s^{-1}. After a further 100 s or so the regular N,S jetting of the premonitory phase recommences.

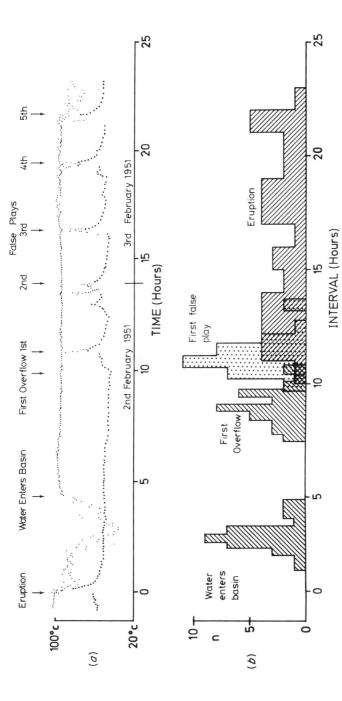

Fig. 17.20. Discharge pattern of the "Great Wairakei" geyser for 38 eruptions from 14 January–10 February 1951 (after McCree, 1957). (*a*) Temperature for one complete cycle, during 2/3 February 1951. Small dots: on the floor of the geyser basin (nearly constant, but falls suddenly when the basin is emptied by the main eruption and remaining low until the basin begins to refill, except for variations caused by being occasionally struck by steam from the vent). Heavy dots: overflow channel (little affected by small continuous overflow but rises immediately at every large overflow). (*b*) Histograms of time intervals measured from the time of an eruption of: water enters basin; first overflow; first false play; next eruption.

The water level rises in the basin, estimated at $0.7\,\mathrm{m\,hr^{-1}}$ corresponding to a recharge of $9\,\mathrm{ton\,hr^{-1}}$, refilling the basin to overflow level in about $1.5\,\mathrm{hr}$. False plays recur on average every $2.7\,\mathrm{hr}$.

(3) Main eruption ($17.5\,\mathrm{hr}$). Some 5 hours before the main eruption the basin water is vigorously stirred and its bottom temperature fluctuates. The eruption starts as does a false play. An N eruption grows to full eruption size of about $15\,\mathrm{m}$ height within a minute. Normally this will be some hours after a preceding false play but may follow as soon as a minute after a false play has stopped (and the basin level is low at $1.2\,\mathrm{m}$ down). The peculiarities of the eruption, in seconds after the start, are:

60s : full eruption established at rate of $0.4\,\mathrm{ton\,s^{-1}}$ to heights of about $15\,\mathrm{m}$;
100s : thumping noise starts;
250s : eruption beings to decrease;
340s : thumping noise louder, eruption decreasing;
385s : water jet small;
420s : discharge of wet steam only, pulsating roaring noise;
490s : discharge of dry steam only, noise greatly diminished.

The eruption is over, the basin is empty and dry. The prelude phase recommences.

The overall behaviour is extremely variable, as indicated in Figure 17.20(b). This is particularly so for the intervals between main eruptions. The average times for the principal events are:

0hr : basin empty and dry;
3hr : water enters basin;
9hr : first overflow;
11hr : first false play;
16hr : main eruption.

Taking the mean discharge time at the maximum rate as $350\,\mathrm{s}$, the total discharge is about 140 ton. Assuming that the recharge rate is the same as that following a false play, namely $9\,\mathrm{ton\,hr^{-1}}$ the time to recharge the volume discharged during the main eruption is $140/9 \approx 15.6\,\mathrm{hr}$. This is sufficiently close to the main eruption interval, $17.5\,\mathrm{hr}$ of the eruption described to suggest that the main discharge is produced from a single reservoir. Taking the mean ground porosity as 0.1, this would be a volume of $1400\,\mathrm{m^3}$.

Identification of the system components

A possible schematic circuit is sketched in Fig. 17.21. There is sufficient data to be able to identify most of the components. Several

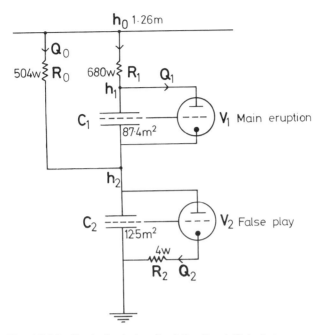

Fig. 17.21. Equivalent circuit of the Great Wairakei geyser.

ways of doing it are possible. Consider the more active false play portion first.

(i) Discharge rate. With valve 2 open, the discharge 0.3 ton s^{-1} is given approximately by $Q_2 = R_2 \Delta h_2$, where h_2 is the change in level, taken as the measured level change of 1.2 m, in the basin. Thus $R_2 = 1.2/0.3 = 4$ w.

(ii) Discharge period. The time constant during discharge with the valve V_2 open is determined by C_2, R_2. Thus $C_2 R_2 = 50$ s, as measured. Hence $C_2 = 50/4 = 12.5 \text{ m}^2$. This is close to the measured area, 13 m^2 of the geyser mouth, so that the reservoir, C_2, for the false plays is the geyser basin itself.

(iii) Recharge period. The time constant of the recharge with the value V_2 closed is $C_2 R_4 = 6300$ s, as measured. Hence $R_4 = 6300/12.5 = 504$ w.

(iv) Available head, h_0. No field data are available for h_0, but it can be deduced from the recharge rate, Q_0 estimated as 9 ton hr^{-1}, since $h_0 \approx Q_0 R_0$ (provided $h_2 \ll h_0$). Hence $h_0 \approx 1.26$ m. This suggests that the basin lip is close to or just below the ambient water table.

Now consider the main eruption.

(v) Discharge rate. Since there are no relevant measurements we can only assume that the discharge is controlled by the same resistance, R_3, which controls the false play discharge rate.

(vi) Discharge period. The time constant during discharge with the valve V_1 open is determined by C_1, R_2. Thus $C_1 R_2 = 350\,\text{s}$, mean value as measured. Hence $C_1 = 350/4 = 87.5\,\text{m}^2$. In this case the main reservoir has a capacitance about 6 times that of the minor reservoirs.

(vii) Recharge period. The time constant of the recharge with the valve V_1 closed is $C_1 R_1 = 59\,400\,\text{s}$, as measured. Hence $R_1 = 59400/87.5 = 680\,\text{w}$.

The precise arrangements of the two reservoirs is not known. They are physically in series since they both use the same mouth.

A geyser complex

One of the best known geysers of New Zealand is Pohutu at Whaka. This is a geyser complex, namely, an elaborate multi-vent structure with complex discharge behaviour. The structure is embedded in a terrace about 6 m above an adjacent stream as sketched in Fig. 17.22. It is constructed from three main elements: (1) the "Cauldron" (C); (2) Pohutu geyser itself (P); and (3) the "Prince of Wales Feathers" geyser (PWF). The eruptive sequence is as follows (data from Grange, 1937).

Phase	Time (hr)	
	0	End of previous eruption of P.
a		Quiescent (May remain inactive for intervals up to 2 yr).
b	10–13	PWF starts (1–4 hr before P, continues till P finishes).
c		C activity increases: water level rises; cyclic violent surface boiling and convective mixing, period 10 min.†
d	13.9	Precursor. C overflows sometimes immediately before P. C not convecting, hot at depth (below BPD but hotter than surface BP)
e	14	P starts: column height about 20 m (part of discharge falls into C in calm weather or with feeble eruptions). C water level gradually falls.
		Near end, spasmodic feeble shots. At end, vent full of water.
		Typical duration 20 min. (Occasional eruptions last for

Phase   Time (hr)

e   14   several hours, the longest recorded being 12hr 10min, 13 May 1920.)
                      PWF also ceases.

          14.3   Eruption finished.

†(i) Normal state. Surface 99.7°C (superheated) and 112.8°C at depth 12 m (BPD 122°C). (ii) Mixing. Strong convection with sudden fall of temperature at 12 m to 99.7°C, surface value.

The discharge during an eruption has not been measured. A crude estimate can be made using the relations for a jet of spray. Pohutu plays to a height of 20 m, corresponding to a discharge velocity of about 20 m s^{-1}. Assuming the water at depth is at 130°C the main discharge density is 10.5 kg m^{-3}. For a throat area of 0.07 m^2 the discharge rate is 14 kg s^{-1}. For a 20 minute eruption this is a total discharge of 17 ton. A similar calculation for PWF with jet height of 2.5 m and throat area 0.025 m^2 requires a discharge rate of 2 kg s^{-1} which for a mean discharge time of about 3 hour gives a total discharge of 20 ton. Altogether the discharge is nearly 40 ton per eruption. An equivalent circuit is shown in Fig. 17.23. This is rather schematic, since sufficient data to identify all the elements are not available.

(1) Long-term behaviour. Unlike geysers such as those in the undisturbed Wairakei geyser valley, these geysers are quite irregular in that more or less periodic activity for intervals of months to a few years will be interspersed with similar times of inactivity. The most likely cause is the response of the system to changes in the local water table. The nearby stream flows continually. Thus there must be a local reservoir draining to the stream of sufficient time-scale to smooth out the effects of rainfall variation and provide a sufficient time lag to explain the observation that there is negligible correlation of activity with local rainfall (Grange, 1937) and indeed sometimes activity is best in a dry year and poor in a wet year. This suggests a time-scale $\gtrsim$ 1 yr. Inspection of the contour map shows a possible local catchment area of about $5 \times 10^4 \text{ m}^2$. With a porosity of 0.1, then $C = 5 \times 10^3 \text{ m}^2$. The local rainfall of 1.5 m yr^{-1}, assuming a 80% immediate run-off, requires a total subsurface discharge rate of 0.5 kg s^{-1}. The ground in the area is fairly flat, indicating a total head to about 10 m. Hence the resistance $R = 2 \times 10^4$ w and $\tau = CR = 3$ yr. Alternatively, we could simply argue that if the time-scale is actually about 1 yr and that if the head has the more extreme possible

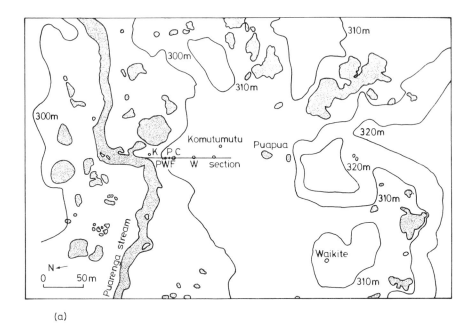

(a)

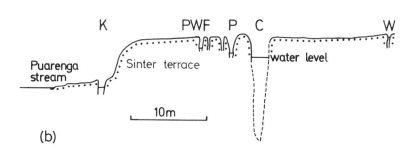

(b)

Fig. 17.22. Pohutu geyser complex. (a) Map showing location near Puarenga
stream, contours in metres above sea level. (b) Vertical section through the vents,
the typical water-level while quiescent is indicated. This section is along the
longest diameter of the vents which are rather narrower perpendicular to this
section. The vent cross sections, dimensions in meters, are: K, Kereru: 1.2 × 0.8
mouth; 0.9 × 0.3 at depth 0.4; PWF, Prince of Wales Feathers; 0.6 × 0.1 mouth;
P, Pohutu: 1.4 × 1.1 mouth; diameter 0.3 at depth 1.7; C, Cauldron: diameter
3.0 mouth; diameter 1.0 at depth 12; W, Waikorohihi: diameter 0.6 mouth;
crack 0.08 wide at depth 0.5. (Circa 1935.)

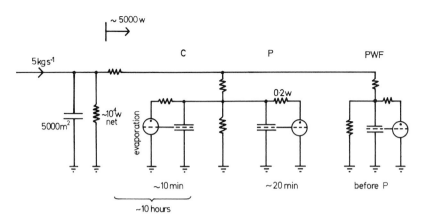

Fig. 17.23. Equivalent circuit of Pohutu geyser complex.

value of 30 m for which $R = 6000$ w, the sub-surface flow is about 5 kg s^{-1}. These figures cannot be taken too seriously.

(2) Average supply. Taking the gross discharge at about 40 ton per eruption and the shortest interval between eruptions of 14 hr, gives a mean supply rate of 0.8 kg s^{-1}. If the fall of the water level in the Cauldron of 3 m is about the working head then the net supply resistance is about 3750 w.

Notice the intriguing possibility that if the long-term time-scale is rather longer than assumed in the above calculation namely 3 years or so, the mean rate of supply is insufficient to maintain the geyser system in permanent operation. After an interval of eruptive activity the main supply reservoir will be drawn down to such an extent that the overall water level is too low to operate the system at all.

18. Bore Discharge

This book is about hydrothermal systems in their natural state: their setting; their mechanism; and their surface hydrology. Where artificial discharges from a borefield have been referred to, our interest has been in identifying the nature of the high-level reservoir and not in the details of how bores and borefields themselves operate. The presentation of these studies in this book would however be incomplete without a description of the mechanics of artificial discharge systems. Our interest is twofold. Firstly, the bulk of our knowledge of hydrothermal systems has been obtained from bores. We use data collected mainly at the well-head to infer conditions in the ground. The state of the fluid changes continuously as it ascends the bore; the state at the well head is different from that in the ground. Furthermore the mechanics of bores are of considerable interest especially as they are artificial fumaroles. Secondly, of course, the major interest in bores is as a practical device to bring geothermal matter and energy to the surface for use: to generate electric power; to provide the working substance for a materials processing plant; to supply hot water for urban and industrial heating and so on.

In this chapter we look at the mechanics of bores themselves. In the following and final chapter the interaction of a borefield and a hydrothermal system is sketched.

18.1. MODEL

The behaviour of a bore can be described in terms of:

(i) conditions in the ambient ground near it;
(ii) the flow in the nearby ground induced by conditions in the bore;
(iii) flow in the bore itself.

Some passing attention is paid to modifications of behaviour arising from the presence of soluble gases, but their role is not emphasized here.

418

Conditions in the ground

In practice, actual ground pressures may be available. Here they are estimated on the basis of some mild assumptions indicated in Fig. 18.1. The temperature with depth in the vicinity of the bore increases

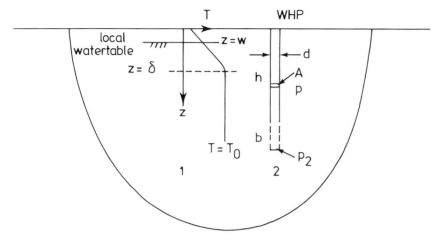

Fig. 18.1. Nomenclature and schematic arrangement for a bore.

in shallow surface zone but is uniform at depth. Thus we define a thermal depth scale δ such that the mean temperature is:

$$T_a = \tfrac{1}{2}T_0 \quad \text{in} \quad 0 \leqslant z \leqslant \delta,$$

$$T_b = T_0 \quad \text{in} \quad z > \delta,$$

corresponding to densities ρ_a, ρ_b. If the local water table is at depth w, then the pressure at depth $z > \delta$ is:

$$P_1 = P_{10} + g[\rho_a(\delta - w) + \rho_b(z - \delta)],$$

where $P_{10} \approx 1$ bar, the ambient atmospheric pressure. In the model calculations below, I will take $w = 0$ and $\delta = 400$ m.

It is important to note that the bore is presumed to operate in a hot ambient and is not affected by conditions of order $10^3 d$ from the bore.

Flow in the ground

The mass flow rate, M, in the ground and the corresponding volumetric flow, Q, in the ground towards the borehole are:

where

$$M = \rho_0 Q,$$

$$Q = Kb\Delta h/\zeta,$$

$$K = kg/\nu, \qquad \zeta = \frac{1}{2\pi}\log(r_1/r_2),$$

and Δh is the difference of pressure head between the distant ground and the wall of the bore. Provided there is no flashing or degassing in the ground, and the flow is sufficiently slow, then k is the permeability. For example, a permeability $k = 1$ darcy $\approx 10^{-12}$ m^2 and a flow of cold water gives $K \approx 10^{-5}$ m s^{-1}.

Thus the pressure drop in the ground is:

$$\Delta p = \zeta\nu M/kb,$$

where ν is the viscosity of the ambient liquid water. Rather then pretend we know all the relevant quantities, rewrite this expression:

$$Q = \Delta h/R,$$

where R is the ground resistance, so that:

$$R = \zeta/bK.$$

In calculations, this is supplied as the value r per unit length of uncased bore for fluid at $20°$C so that $R = r\nu'/b$ where $\nu' = \nu/\nu(20)$ is the ratio of the kinematic viscosity of the ground water at temperature T_0 to the kinematic viscosity of water at $20°$C. For example with a permeability of 1 darcy, and uncased length of 100 m for cold water, $R \sim 10^3$ w.

The pressure at the bore face as a result of the loss of pressure due to the flow, noting that $\Delta P = \rho_0 g\Delta h$, is:

$$P_2 = P_1 - \Delta P,$$

where

$$\Delta P = gRM.$$

A note on the choice of outer radius, r_1

It is a well known and perennial bugbear of a cylindrical geometry that a steady state is not possible. In steady state problems we usually get around this difficulty, somewhat arbitrarily by specifying conditions not at infinity but at some large but finite outer radius, r_1. This leads to a factor $\log(r_1/r_2)$ in the analysis. Fortunately this is a slowly varying function of r_1. In the illustrative cases chosen here, I have chosen r_1 such that $\log(r_1/r_2) = 2\pi$ so that $\zeta = 1$ and for single-phase flow the ground resistance is $R = \nu/gkb$. Thus, for a bore of radius 0.1 m the "zone of influence" extends to a radius of 53.55 m.

In reality, r_1 is the outer limit of nearly axisymmetric horizontal flow towards the bore. Thus for example if a bore penetrates a rising column of fluid, r_1 would be somewhat less than the radius at which the vertical and horizontal mass fluxes were equal.

In practice, as already mentioned elsewhere, it is the ground resistance, R, or the specific resistance, $r = Rb$, which is measurable and is used in most of the models discussed here.

Permeability and resistance

Permeability is related to the ground resistance by $R = \nu/kgb$. For the examples below, we have in mind a typical permeability in practice of about 0.1 darcy $\approx 10^{-13}$ m^2. Hence at, typically, $T_0 = 250°$C with $\nu \approx 10^{-7}$ m^2 s^{-1} and, say, $b = 100$ m, we have $R \approx 10^3$ w, or per unit length of uncased bore about 10^5 w m and a cold fluid value of 10^6 w m.

Flow in the bore

Conditions within the bore for a system in a steady state are determined by the conservation of mass, momentum and energy.

(i) The constant mass flow rate up the pipe is:

$$M = \rho Aw; \qquad A = \tfrac{1}{2}\pi d^2,$$

where ρ is the density of the fluid and w its mean velocity.

(ii) The pressure gradient in the bore,

$$dP/dz = \rho g + \tfrac{1}{2}f\rho w^2/d,$$

arises from the weight of the fluid column together with the effects of hydrodynamic drag. Since for most bores the flow Reynold's number is high the friction factor f is determined by the roughness of the bore wall; values of f in the range 0.01 to 0.02 are typical.

(iii) The fluid density is estimated from thermodynamic considerations, but in effect, for a given P, then from the SVP relation we obtain T, from enthalpy conservation we obtain the steam fraction and hence ρ.

The pressure equation is numerically integrated from $P = P_2$ at $z = h + b$ to $z = 0$ giving the WHP at $z = 0$ for a given M.

18.2. CHARACTERISTIC FEATURES

Example of a hypothetical bore

The following figures illustrate the behaviour of a discharging bore with: $d = 0.2\,\text{m}$, $h = 900\,\text{m}$, $b = 100\,\text{m}$, $r = 10^6$ w m, $T_0 = 250°\text{C}$, $\lambda = 0$ (no gas).
The discharge characteristic is shown in Fig. 18.2.

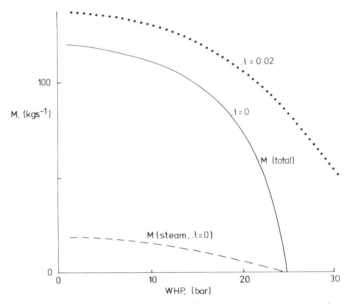

Fig. 18.2. Bore hole output characteristic. Total mass rate, M, as a function of well-head pressure, WHP. Also shown, the steam mass rate, ξM. In addition, the output characteristic for gas mass fraction, $\lambda = 0.02$, is shown for comparison.

(i) WHP > WHP(max) $\approx$ 25 bar. Above a certain WHP, no discharge occurs. The WHP(max) determined here corresponds to the limit $M \to +0$, namely the well-head valve being closed slowly. In practice, the flow regime will collapse once the valve is closed: the upper part of the bore fluid will cool and the ultimate closed in pressure will be quite different from WHP(max), being determined by ambient ground conditions and convection in the bore.

(ii) 1 bar $\lesssim$ WHP $\lesssim$ WHP(max). For a range of pressures, below WHP(max) to the well-head valve fully open, the output increases with decreasing WHP reaching a maximum when the WHP is atmospheric.

Various profiles for a particular discharge, $M = 100 \, \text{kg s}^{-1}$, WHP = 15 bar, are shown in Fig. 18.3.

(a) T. Below z_* $\approx 600 \, \text{m}$ there is no boiling and the fluid is solely liquid water at the inlet temperature, $250°\text{C}$. Above z_*, in the boiling zone, the temperature falls nearly linearly to its value of $200°\text{C}$ at the well-head. This fall arises because an increasing proportion of fluid is boiling off to high-enthalpy vapour.

(b) P. Below z_* the pressure varies linearly, being controlled almost entirely by the fluid weight. Above z_* the pressure also varies nearly linearly, the decreasing weight of the lower density boiling fluid is compensated by the increase in hydrodynamic drag.

(c) ρ. Below z_*, the fluid density is constant, $800 \, \text{kg m}^{-3}$. Above z_* there is an initially rapid drop of density. In this case, at the well-head the fluid density is $73 \, \text{kg m}^{-3}$, about 9% of its original value.

(d) ξ. Below z_* there is no boiling point and $\xi = 0$. Above z_*, the steam fraction rises nearly linearly to its well-head value, in this case 0.112.

(e) w. Below z_*, the density is high and the vertical velocity, w, low. Above z_*, the velocity increases nearly linearly at first and then more rapidly as the well-head is approached. In this case the velocity ranges from $4 \, \text{m s}^{-1}$ to $41.7 \, \text{m s}^{-1}$ at the well-head. For the stronger bores, particularly with low WHP and especially at throats or constrictions, in the uppermost part of the bore sonic conditions will be met. It is not normally necessary to include compressibility effects for consideration of the discharge characteristics of bores in current practice.

The pressure gradient

The vertical pressure gradient within the bore is balanced by two terms: the static term, ρg, arising from the weight of the fluid; the dynamic term, $\frac{1}{2} f \rho w^2 / d$, arising from the hydrodynamic drag. The relative importance of these terms for our hypothetical standard bore is illustrated in Fig. 18.4.

(i) Below z_*, where the fluid is entirely liquid, $\text{d}P/\text{d}z \approx 0.0816$ bar m^{-1}, of which $\rho g = 0.0784 \, \text{bar m}^{-1}$ and $\frac{1}{2} f \rho w^2 / d = 0.0032$ bar m^{-1}. In the region there would be little error in taking $\text{d}P/\text{d}z \approx \rho g$ (and this is a fairly strong bore). Frictional effects in the liquid zone are negligible.

(ii) Above z_*, where the fluid is a two-phase mixture, the pressure

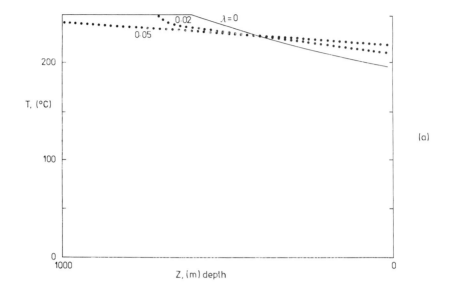

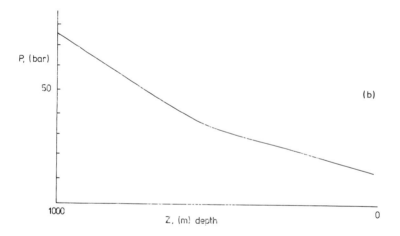

Fig. 18.3. Bore profiles (standard hypothetical bore $h = 900$ m, $b = 100$ m, $f = 0.01$, $T_0 = 250°$C, discharge 100 kg s^{-1}) as a function of depth, z. Above: (a) Temperature, T; (b) pressure, P. Facing page: (c) Density, ρ; (d) steam mass fraction, ξ; (e) mean vertical velocity, w.

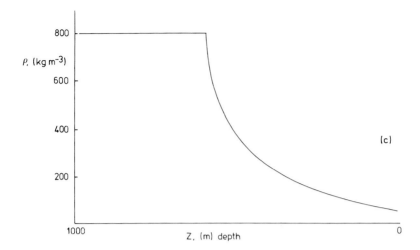

(c)

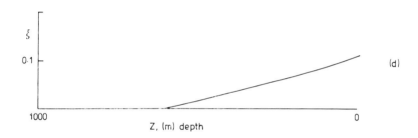

(d)

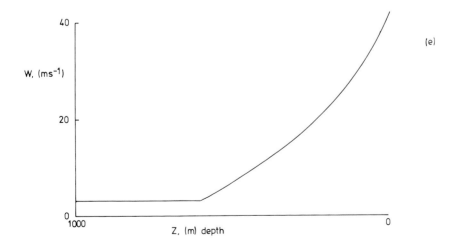

(e)

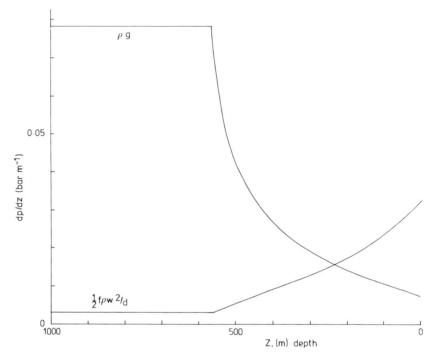

Fig. 18.4. Pressure gradients in a bore (standard hypothetical bore) as a function of depth, z. Static component, ρg; drag component, $\frac{1}{2}f\rho w^2/d$. Discharge 100 kg s^{-1}.

gradient falls rapidly, largely through the rapid drop in fluid density. Frictional effects increase and become dominant in the uppermost part of the bore, here above 240 m deep.

In this example the net effect, a total pressure drop from the bottom of the bore to the well-head (at WHP = 14.6 bar) of 56 bar arises from a column weight contribution of 46 bar and total drag of 10 bar. For bores of this class, the bulk of the pressure is used in the mere lifting of the fluid to the well-head.

Bore data

Measurements on some Wairakei bores are shown in Fig. 18.5.

(a) The assumption of constant enthalpy for the bore fluid is directly tested and is valid within experimental error.

(b) The distribution of temperature follows the nearly linear trend in the flashing region.

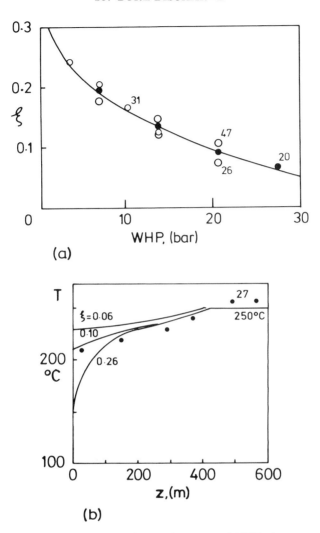

Fig. 18.5. (a) Well-head dryness, ξ as a function of WHP, in bar: theoretical curve for 250°C; field values for Wairakei bores 20, 26, 31, 47. (b) Temperature distribution with depth in a discharging bore. Model curves. Field data for Wairakei bore 27, $T_3 = 267$°C, WHP = 17 bar. (Data from Smith, 1958.)

18.3. ROLE OF BORE PARAMETERS

Role of bore diameter

The relative importance of the flow resistance produced by the ground, R, and the bore, R', can be illustrated by varying the bore

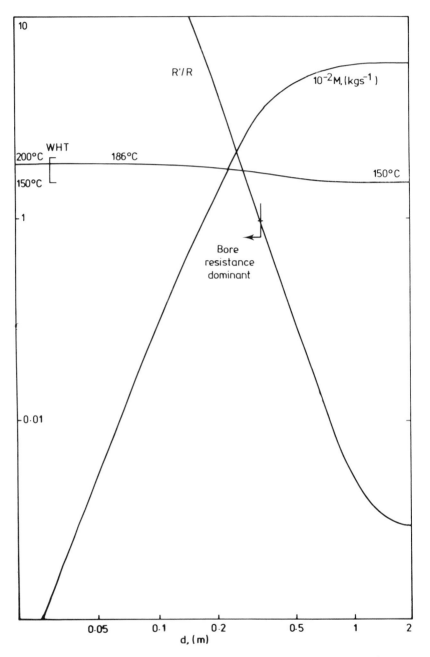

Fig. 18.6. Role of bore diameter. Resistance ratio bore to ground, R'/R, mass output rate, M, and well-head temperature as a function of bore diameter, d.

diameter, d. Thus for the standard hypothetical bore with: $h = 900$ m, $b = 100$ m, $r = 10^6$ w, $T_0 = 250°$C and WHP $= 5$ bar, various data are shown in Fig. 18.6 as a function of d. In this case the ratio $R'/R \gtrsim 1$ for $d \gtrsim 0.35$ m. For large d the output, M is limited by the ground resistance, in this case for diameters greater than about 0.7 m the discharge is nearly independent of diameter. For small d the output is limited by the bore resistance, in this case for diameters less than about 0.1 m the discharge is determined solely by the bore, the loss of pressure in the ground being relatively small.

If the choice of d in practice were based on the requirement of greatest output/cost then $R'/R \approx 2$ is suggested. This would not necessarily be a good choice. Considerations of local drawdown suggest advantages of a more distributed withdrawal. In any event ground permeability is far too variable to allow criteria of this kind to be anything other than a guide.

Role of drag coefficient, f

The effect of f on the output characteristic is illustrated for our hypothetical standard bore in Fig. 18.7 for values of f in the range 0—0.1. The output characteristic for $f = 0$ is of theoretical interest only, but is also what would be obtained as $d \to \infty$. The output characteristic for $f = 0.1$ is for a value which is probably typical of natural systems (fumaroles and the like). We are interested in values near $f = 0.01$. It is important to note that the output is affected mildly by changes in f — of course, because the drag is not the dominant effect in the net pressure drop in the bore.

Role of temperature, T_0

Of all the factors which affect the output characteristics of bores that of the ambient ground temperature, T_0, is the most dramatic. For a hypothetical bore Fig. 18.8 shows M (WHP) for $T_0 = 150$ (25) $300°$C. A number of features are of interest.

(1) The pressure, P_0, obtained as $M \to 0$, increases rapidly with T_0. The characteristics near $M = 0$ are increasingly spaced at higher T_0. This is simply because of the dominant role of SVP (T_0). In effect the bore is driven by the SVP. (For this arrangement a bore with $T_0 = 125°$C cannot run.)

(2) The mass rate, M_0, obtained as WHP $\to 1$ bar increases with T_0, but at a rapidly decreasing rate. The characteristics near WHP $= 1$ bar

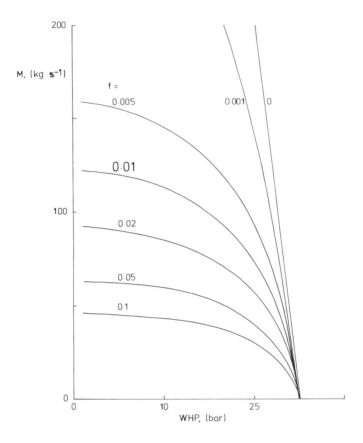

Fig. 18.7. Output characteristics for various values of friction factor, f.

are more closely spaced at higher M_0. This is because of the increasing dominance of flow resistance, in this illustration particularly of the bore resistance.

(3) Otherwise, all the characteristics have the same qualitative form of roughly elliptical shape.

The behaviour of P_0 emphasizes the important point that the operating range of WHP is largely determined by ground conditions. In practice we need to select at any given moment a well-head pressure: suppose we decide to set WHP $= \xi P_0$ with $0 < \xi < 1$. Plant calculations, described below, suggest that $\xi \approx 0.5$ is a suitable value for the optimum WHP for the generation of electric power.

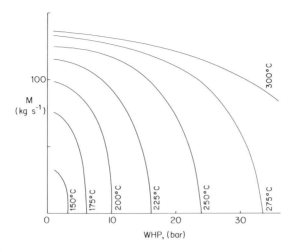

Fig. 18.8. Output characteristics for various values of reservoir temperature, T_0: $h = 800\,\text{m}$; $b = 200\,\text{m}$; $r = 10^6$ w m (300°C curve for $h = 1200\,\text{m}$; 250°C above $100\,\text{kg s}^{-1}$ for $h = 1200\text{m}$).

If therefore throughout the exploited life of a field there is a change of T_0, then for no other reason than that there will be corresponding changes in P_0, we must make changes to the well-head pressure.

Role of dissolved gas

When the deep fluid has dissolved gas, $\lambda > 0$, there are two critical pressures: P_s^* and P_c^*. For liquid of given temperature, T_0 and $\lambda = 0$, boiling will not occur when $P > P_s^* = \text{SVP}(T_0)$. For liquid of given total gas concentration λ, degassing will not occur when $P > P_c^* = \lambda/\alpha$ (all the gas is in solution). Thus there are two distinct possibilities:

(i) $P_c^* < P_s^*$. With a given T_0 there will always be a value of λ sufficiently small for this arrangement to occur. Whereupon, as a fluid particle rises and its pressure falls, the pressure first drops below P_s^* and the fluid begins to boil.

(ii) $P_c^* > P_s^*$. In this case the pressure first drops below P_c^* and the fluid begins to degas. At the liquid water—gas interfaces of the bubbles, water vapour will also be produced by "evaporation" so that the bubbles will contain a mixture of gas and water vapour. Strictly this is not a boiling system (we do not regard the effervescent discharge from a bottle of champagne newly opened as boiling).

The distinction between these two possibilities is not of significance in a numerical model since the same thermodynamics apply once two phases are present. Nevertheless, for values of $\lambda > \alpha P_s^*$, degassing dominates and can have strong effects, notably in producing thick two-phase zones; for example with water at $250°C$, $P_s^* \approx 40$ bar, so that degassing occurs first for $\lambda \gtrsim 0.016$ for CO_2. Hence, for gas concentrations of only a few percent strong effects are possible.

This is illustrated in Fig. 18.9 which shows WHP, WHT for a bore like that in Fig. 18.2 except for a non-zero gas concentration. In this case, even a concentration $\lambda = 0.01$ has a pronounced effect particularly on the WHP. The gas provides additional lift to the bore fluid and the temperature drop as the flashing mixture rises in the bore is less.

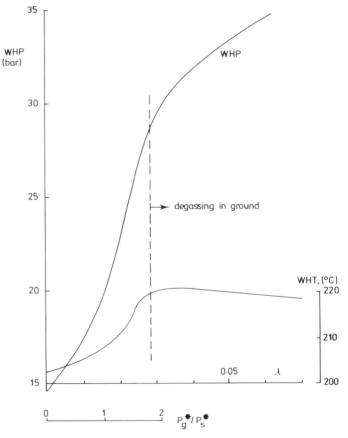

Fig. 18.9. Well-head pressure, WHP, and temperature, WHT, as a function of gas mass fraction, λ. Discharge rate set at $100 \, \text{kg s}^{-1}$, standard hypothetical bore.

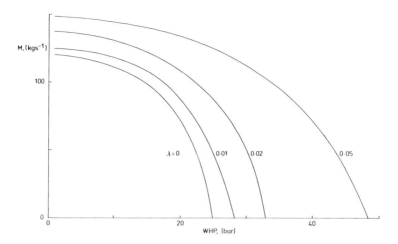

Fig. 18.10. Output characteristics for various values of gas mass fraction, λ ($d =$ 0.2 m, $h = 900$ m, $b = 100$ m, $r = 10^6$ w m, $T_0 = 250°$C).

The output characteristics of our standard bore for various values of λ are shown in Fig. 18.10. The gross elliptical form is the same for all values of λ but the pressure of soluble gas enhances both the discharge and the operating range.

The role of water table level, w

A bore interacts with the ground in which it is embedded in two ways: (i) through processes which control the operation of the bore itself and are localized to the immediate vicinity of the bore; (ii) through the collective effect of the borefield in producing a gross drawdown or pressure fall in the reservoir.

The role of w is illustrated in Figs. 18.11 and 18.12 for two hypothetical bores. (1) For the case of a set WHP, we notice that output falls dramatically with w. The temperature at the well-head, WHT, also falls. (2) For the case of a set discharge rate the WHP falls dramatically with w. In this case, where we are near the bore output maximum (120 kg s^{-1}), the operating range is small.

18.4. TWO-PHASE EFFECTS

So far we have considered the case in which boiling or degassing does not occur in the intake region of the bore. In other words, the

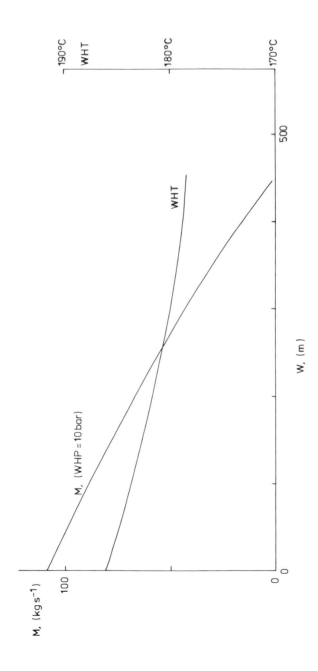

Fig. 18.11. Role of reservoir pressure drop, w, on bore discharge rate, M and well-head temperature, WHT: $d = 0.2\,\mathrm{m}$, $h = 460\,\mathrm{m}$, $b = 120\,\mathrm{m}$, $r = 10^6\,\mathrm{w\,m}$, $T_0 = 250^\circ\mathrm{C}$, WHP $= 10\,\mathrm{bar}$.

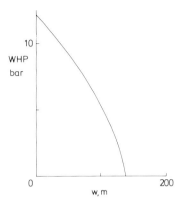

Fig. 18.12. Role of reservoir pressure drop, w, on well-head pressure, WHP: $d =$ 0.2 m, $h = 900$ m, $b = 200$ m, $r = 10^6$ w m, $T_0 = 250°C$, discharge rate, $M = 100$ kg s^{-1}.

bore intake is solely liquid water. This will not always be the case and three distinct situations can arise:

(i) Ground fluid is liquid and remains liquid over the entire operating range of the bore — the case considered so far.

(ii) Ground fluid is liquid in the undisturbed state, but partial boiling or degassing occurs within part of the intake region over part of the operating range of the bore — at the higher mass rates when pressures within the bore are lowest.

(iii) Ground fluid boiling or degassing in the natural state and boiling or degassing becoming further advanced as bore mass rates are increased.

All these effects can be handled and are reasonably well understood. In practice, however, many of the parameters specifying conditions in the group will not be known. It is therefore desirable to look for a model which is simple and realistic, which allows incorporation of the most pronounced effects of the above conditions. After some illustrative examples such a model is presented. Undoubtedly, other approaches are possible.

This is a good place to make an important point. In designing a system to utilize a geothermal resource, where possible, we choose operating conditions within acceptable regions of certainty. For example in designing a steam or two-phase pipeline we select a suitably large diameter so that choking will not occur within the pipeline's operating range. In this case, the details of the choking mechanism are irrelevant, all we need to know is when they occur. Of course, a more sophisticated design may be possible, but only if the relevant information is available. In effect, then, the simple model defines a "drawdown limit" and I propose that the system be designed to avoid this limit and, if that is not possible, to operate the system on the drawdown limit.

Ground resistance for two-phase conditions

In order to illustrate the gross effects of the mutually reduced relative permeabilities of the two phases, consider the following situation. The ambient ground at temperature, T_0, and local pressure, p_1, supplies a bore through a thin homogenous stratum. For a given mass flow rate per unit thickness of stratum, m', the flow equations can be integrated directly. Typical results are shown in Fig. 18.13 for: $T_0 = 250°C$; various p_1; $\log(r_1/r_2) = 5$; $k = 0.1$ darcy; where the resistance per unit thickness of stratum, $R = g\rho_0(p_1 - p_2)/m'$ is given as a function of m'.

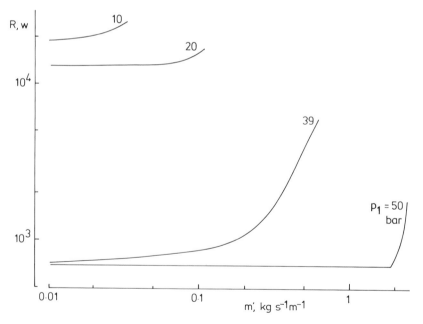

Fig. 18.13. Resistance of bore input region of ambient ground with local boiling present. Resistance, R, as a function of mass rate per metre section for various ambient ground pressures, p_1.

Various points can be noted:

(i) $p_1 \gg \text{SVP}(T_0)$. Below a critical flow rate, boiling does not occur in the ground and R is a constant equal to its value for liquid water. Above a critical flow rate ($1.85 \text{ kg s}^{-1} \text{ m}^{-1}$ here at $P_1 = 50$ bar), boiling occurs in the ground and the resistance increase rapidly with flow rate, in this case by a factor of nearly 3 before "choking" ($p_2 \lesssim 1$ bar) occurs in the ground.

(ii) $p_1 \sim \text{SVP}(T_0)$. Boiling in the ground occurs over the full range of m', the resistance for low values of m' being near the one-phase value but at higher values of m' increasing rapidly with m'. For the data shown, the resistance increases by a factor of 8 before choking occurs in the ground.

(iii) $p_1 \ll \text{SVP}(T_0)$. Not only does boiling occur in the ground over the entire range of m' but it is also in an advanced state. The ground resistance is high for all values of m' and shows only moderate variation with m' — indeed, for p_1 sufficiently low the flow is largely steam and the variations of resistance are dominated by the compressibility of the steam.

Pressure profiles $p(y)$ where $y = \log(r/r_2)$ are shown in Fig. 18.14 for the case $T_0 = 250°\text{C}$, $p_1 = 39$ bar $\approx \text{SVP}(T_0)$. For small values of m', $\lesssim 0.2 \text{ kg s}^{-1} \text{ m}^{-1}$ here, the profiles are nearly straight: the resistance is not very different from its single-phase value. For higher values of m', the profiles are increasingly curved as boiling in the ground becomes more advanced.

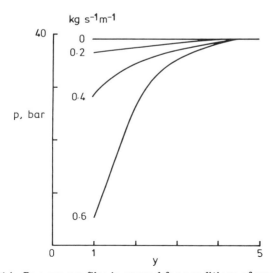

Fig. 18.14. Pressure profiles in ground for conditions of previous figure.

Single-slice two-phase bore model

For the case in which fluid reaches a bore through a single thin stratum, the bore equations can be integrated directly (as for a single-phase ground fluid), provided the mass flow rate is specified. The only difference from the case of entirely single-phase ground fluid, where the pressure drop through the ground is given by a simple algebraic relation, is the need to integrate the two-phase flow equations through the ground.

Mass output characteristics for two hypothetical bores are shown in Fig. 18.15 based on: $T_0 = 250°C$, $k = 0.1$ darcy.

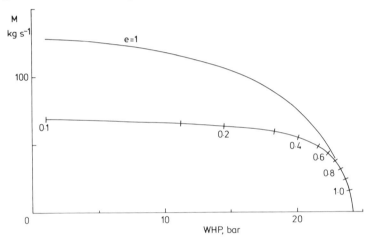

Fig. 18.15. Bore output characteristic with local boiling in ambient ground, single-slice model. $h = 400$ m, $b = 100$ m, $T_0 = 250°C$, $k = 0.1$ darcy. Values of volumetric liquid fraction, e, are indicated. For comparison, a characteristic for $h = 600$ m, $b = 100$ m and no boiling ($R = 701$ w, $e = 1$ throughout).

(i) For the deep bore there is no boiling in the ground and $M(\text{WHP})$ has the familiar elliptical shape.

(ii) The shallow bore has boiling in the ground at all but the smallest mass flow rates. For rates such that the liquid saturation at the bore wall is $e \lesssim 0.5$, the characteristic is little affected by the boiling in part of the ground. For higher flow rates liquid saturations fall rapidly with flow rate, the resistances rise correspondingly and the characteristic is very flattened. The maximum output is considerably reduced.

This model is not very realistic, except for bores fed through a thin permeable stratum. Once we consider the case of flow into the bore through a thick zone, the calculations are much more complex.

Multi-slice two-phase steady state bore model

Consider now a bore of cased depth, h, and length of uncased section, b, in ground of uniform permeability but not necessarily uniform fluid. Make the approximation that flow near the bore is axisymmetric and nearly horizontal. Split the cylinder of ground height, b, into n slices (annuli) of thickness, b/n, and consider the ground flow from an outer radius, r_1, to an inner radius r_2, that of the bore as shown in Fig. 18.16.

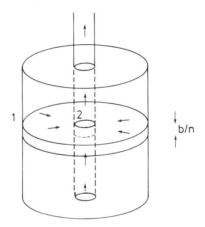

Fig. 18.16. Schema and nomenclature for multi-slice bore model.

Within the uncased section of ground, fluid flows radially towards the bore and in the bore vertically upwards. Each slice adds its contribution to the bore flow, so that the mass flow rate increases up the uncased part of the bore, but no further increment arises in the cased section in which the mass flow rate is constant. The numerical model therefore treats the two sections of the bore, uncased and cased, differently. Furthermore, within each slice we in effect know the outer pressure, p_1, from the given ground conditions, and the inner pressure, p_2, from integrating upwards within the bore, but the mass flow rate in the slice m' must be found by iteration.

Some typical results are shown in Fig. 18.17 for various hypothetical bores in uniform ground at $250°$C.

(i) The data for three bores all with $b = 200$ m, and $h = 300, 400, 500$ m.

(a) The deepest bore differs insignificantly from single-phase ground fluid cases. There is a little boiling in the ground near the bore confined to the upper 40 m of the uncased section.

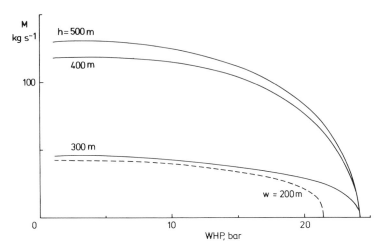

Fig. 18.17. Bore output characteristic with local boiling in ambient ground, multi-slice model. $b = 200$ m, $k = 0.1$ darcy, $T_0 = 250°$C and various values of h: 300, 400, 500 m. Also shown characteristic for $h = 500$ m, $b = 200$ m but with system pressure drop $w = 200$ m.

(b) With $h = 400$ m, the effects of boiling in the ground are apparent but an effect of this magnitude would be unimportant in practice.

(c) With $h = 300$ m, the effects of boiling in the ground are pronounced with ground saturation falling well below 0.1 in the upper part of the characteristic. As with the single-slice case the characteristic has a strongly flattened form.

(ii) A bore with $h = 500$ m, $b = 200$ m with water drawn down 200 m. This differs little from the above bore with $h = 300$ m except for the shift in P_0.

Role of non-uniform ground fluid

The effect on bores which draw from this two-phase zone is pronounced. The output characteristics of deep bores which draw solely from the deep single-phase zone are little affected by the presence of the two-phase surface zone. The discharge characteristics of two hypothetical shallow bores drawing exclusively from the two-phase zone are shown in Fig. 18.18. A number of features are apparent.

(i) The outputs are considerably reduced compared to that of the deep bores. Much of the reduction in output arises from the mutually increased relative resistances of the flow of the two phases in the ground.

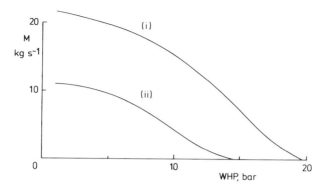

Fig. 18.18. Bore output characteristic for ground in a naturally boiling state. Curve (i), $h = 300$ m, $b = 200$ m, $w = 200$ m; curve (ii), $h = 100$ m, $b = 200$ m, $w = 400$ m.

(ii) The shape of the characteristic curve is no longer so closely elliptical. It is much more nearly straight with minor curvature and a rather distinct "toe" near $M = 0$, particularly for the shallower bores.

Effect in the ground of flashing below the casing

This is a complex subject. For the present purposes we make the following strong simplifications — see Fig. 18.19.

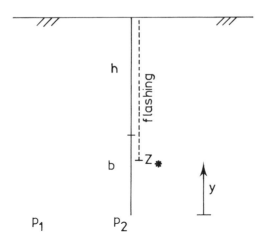

Fig. 18.19. Schema and nomenclature for simple drawdown limit model.

(a) The elevation, y above the base of the bore at which flashing or degassing commences is:

$$y = (p_2 - p_*)/\rho_1 g,$$

where

$$p_* = \max(P_s, P_g); \quad \text{or if } \lambda = 0, \text{ set } p_* = P_s$$

This is obtained on the assumption of negligible hydrodynamic pressure loss below the flashing level.

(b) Any part of the ground near the uncased hole in the flashing zone is completely choked. Hence the net ground resistance, otherwise R, becomes:

(i) $y \geqslant b$, R;   (ii) $y \leqslant 0$, ∞;   (iii) $0 < y < b$, $Rb/y \equiv Ru$,

where $u = b/y$. This is a rather fierce approximation.

All this has the advantage that the blocking factor can be calculated explicitly independent of the details of the flow in the bore and most important it is not a function of the bore diameter. For a given mass flow rate we have:

$$p_2 = p_1 - \Delta p,$$

where

$$\Delta p = u \alpha R, \quad \alpha = gM,$$

and with no blocking the blocking factor $u = 1$.

(i) Hence we find y assuming $u = 1$ and if $y/b > 1$ there is nothing more to do.

(ii) Otherwise writing $y' = y/b$ and $u = 1/y'$ we have

$$y'^2 - \beta y' + \gamma = 0,$$

where

$$\beta = (p_1 - p_*)/\rho_1 gb; \quad \gamma = \alpha R/\rho_1 gb,$$

and hence

$$y' = \tfrac{1}{2}\beta[1 \pm (1 - 4\gamma/\beta^2)^{1/2}]$$

(and as we shall see the larger root is appropriate).

The net effect is illustrated for a particular case in Fig. 18.20 (and 18.21) which shows the well-base pressure, p_2, as a function of mass flow rate, M. In the case shown, the flashing level is in the uncased part for all M. The quadratic form of $p_2(M)$ is seen.

Consider the well-head valve being gradually opened and thereby M gradually increased. The initial $p_2 = p_1$ and the bore follows the upper branch (positive sign above). The pressure p_2 falls more rapidly than proportionally to M, since the flashing level is descending and less of the uncased section is unchocked. The values of p_2 for no

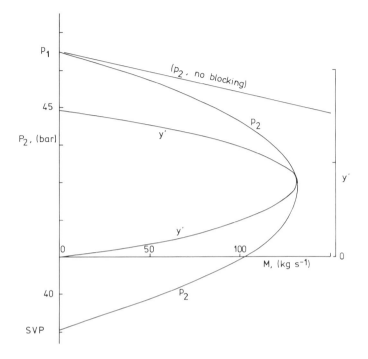

Fig. 18.20. Drawdown limit model: bore base pressure, p_2, and dimensionless length of unblocked uncased zone, y', as a function of mass discharge rate, M. Blocking over the entire operating range.

blocking are shown for comparison. There is a maximum discharge, in this case $M = 131\,\mathrm{kg\,s^{-1}}$ at $p_2 = 43\,\mathrm{bar}$ and $y' = 0.40$: the discharge is maintained from the lower 40% of the uncased well.

The lower branch of the function $p_2(M)$ is inaccessible. Note that it terminates at $M = 0$, $p_2 = \mathrm{SVP}\,(T_0)$.

Figure 18.22 shows a case in which blocking does not occur from the outset. In the case shown for M less than $28.7\,\mathrm{kg\,s^{-1}}$ flashing occurs above the uncased portion and there is no blocking and the fall of p_2 is proportional to M. Otherwise, for larger values of M, blocking occurs and $p_2(M)$ has the quadratic form of the blocked state.

The drawdown limit

In this model the maximum discharge defines what I call the "drawdown limit". This occurs, as seen from the solution for y' given

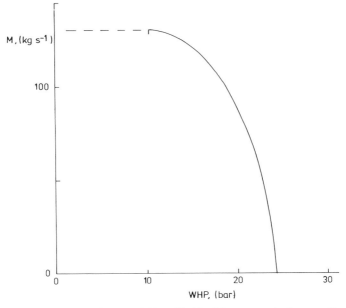

Fig. 18.21. Bore output characteristic with blocking over whole operating range. Characteristic to drawdown only.

above when $\gamma = \frac{1}{4}\beta^2$ giving $y' = \frac{1}{2}\beta$. The corresponding maximum discharge is:

$$M(\text{max}) = (p_1 - p_*)^2 / 4\rho_1 gb(gR).$$

Beyond this limit the model requires a rapidly falling output. The detailed studies of two-phase conditions show, however, that once boiling or degassing in the ground is advanced, the mass rate is nearly independent of well-head pressure. In this model, therefore, if WHP(max) corresponds to $M(\text{max})$, I quite simply take for WHP $\leqslant$ WHP(max), $M = M(\text{max})$.

This is rather arbitrary (we are trying to get something for nothing). Until more sophisticated models for the region WHP $\lesssim$ WHP(max) can be used in the discussion of field-plant models we proceed as follows. Choose a management policy — for example maintain the WHP at a preset value. Run the simulation. If, as a consequence of the fall of reservoir pressure, etc., the drawdown limit is reached, abandon the management policy and run the system down sitting on the current drawdown limit.

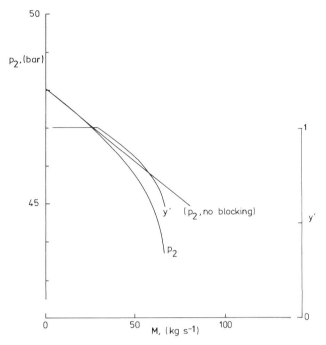

Fig. 18.22. Drawdown limit model: bore base pressure, p_2, and dimensionless length of unblocked uncase zone, y', as a function of mass discharge rate, M. In this case blocking occurs only for $M \gtrsim 30 \text{ kg s}^{-1}$.

The drawdown catastrophe

Sufficiently deep bores, in a given hydrothermal system, can be operated over the entire range of their output characteristics. The discharge can be set anywhere in the range $0 < M < M(\text{max})$ for WHP (max) $>$ WHP > 1 bar. As a further illustration, Fig. 18.23 is for a set of bores which differ only in the depth of the solid casing, h. All these bores behave similarly — the deepening of the bores has little effect, since that merely increases the length of liquid water-substance below the flashing level, z_*. The small differences arise from the small hydrodynamic drag in the region below z_*.

On the other hand, for bores that are relatively shallow, their behaviour is strongly affected by their depth. These effects occur for bores less than a critical depth — for the above hypothetical bores with $h \lesssim 575$ m. Output characteristics for a set of shallow bores is shown in Fig. 18.24. As the depth is reduced there is a progressively narrow operating range. In the examples shown here, this effect is non-existent for $h \lesssim 575$ m and so extreme for $h \lesssim 375$ m that the

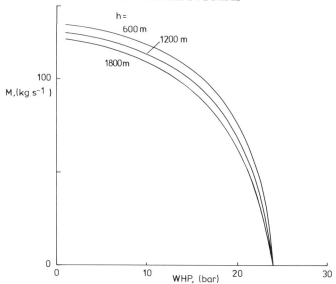

Fig. 18.23. Bore output characteristics: $b = 200$ m, $r = 10^6$ w m, $T_0 = 250°$C and various h: 600, 1200, 1800 m. Drawdown limit not reached with these bores. Compare Fig. 18.24.

bore cannot be operated satisfactorily. We notice in particular that for bores in this range, a small increase in depth has a large effect on the operating range. (Clearly a system which operates in this region can be greatly improved by deepening wells and, furthermore, the effect on possible output would be immediate.)

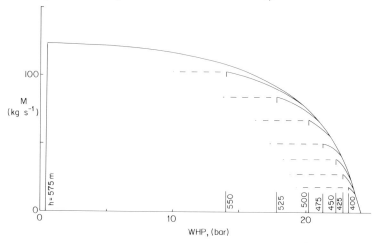

Fig. 18.24. Bore output characteristics: $b = 200$ m, $r = 10^6$ w m, $T_0 = 250°$C and various $h = 400$ (25) 575 m. Characteristics drawn to drawdown limit only.

Similarly, it is of interest to compare a hypothetical set of bores of given cased depth but different uncased depths. As illustrated in Fig. 18.25 with fixed $h = 500$ m for values of $b \gtrsim 300$ m, the effects of blocking are negligible, but at $b \lesssim 100$ m they are extreme.

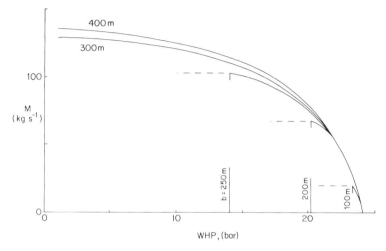

Fig. 18.25. Bore output characteristics: $h = 500$ m, $r = 10^6$ w m, $T_0 = 250°$C and various $b = 100, 200, 250, 300, 400$ m.

If in practice when this situation were encountered, it was desired to improve the bore performance by deepening the bore — since in this model there is no essential difference between the two cases described above — mere deepening of the uncased section would be sufficient. A small deepening can produce a dramatic improvement in output (subject to ground permeability being maintained at the greater depth).

EPILOGUE

PLATE. Part of the multiple steam pipeline at Wairakei, New Zealand. In the background, steam being discharged from cyclone separators and silencing towers. (November, 1978).

19. Epilogue: Exploitation of Hydrothermal Systems

There are a great variety of possible uses of the matter and energy of a hydrothermal system. Where exploitation is intense, with artifical discharge comparable or greater than the natural output, there are gross changes in the hydrothermal system. Understanding and control of the long-term effects are a vital feature of the strategy of management of these resources.

The "moment of truth" in the design of a geothermal power project occurs when the decision "how much" is finalized. At first sight this is an easy question. Given, say, the area of the field, perhaps from a resistivity survey, the output from a single proto-production bore and a logistically suitable mean bore spacing we obtain the total output. Such an approach, more or less the norm till now, treats the system as a heat mine. The object is to extract the deposit of heat as quickly as possible fully conscious that after a finite number of years the plant will be inoperable. Current experience shows this to be an economically feasible strategy. On the other hand, in the undisturbed natural system the surface heat output must be close to that supplied to the deep parts of the system. Current models suggest that the time-scales of the heat supply are at least of order 10^3 yr. Thus in the short run of 10^2 yr or so the heat supply is fixed. Further, laboratory model studies show that major alterations of the upper parts of a hydrothermal system make no change in the power entering the system at depth. Hence the most that can be permanently extracted (over intervals of order 10^2 yr) equals the natural heat discharge.

The resource has two components: a deposit in a high-level reservoir and a permanent deep recharge. Now the question "how much" becomes more subtle. Should one be greedy and immediately mine the deposit or be timid and just take the recharge. Perhaps there is an intermediate strategy. Of course, in the long run there is no intermediate strategy, for sooner or later all that will remain will be the recharge. Clearly two key factors concern us here. Firstly, when do we need this heat, and secondly, to what extent can we draw off the deposit and collapse the high-level reservoir and still in practice draw on the recharge.

19.1. IDENTIFICATION OF THE SYSTEM

It is frequently stated unequivocally that geothermal systems are a resource which is renewable. This is not so! A hydroelectric system utilizes a more or less renewable resource since the high country water supply is more or less continually resupplied by precipitation. A mine or an oil or gas field is a non-renewable resource. The utilization of a geothermal system lies between these two extremes. Current practice in effect treats the geothermal system as a mine by rapidly extracting water and energy from the high-level reservoir. The rate of extraction of water and energy can be very much greater than the natural output, since it is determined by the bore field. If after an interval mining ceased, say because with the installed technology the plant was no longer economically viable, then the matter and energy will again accumulate in the reservoir until the system returns to its natural state, in a time of order 10^2 yr. During this time further mining would not be viable. On the other hand, if utilization proceeded at a rate roughly the same as the natural output the plant could continue in use indefinitely.

The mining concept

For the want of an appropriate expression, I refer to the "mining" phase etc., to describe graphically an interval during which the particular hydrothermal system has matter and energy extracted artificially at rates considerably greater than in the natural undisturbed system. We have two detailed sets of measurements of mass discharge rate and change of system pressure, for Wairakei and Ohaki—Broadlands, together with measurements of the areal distribution of the local changes of gravity. This data can be interpreted, within experimental error, in terms of a fixed container — the reservoir — of constant resistance, supplied at a nearly constant rate from depth. The effect is simply that of a net withdrawal of fluid from the reservoir. If, however, we study this process in laboratory model hydrothermal systems we find that there is a further slower hydrological adjustment — the contraction of the reservoir volume to a new smaller equilibrium shape. The effect is simply that of a dominantly lateral inward net invasion of the reservoir by cold ground water. This aspect of the hydrological change is barely apparent at Wairakei after 20 years of operation. At the moment, then, this process is largely a speculation. It suggests however that long-term changes in the reservoir shape will occur.

Now consider the very long-term possibility of exploiting the matter and energy from deep in, for example, the Taupo—Rotorua system. A first sight there would appear to be a more or less continuous region of hot rock at depth. This is probably true. Nevertheless, if the natural hydrothermal systems are at all like those of laboratory scale models then the total amount of energy in the entire system is very much smaller than in a corresponding non-volcanic zone (the recharge systems are largely of cold water), and moreover the bulk of the energy is in the high-level reservoirs. If this interpretation is correct then the phrase "mining" takes on almost its full common meaning: namely that the excess matter and energy, over the natural replacement, is available in and can be extracted from the high-level reservoirs only.

Artificial recharge

Artificial withdrawal of fluid from a reservoir without artificial replacement will lead to the greatest rate of loss of reservoir fluid and to the most rapid and intense degradation of the reservoir pressures. These effects can be muted by artificially replacing some or all of the artificial discharge. This artificial recharge may be with local surface water or with effluent from the plant. In the latter case this is called reinjection. The artificial recharge rate could be greater or less than the discharge rate.

Experience with artifical recharge is as yet very limited. It is straightforward to actually do it and at a sufficient rate it can have powerful effects on the system pressures, but the long-term effects are not known. Undoubtedly however it provides a key feature of a resource management strategy.

In the illustrative examples described below there is no artificial recharge.

Short term resource definition

The magnitude of a hydrothermal resource can be crudely specified by two quantities, U, F. The total energy of the reservoir, U is obtained from a knowledge of the reservior volume and the mean reservoir temperature, θ. Since, for most of the systems of current interest, the reservoir temperatures are much of a muchness, in New Zealand for example $270 \pm 30°C$, it is usually sufficient to refer to the thermal volume, Φ, such that $U = \rho_0 c_0 \Phi \theta$. The natural heat flow, F, which occurs directly or indirectly through the surface discharge is the best measure of the deep natural energy recharge and the "renewable" exploitable power.

It is misleading to refer to the resource in terms of the discharge available from an existing bore field. This is largely a matter, in a given field, of the number of bores drilled. For example, as at Wairakei with bores of 0.2 m diameter in ground of mean net permeability about 50 millidarcy, of typical outputs $50 \, \text{kg s}^{-1}$, non-interfering bores could be spaced at least as close as 0.1 km; say 100 bores per km^2 giving about $5000 \, \text{kg s}^{-1} \, \text{km}^{-2}$, which as generated electricity is about $1000 \, \text{MW(e)} \, \text{km}^{-2}$.

Of course at Wairakei such a drawoff would have led to a very rapid depressurization of the field and in practice bore spacing is not usually so dense. The point is that the size of a project is determined by considerations of the acceptable rate of mining and this, together with the mean field permeability, defines the number of bores needed (subject to the obvious constraint that if the total bore field cost is significantly more than the value of the exploited discharge then development is economically unjustified).

Field development

With existing or forseeable knowledge of the behaviour of hydrothermal systems and techniques for exploiting them, there are five key parameters needed to predict overall short-term behaviour. These parameters are all directly measurable. They are: θ, mean reservoir temperature; F, natural heat output which occurs directly or indirectly as flow of water-substance; C, the reservoir capacitance; R, the reservoir resistance; Φ, the reservoir thermal volume. Of these only θ and F can be measured immediately. The others C, R and Φ need an existing bore field which discharges for an interval of time, of order 10 years, and at sufficient rate to change measurably pressures and temperatures within the reservoir.

In these circumstances there are two distinct approaches: (i) wildcat development; (ii) adaptive development. Wildcat development, the system used in New Zealand (and elsewhere) up to now, is based on drilling wells until the total output reaches some arbitrary level. This is a crude but successful method in the short run largely because the time-scale of hydrological adjustment, RC, is typically of order

10 years. Adaptive development, is a step-by-step development in which there is no final design frozen at the outset. Existing plants have, of course, been changed but through necessity and expediency. The essence of adaptive development, however, is that the changes and their inevitability are planned from the start.

In order to illustrate the problems which arise, consider the exploitation of a liquid-filled system for the generation of electricity. The data are for hypothetical systems, but the system parameters have been chosen to be somewhat similar to those of Wairakei. An electricity plant is most suitable for this description since the plant is straightforward to describe quantitatively. If, however, the plant was designed for materials processing or direct use in an urban or industrial heat reticulation system, only relatively minor details of the description would need to be changed. The major conclusions would be the same.

Preliminary estimates of the most important quantities can be made at an early stage. In the early stages of exploitation (namely over time intervals of order RC), the important quantities are R and C. A fair estimate of the capacitance of a liquid system is given by $C = \epsilon A$ where ϵ is the mean connected porosity of the surface zone rocks and A is the area of the reservoir cross-section, obtained in the first instance from, say, the area of low resistivity. The resistance can be estimated, not very reliably, from the mean *in situ* permeability, k, measured in boreholes from $R \approx \nu l/gkA$, where ν is the kinematic viscosity of the reservoir fluid and l is the high-reservoir depth (2 km would be a good first guess). Similar estimates can be made for reservoirs dominated by two-phase or vapour conditions.

If it is now decided to pursue adaptive development the stages might be as follows for the exploitation of a single hydrothermal system(of say natural output $F = 200$ MW).

(1) Measurement of R, C. A small pilot plant produces 20 MW(e) and discharges its waste liquid water outside the field (in a stream, lake or by injection). The time interval for this phase is initially unknown but needs to be of order RC. A lower bound for Φ is also measured.

(2) Test of proposed reinjection scheme using the existing plant. Modification may be needed, whereupon a further interval for measurement will be needed. The time interval for this step needs to be of order RC.

(3) Initial mining step. The scale and duration of this step will be determined by methods suggested below. Perhaps. a 50 MW(e) plant is installed. The scale should be sufficiently large, however, to allow

measurement of Φ. The time interval for this step needs to be of order Φ/D where D is the total equivalent volumetric discharge rate.

(4) Production mining step. The scale and duration of this step will be determined by methods suggested below. Perhaps a 100 MW(e) plant is installed. This step may itself involve several stages. Continued measurement together with 2 and 3 dimensional reservoir and detailed plant models will permit the controlled collapsing of the reservoir.

(5) Permanent stage. With the reservoir now suitably collapsed a permanent station in equilibrium with the natural input is run. Surplus equipment may be used elsewhere.

In some fields there may be a number of hydrologically distinct parts. If this is the case, separate determinations of the parameters for each part which will contribute significantly to the total mined output is needed.

Clearly there are many variations on this scheme. Nevertheless it cannot be emphasized enough that for satisfactory management of the exploitation of the resource it is necessary to measure the key reservoir parameters before production mining is begun.

19.2. THE PLANT

The variety of components which take part in the utilization of geothermal energy can be considered as being in three groups:

(i) above-ground plant — all the equipment downstream of the well-head;

(ii) bores — the well up to the well-head equipment and including the nearby ground in which it is embedded;

(iii) reservoir — the hydrothermal system itself, but in the case of mining, predominantly the upper part of the hydrothermal system.

The distinction between these groups is somewhat arbitrary and is used here largely for convenience of description. We note however that they are in sequential reverse order with respect to fluid flow, in order of difficulty of manipulation, and in order of ignorance of behaviour.

The clearest distinction is between the ground system (iii) with time-scales typically of order 10 yr and the "plumbing" system (i) and (ii) with time-scales typically of order $10^2 - 10^3$ s. Thus, as far as the analysis of the behaviour of the plumbing system is concerned the reservoir can be considered as in a quasi-steady state. Further-more, the response of the reservoir to alterations in the state of the plumbing system will be negligible unless those alterations persist for times of order 10 yr.

Plant components

For the purpose of this sketch the system shown in Fig. 19.1 is considered. The above-ground plant is composed of a number of

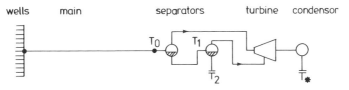

Fig. 19.1. Schematic arrangement of hypothetical plant.

identical wells which at a common point feed a single main; there is no separation of water and steam near the wells so that the flow in the main is of a two-phase mixture; a two-stage separation is per-formed and steam at two pressures fed to the turbine which exhausts into a condenser. The reservoir is filled with liquid water of zero gas content and the uncased portion of the bores is sufficiently deep to be fully wet at the start of exploitation.

The quantitative analysis of the behaviour of the individual components is straightfoward and is adequately described in books about the applications of thermodynamics (for example, Van Wylen and Sonntag, 1976). In all the calculations referred to below we assume that the fluid is saturated water-substance and wherever appropriate that enthalpy is conserved. Only a few comments are necessary here.

Pipelines

The flow of fluid along the pipelines between the well-head equipment and the fluid processing equipment near the generators can be described by the same relations used for bores, with the simplication that, since the flow is horizontal, the pressure gradient does not involve that arising from the mere weight of fluid.

There is, however, an important aspect of the flow which should not be overlooked. We are considering the flow of a flashing mixture of liquid water and steam (as in a bore, to which these remarks also apply). As the pressure falls along the pipe, the steam fraction increases, the density falls and the volumetric flow rate increases so that the hydrodynamic friction increases. In an extreme case the flow is in effect choked. For a given pipe this will occur when the pipe exit pressure equals the condenser pressure; and for a given mass flow rate will occur at WHP less than a critical value. This condition is undesirable and can be avoided by operating at sufficiently high WHP with pipes of sufficiently large diameter — alternatively a portion of the pipeline near the plant could be of larger diameter.

Separators

The mechanical separation of a flashing mixture of water-substance into liquid water and steam is very efficient and the mass proportion of steam produced is determined by the chamber pressure (or corresponding saturated state temperature). A greater quantity of steam can be produced by successively separating the liquid effluent. A dual separation is adequate in practice.

Turbine and condenser

The power, P, produced by a turbine-condenser arrangement of efficiency, η, is given by $P = \eta W$, where the "available work", $W = \Delta \psi$, is obtained from the change of $\psi = h - T_* s$ between inlet and outlet. Here h, s, are the specific enthalpy and entropy of the fluid (saturated steam) and T_* is the condenser temperature.

It is readily shown that for a given throughput of water-substance that the greatest power produced is obtained close to (T_0, T_1, T_2, T_*) equally spaced in temperature. I have used this result in the calculations below.

A note on overall plant efficiency

Having optimized the separator conditions, the remaining free variable is the condenser temperature, T_*. The available power is shown in Fig. 19.2 as a function of T_*. In addition, there is shown a scale of net efficiency for a plant efficiency of 0.8. We see that the overall efficiency lies in the range 0.05 to 0.15: near 0.05 for a direct exhaust turbine; near 0.1 for recirculated cooling tower water to the condenser; near 0.15 for ample ambient cooling water for the condenser.

While there appears everything to gain from the lowest possible, T_*, this ignores the correspondingly greater gross expansion ratio in

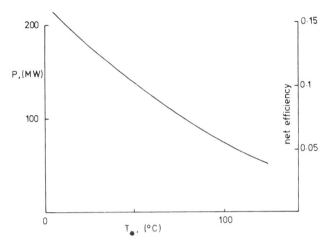

Fig. 19.2. Available power, P, per $1\,\text{ton}\,\text{s}^{-1}$ input, as a function of condenser temperature, T_* for a dual stage separator, turbine, condenser. Also shown, net efficiency for a turbine efficiency of 0.8.

the turbine and its effect on its size, cost and maintenance. $T_* = 50°\text{C}$ is used in the models below.

The above-ground plant

For the plant operated at near optimum conditions at a chosen condenser temperature, the generated power for a given mass rate is determined by the well-head pressure (controlled by the well-head valve) and the temperature of the fluid at the well-head. This is illustrated in Fig. 19.3. The available power is a strongly varying function of temperature, but a weakly varying function of well-head pressure, except at low well-head pressure when the main begins to chock. In other words for a given plant its output is largely determined by its input, namely the mass rate and fluid enthalpy.

The entire plumbing system

The input to the above-ground system comes from the wells. When we consider the combined effect of above-ground plant and wells the performance is as illustrated in Fig. 19.4. For a particular system and throughput, the available power is controlled by the well-head pressure. The behaviour at high WHP is determined by the well characteristics: $M \to 0$ as WHP $\to P_0$, where P_0 is determined by conditions in the ground and principally by its temperature. The behaviour at low

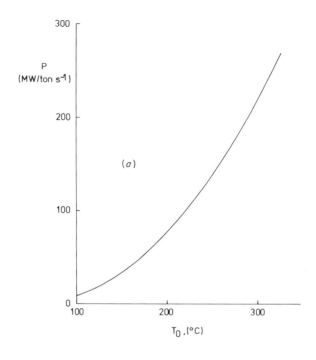

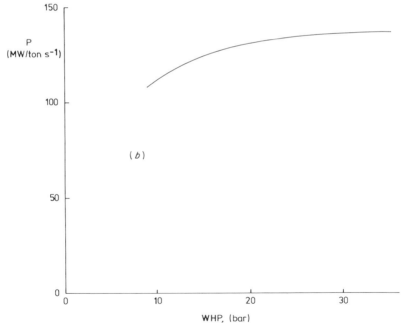

Fig. 19.3. Above-ground plant, available power, P, per $1\,\mathrm{ton\,s}^{-1}$ throughput, as a function of: (a) equivalent temperature, T_0; (b) well-head pressure, WHP.

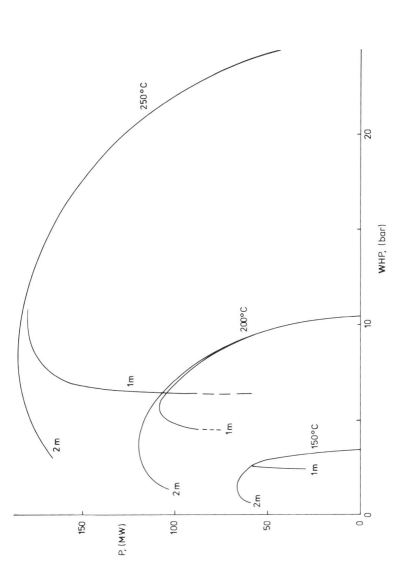

Fig. 19.4. Available power, P, from entire standard hypothetical plant as a function of well-head pressure, WHP, for main diameter 1 m and 2 m, for various reservoir temperatures, T_0, 150, 200, 250°C.

WHP is determined by the above-ground plant characteristics; namely, $M \to 0$ as WHP $\to$ (critical chocking pressure). In between these two extremes there is a single broad maximum available power.

It is presumed, in subsequent calculations, that the system will be operated at that WHP which produces the maximum available power, as we now consider the interaction of the plumbing system and the reservoir.

19.3. THE FIELD—PLANT SYSTEM

The behaviour of the combined field—plant system will now be described by means of some simple examples. Note that these are all for initially fully liquid reservoirs. The role of flashing in the ground near the uncased parts of bores, should it occur, is dealt with in the manner described in Chapter 18. Furthermore, any system which reaches the drawdown limit is thereafter presumed to be run with the system at the drawdown limit.

We assume that all natural activity ceases once exploitation begins. This is plainly a very artificial assumption and, for example, would be quite invalid at Wairakei. In the case of applying these considerations to actual systems, the level of continuing natural activity would need to be taken into account.

The system data chosen here are quite arbitrary but have been selected to be typical of fairly large wet systems like those found in New Zealand — see Table 19.1.

Fixed plant — maximum power strategy

Let us first consider the behaviour of a fixed plant, including its borefield, with the management strategy of running it for maximum generated power. As the field responds to the increased discharge, the operating conditions are set at their optimum by adjusting well-head pressures, separator pressures and so on. This is a case in which the system is designed and built in its final form prior to productive exploitation.

Simple exploitation

The simplest type of behaviour is for a system with a very large thermal volume, so that the reservoir is not thermally degraded, and bores sufficiently deep and with sufficiently small discharge so that no flashing occurs in the ground or uncased sections of the bores.

Table 19.1. *Hypothetical field–plant parameters.*

Symbol	Value†	Name
θ_0	(250°C)	Initial reservoir temperature
λ	(0)	Soluble gas content mass fraction
R	$200\,\mathrm{s\,m^{-2}}$	Resistance of reservoir
C	$1\,\mathrm{km^2}$	Capacitance of reservoir
Φ	$(4\,\mathrm{km^3})$	Thermal volume of reservoir
F	(500 MW)	Natural power input to reservoir
n	(10)	Number of equal bores
r	$10^6\,\mathrm{s\,m^{-1}}$	Specific bore input resistance per unit length of uncased hole
f	0.01	Friction factor
d	0.2 m	Internal diameter of bore
b	200 m	Uncased length of bore
h	(600 m)	Cased length of bore
WHP	($\sim$ 10 bar)	Well-head pressure, optimum value
d_m	1 m	Diameter of single pipeline
l_m	1000 m	Length of pipeline
T_1	($\sim$ 180°C)	Temperature of first-stage separator, optimum value
T_2	($\sim$ 120°C)	Temperature of second-stage separator, optimum value
T_*	50°C	Condenser temperature

† Typical value; otherwise, value stated in text.

This is illustrated by the case shown in Fig. 19.5. The system makes a hydrological adjustment, a pressure fall of about 9 bar in a time of about 10 yr, and a corresponding small fall in available power and is thereafter in equilibrium. This is the case of minor or minimal exploitation of the resource.

Role of number of bores, n

For a given field and bore geometry, the overall power of the system is determined by the choice of the number of bores, n. Data for our standard hypothetical field and bore is shown in Fig. 19.6 for various values of n. We notice a number of features.

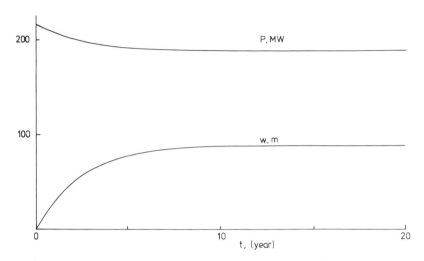

Fig. 19.5. Simple exploitation. Available power, P, and loss of reservoir pressure, w, as a function of exploitation time, t. $R = 100\,w$, $C = 1\,km^2$, $\Phi = \infty$, $r = 10^6\,w\,m$, $H = 700\,m$, $b = 200\,m$, $T_0 = 250°C$.

(1) For n small, here with $n \leqslant 4$, the drawdown limit is not reached, and total output is proportional to n, except for the moderate drop arising from drawdown.

(2) For n large, here with $n \geqslant 10$, the drawdown limit is reached early in the exploited life, and although total output continues to increase with n it is at a diminishing rate.

(3) The change in power as a function of time is large for n large, as in the case shown of $n = 20$.

Field plant: role of T_0

During the exploited life of the system the reservoir temperature T_0 may change. It is therefore of interest to consider the role of T_0.

First, let us consider the rather artifical situation of a given field, given plant, given WHP and change only T_0. (The situation is artificial only in the sense that the plant would be designed and operated for the anticipated range of T_0 rather than the case first considered here). We do this in two illustrative cases:

(i)   with sufficiently deep bores so that the drawdown limit is never reached;

(ii) with bores of intermediate depth so that the drawdown limit is reached for the higher ground temperatures.

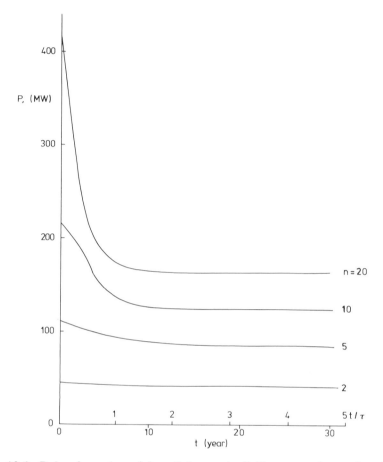

Fig. 19.6. Role of number of (equal) bores. Available power, P, as a function of exploitation time, t, for various numbers of bores, n. $R = 200$ w, $C = 1$ km^2, $r = 10^6$ w m, $h = 600$ m, $b = 200$ m, $T_0 = 250°$C.

Case (i). *Deep bores, given reservoir temperature.* An illustrative example is given in Fig. 19.7, which shows the output of a given plant for various reservoir temperatures (fixed). After the effects of the simple hydrological adjustment (here $RC = 7$ yr), the system is in a steady state. This system is virtually inoperable below $T_0 = 150°$C, but above this temperature the power output increases dramatically because of bore performance.

Case (ii). *Shallow bores, given reservoir temperature.* An illustrative example is given in Fig. 19.8. As before there is an early interval of

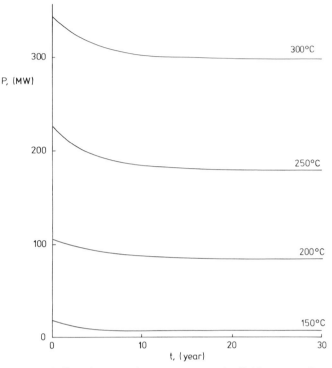

Fig. 19.7. Role of (fixed) reservoir temperature. Available power P, as a function of exploitation time, t, for various reservoir temperatures. $R = 200$ w, $C = 1\,\mathrm{km}^2$, $r = 10^6$ w m, $h = 800\,\mathrm{m}$, $b = 200\,\mathrm{m}$, number of bores, $n = 10$.

simple hydrological adjustment. Here, however, there is a pronounced peak system performance near $245°\mathrm{C}$. This behaviour is produced because of the nature of the blocking of the inlet to the bore. As an aside, this indicates the interesting possibility of a particular system with shallow bores and sufficiently high initial temperature first increasing its output as reservoir temperatures fall.

Field—plant transition

We have considered systems of different T_0, but in which for a particular system T_0 is fixed. Now we consider the effects which arise when the reservoir temperature itself depends on the operation of the system. As we saw above, the reservoir temperature, θ, satisfies:

$$\frac{\mathrm{d}\theta}{\mathrm{d}t} = (Q_0 \theta_0 - Q\theta)/\Phi,$$

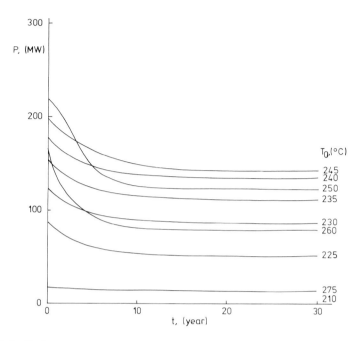

Fig. 19.8. Role of reservoir temperature (fixed). Available power, P, as a function of exploitation time, t, for various reservoir temperatures, T_0. $R = 200$ w, $C = 1 \, \text{km}^2$, $r = 10^6$ w m, $h = 600$ m, $b = 200$ m, number of bores, $n = 10$.

where Q_0, θ_0 are the natural or initial equivalent volumetric discharge rate and reservoir temperature; and Q, θ are the total rate discharge rate (natural plus artificial) and reservoir temperature. In this relation, the thermal volume Φ need not be constant, but will be taken as a constant here. If Q tends to a steady state, $Q \sim Q_\infty$ then θ also will tend to a steady state, $\theta \sim \theta_\infty$. Thus, we can think of the system as simply making a transition from one steady state to another, from (Q_0, θ_0) to $(Q_\infty, \theta_\infty)$. The time-scale of the interval during which this change occurs will be of order Φ/Q_∞.

The relation above assumes that the total power. F, supplied to the reservoir is constant. We have $F/\rho_0 c_0 = Q_0 \theta_0 = Q_\infty \theta_\infty$.

A number of illustrative cases are shown in Figures 19.9 and 19.10. The system makes a transition from an initial to a final state. In example 19.9(b) the transition lasts about 200 yr. After the transition, the reservoir temperature has fallen from $250°C$ to $150°C$, the generated power has fallen from 225 MW to 20 MW. The most striking feature of the transition is the early peak in the system pressure

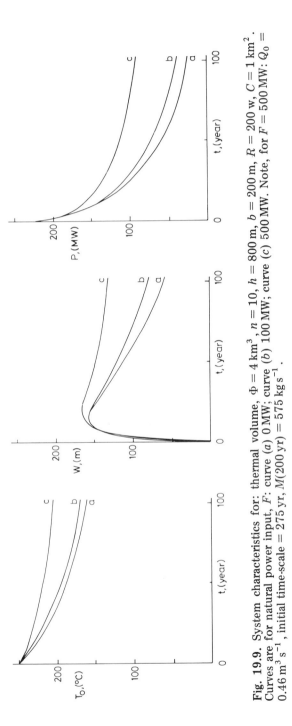

Fig. 19.9. System characteristics for: thermal volume, $\Phi = 4\,\mathrm{km^3}$, $n = 10$, $h = 800\,\mathrm{m}$, $b = 200\,\mathrm{m}$, $R = 200\,w$, $C = 1\,\mathrm{km^2}$. Curves are for natural power input, F: curve (a) 0 MW; curve (b) 100 MW; curve (c) 500 MW. Note, for $F = 500\,\mathrm{MW}$: $Q_0 = 0.46\,\mathrm{m^3\,s^{-1}}$, initial time-scale = 275 yr, $M(200\,\mathrm{yr}) = 575\,\mathrm{kg\,s^{-1}}$.

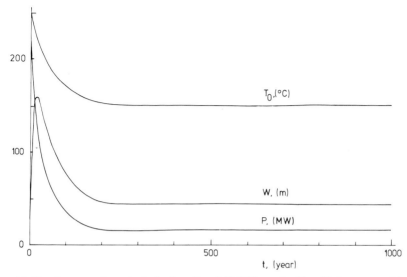

Fig. 19.10. System characteristic for $F = 100$ MW, as in Fig. 19.9, curves (b), but for a longer time.

loss, w. During the interval, near this pressure minimum, say from 10—30 yr, the system could appear to be in hydrological equilibrium — but this would be a mere illusion.

Variable bore field — constant power strategy

With intense exploitation, the system changes dramatically through time in a manner dependent in part on the details of the exploitation. The maximum power strategy has the impractical consequence of a rapid fall in generated power. A more acceptable management policy would be to operate at a constant generated power. Consider a fixed plant and a bore field (of hypothetical equal bores) to be run at nearly constant generated power output by responding to changes in the system solely by making changes in the bore field either by closing some bores or drilling extra bores. Let the required number of bores be reassessed at regular intervals (less than the hydrological time-scale, RC), for the example here every 2 years.

Consider first a field with natural thermal power input, $F = 500$ MW and a 100 MW(e) plant. The behaviour of the system is indicated in the data of Fig. 19.11(a). This is about the simplest possible situation. There is an early hydrological adjustment, but after about 20 years, further changes are small. In the first 100 years there is negligible change in reservoir temperature and the required expansion of the

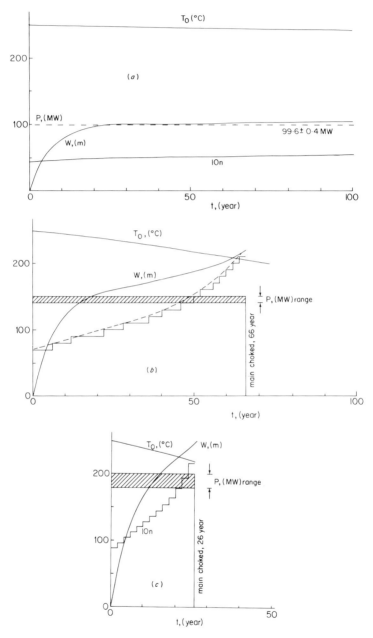

Fig. 19.11. Constant power strategy. Natural power 500 MW. System characteristics and number of bores, n (bore hole reassessment interval, 2 years), for designed available power: (a) 100 MW, (b) 150 MW; (c) 200 MW.

bore field is minor. The behaviour of this system is determined by wear and tear on plant and not through gross changes in the reservoir.

In the same field, consider now a 150 MW(e) plant illustrated in (b). There is now no longer approach to a quasi equilibrium. Reservoir pressure and temperature fall continuously. The size of the bore field needs continual expansion at an accelerating rate. The system becomes inoperable after 66 years when the pipeline chokes. In practice, of course, a new bigger diameter pipeline could be installed somewhat earlier.

In the same field, consider, finally, a 200 MW(e) plant illustrated in (c). Changes in reservoir pressure and temperature and bore field size are now rapid. The pipeline chokes after 26 years.

If we compare the performance of these three hypothetical systems we see that there is a relatively narrow range of exploitation levels of intense exploitation in which there are increasingly dramatic changes in the performance of the field—plant system. If the ratio of rate of energy removed to rate supplied is about 2:1 (a), effects are minor, but at ratio 4:1 (c), effects are dramatic.

Appendix

Much of the description of hydrothermal systems is dominated by the properties of water-substance near saturation. Most of the quantities of direct interest for saturated water and steam are indicated in the tables and figures below.

A1. THERMODYNAMIC PROPERTIES OF WATER

Numerical representation of properties of water-substance

A ubiquitous and central feature of models involving hydrothermal systems is an appropriate representation of the properties of ordinary water-substance. The forms used in this work are based on the *UK Steam Tables in SI units* 1970. An extract is shown in Table A1. These tables are a particularly good example of the modern approach to this problem. Internal consistency is achieved by first constructing a pair of related functions; in this case, the Gibbs function (specific free enthalpy), $g = g(p, T)$; and the Helmholtz function (specific free energy), $f = f(v, T)$; the other functions being derived directly from them. The representations are very elaborate. As a consequence they are not very suitable for anything but the most detailed computations.

Simpler forms are often desirable. Two approaches can be used. One is to find a very simple approximate relationship, such as $T_s \sim 100 P_s^{1/4}$. Another is to find a suitable finite power series expansion. Suitable forms, used in this work are given below. Simple forms can be chosen in a variety of ways. Those quoted in the literature are usually disappointingly crude. Nevertheless if overall accuracy of say $\pm 20\%$ is adequate — and other key parameters of a model may be far more uncertain — then all is well.

Power series of the form:

$$y = \sum_{i=0}^{m} a_i x^i$$

475

Table A1. Saturated water and steam values

T (°C)	P_s (1 bar)	v_f (10^{-3} m³ kg⁻¹)	v_g (10^{-3} m³ kg⁻¹)	K_f (10^{-3} W m⁻¹ K⁻¹)	K_g (10^{-3} W m⁻¹ K⁻¹)	μ_f (10^{-6} N s m⁻²)
0	0.01	1.000	206163	569	17.6	1786
50	0.12	1.012	12046	643	20.9	547.8
100	1.01	1.044	1673	681	24.9	283.1
150	4.76	1.091	392	687	30.0	179.8
200	15.55	1.157	127.2	664	37.4	133.9
250	39.78	1.251	50.0	616	49.5	108.7
300	85.93	1.404	21.6	541	71.9	90.7
350	165.35	1.741	8.80	434	134	69.4
374.15	221.20	3.17	3.17	240	240	41.4

T (°C)	μ_g (10^{-6} N s m⁻²)	h_f (10^6 J kg⁻¹)	h_{fg} (10^6 J kg⁻¹)	h_g (10^6 J kg⁻¹)	u_f (10^6 J kg⁻¹)	u_{fg} (10^6 J kg⁻¹)
0	8.1	0.000	2.502	2.502	0.000	2.375
50	10.1	0.209	2.383	2.592	0.209	2.234
100	12.1	0.419	2.257	2.676	0.419	2.088
150	13.9	0.632	2.113	2.745	0.632	1.928
200	15.7	0.852	1.939	2.791	0.851	1.745
250	17.5	1.086	1.715	2.800	1.080	1.522
300	19.7	1.345	1.406	2.751	1.332	1.231
350	24.2	1.672	0.896	2.568	1.642	0.777
374.15	41.4	2.107	0	2.107	2.030	0

T (°C)	u_g (10^6 J kg^{-1})	s_f (10^3 J kg^{-1} K^{-1})	s_{fg} (10^3 J kg^{-1} K^{-1})	s_g (10^3 J kg^{-1} K^{-1})	c_{pf} (10^3 J kg^{-1} K^{-1})	c_{pg} (10^3 J kg^{-1} K^{-1})
0	2.375	0.000	9.158	9.158	4.22	1.86
50	2.444	0.704	7.374	8.078	4.18	1.91
100	2.507	1.307	6.049	7.355	4.22	2.03
150	2.560	1.842	4.994	6.836	4.31	2.32
200	2.595	2.331	4.097	6.428	4.49	2.88
250	2.602	2.794	3.277	6.071	4.87	3.92
300	2.563	3.255	2.453	5.708	5.79	6.15
350	2.418	3.780	1.438	5.218	10.07	15.80
374.15	2.030	4.443	0	4.443	∞	∞

Table A2. *Polynomial coefficients, $m = 4$.*

Symbol	Unit	Value at 100°C	rms	a_0	a_1	a_2	a_3	a_4
T_s	10^2 °C	1.00	0.0005	−0.333450	1.614960	−0.368806	0.091327	−0.008632
P_s	1 bar	1.01	0.04	0.085	−0.794	2.203	−1.976	1.503
v_f	10^{-3} m^3 kg^{-1}	1.04	0.005	1.0078	−0.0708	0.1769	−0.0846	0.0161
$v_g(P_s/\widehat{T}_s)$	(10^{-3} m^3 kg^{-1})	4.54	0.008	4.5020	0.3479	−0.3990	0.1204	−0.0242
K_f	1 W m^{-1} K^{-1}	0.681	0.0006	0.5685	0.1915	−0.0899	0.0119	−0.0014
K_g	10^{-3} W m^{-1} K^{-1}	24.9	1.2	19.575	−11.701	31.553	−17.531	3.435
$1/\mu_f$	10^3 m^2 N^{-1} s^{-1}	3.53	0.07	0.702	0.594	3.710	−1.650	0.244
μ_g	10^{-6} N s m^{-2}	12.06	0.08	8.230	2.846	2.056	−1.272	0.233
h_f	10^6 J kg^{-1}	0.42	0.0015	0.0027	0.3957	0.0382	−0.0219	0.0050
h_{fg}	10^6 J kg^{-1}	2.26	0.004	2.4941	−0.1698	−0.1115	0.0558	−0.0134
h_g	10^6 J kg^{-1}	2.68	0.003	2.4968	0.2258	−0.0730	0.0338	−0.0084
u_f	10^6 J kg^{-1}	0.42	0.002	0.0024	0.3980	0.0341	−0.0193	0.0043
u_{fg}	10^6 J kg^{-1}	2.09	0.003	2.3693	−0.2279	−0.0905	0.0465	−0.0111
u_g	10^6 J kg^{-1}	2.51	0.002	2.3718	0.1696	−0.0558	0.0270	−0.0067
s_f	10^3 J kg^{-1} K^{-1}	1.31	0.003	0.0054	1.4856	−0.1966	0.0103	0.0035
s_{fg}	10^3 J kg^{-1} K^{-1}	6.05	0.01	9.1362	−3.9916	1.0663	−0.1735	0.0045
s_g	10^3 J kg^{-1} K^{-1}	7.36	0.007	0.1416	−2.5063	0.8701	−0.1633	0.0081
c_{pf}	10^3 J kg^{-1} K^{-1}	4.22	0.13	4.411	−1.904	3.150	−1.710	0.309
c_{pg}	10^3 J kg^{-1} K^{-1}	2.03	0.3	2.324	−3.954	6.753	−3.661	0.667

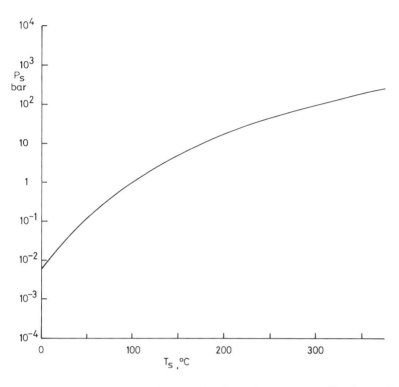

Fig. A1. Saturation pressure, P_s, as a fuction of teperature, T_s, for water-substance.

fortunately provide a good fit to most functions over a suitably small range. The calculation is very straightforward with, in an obvious notation:

set $y = 0$; for $i = m, -1, 0$;   set $y = a_i + yx$

It is computationaly desirable to have a uniform approach. Since the important function $P_s = P_s(T)$ shown in Fig. A1, is dominated by $P_s \sim (T/100)^4$ as indicated in Fig. A2, I have chosen $m = 4$ and $x = T/100$, where T is in °C, for functions of T.

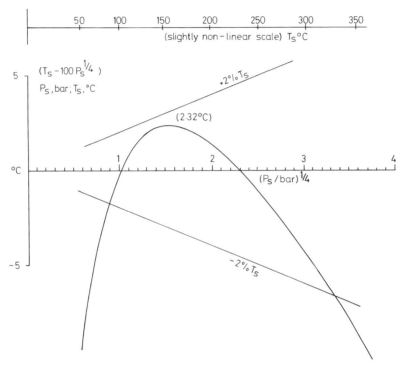

Fig. A2. Correlation of saturation pressure, P_s, and temperature, T_s, for water-substance with the variable, $P_s^{1/4}$.

Saturated liquid water and steam

The procedure for obtaining the power series representation of the functions of interest has been as follows:

(1) List the quantities, q_j, $j = 0,n$, to tabulated precision, at temperatures $0(10)370°C$.

(2) Solve for the a_j by the method of least squares. This involves computing the rectangular matrix of the cross products of the powers of x and the q_j and solving by the method of Gaussian elimination (or otherwise).

(3) Compare the original quantities q_j with the corresponding y_j given by the series and obtain the distribution of error over the range, and the mean error, ϵ.

(4) Progressively round (by retaining one less significant digit at each step) the coefficients a_j until the mean error begins to increase to no more than 2ϵ.

The quoted values of the coefficients and the mean error are for the rounded quantities — see Table A2.

All quantities of interest, except T_s, are given as functions of T; see also Fig. A3. Only one function of P_s is given, namely $T = T(P_s)$. Should a quantity be needed as a function of P_s then first obtain $T = T(P_s)$ and then the quantity as before. It is fortunate in this respect that the function $T(P_s)$ has a very precise series representation.

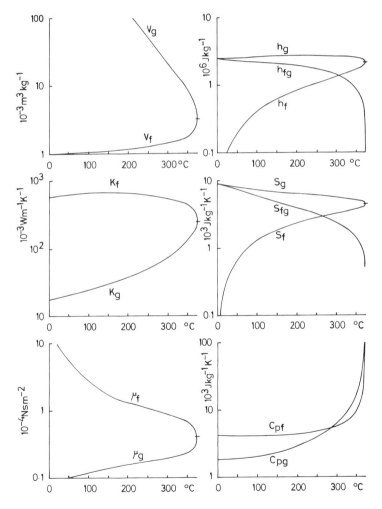

Fig. A3. Frequently used (in this book) thermodynamic properties of saturated water-substance as a function of temperature in °C.

Remarks on individual functions

$T = T(P_s)$. In the range $T = 100-350°C$ the saturated vapour temperature closely follows $T \sim P_s^{1/4}$. Thus, writing

$$T/°C = 100(P_s/\text{bar})^{1/4} \xi$$

we find ξ in the range 0.976–1.016. Note that the choice of the coefficient 100 is purely a matter of convenience. A 4-order fit to ξ has a mean error of 2.7×10^{-4}, that is 0.027%. Corresponding errors of T are at most about 0.13°C, with mean error of 0.06°C.

Over the restricted range of $P_s^{1/4} = 0.5-1.5$ (T_s from 37–152°C), useful for natural discharge systems, the polynomial coefficients:

$$(-0.52390, \quad 2.22141, \quad -1.07313, \quad 0.447037, \quad -0.074895)$$

provide a fit to $T_s(P_s)$ with the variables $(T_s/100°C)$, $(P_s/\text{bar})^{1/4}$ with mean error 0.02°C.

P_s. Thermodynamic considerations of a one-component, two-phase system give the Clausius–Clapeyron relation:

$$d\tilde{T}/dP = \tilde{T}(v_1 - v_2)/L,$$

where $L = h_{fg}$, is the latent heat absorbed by the system in passing from phase 2 to phase 1. If phase 2 is liquid water, $v_1 = v_f$; if phase 1 is steam, $v_2 = v_g$; and distant from the critical point $v_g \gg v_f$; with $Pv_g = RT$, where R is the gas "constant" (8.31434 kJ/kg mole K, if steam were a perfect gas, with 1 kg mole steam = 18 kg) then

$$d \log P/d\tilde{T} = L/R\tilde{T}^2,$$

so that

$$P = P_0 \exp\left[\frac{L}{R}\left(\frac{1}{\tilde{T}_0} - \frac{1}{\tilde{T}}\right)\right],$$

where at a reference point $P = P_0$, $T = T_0$. For example, choosing $T_0 = 100°C$ $P_0 = 1.01325\,\text{bar}$ with $L/R = 4886\,\text{K}^{-1}$ (steam at 100°C and perfect gas R), gives a mean error of about 0.01 bar from 0 to 150°C (spot on at 100, 150°C). A more accurate fit in the range 100–110°C is given with $L/R = 4953\,\text{K}^{-1}$.

Unfortunately neither L or R is a constant.

Over the range 0 to 100°C, a 5-order polynomial provides a very good fit, with mean error 0.0002 bar, with variables $(T/100°C)$, (P/bar) and coefficients:

$$(0.0060, \quad 0.0424, \quad 0.1500, \quad 0.2566, \quad 0.2893, \quad 0.2685)$$

Table A3. *Saturated steam gas constant* $R_g = v_g P_s / \widetilde{T}_s$.

T ($^\circ$C)	R_g (10^2 J/kg K)	β_g (kg m^{-3} (K/bar))
0	4.53	220.8
50	4.59	218.1
100	4.54	220.2
150	4.41	226.5
200	4.19	239.3
250	3.80	262.9
300	3.25	308.1
350	2.33	428.3
374.15	1.08	926

$(T \to \infty, P \to 0; R = 4.615 \times 10^2 \text{ J kg}^{-1} \text{ K}^{-1})$

v_g. The specific volume of saturated steam behaves roughly like that of a perfect gas, so that $R_g = v_g P_s / \widetilde{T}_s$ is a slowly varying function of $\widetilde{T}$ (certainly if compared to the variation of v_g itself) — see Table A3. Thus given a polynomial representation of R_g we obtain $v_g = R_g \widetilde{T}_s / P_s$, where, if it has not already been done, it will be necessary also to evaluate $P_s(\widetilde{T}_s)$.

It is often rather convenient to work with $\beta_g = 1/v_g$ so that the density $\rho_g = \beta_g P_s / \widetilde{T}_s$. The 4-order polynomial for β_g in the independent variable $(T/100^\circ C)$ is given by:

$$a = (225.2, \ -45.0, \ 68.1, \ -34.2, \ 6.6) \text{ kg m}^{-3} (\text{K/bar})$$

with mean error of about 1%.

Values of v_g for super-critical steam are given in Table A4.

μ_f. The power series for the viscosity of the liquid is a very poor fit, with errors up to about 50%. As was first noted by Dorsey,

$$1/\mu_f \approx a_0 + a_1 T,$$

where a_0 and a_1 are constants. The power series for $1/\mu_f$ gives quite a good fit with errors less than 1% above 100°C, but not so good below 100°C, reaching about 6% near 50°C.

μ_g. For a gas assumed to be composed of small, elastic independent molecules, Maxwell showed that

and
$$\mu_g = K\widetilde{T}^{1/2}, \qquad (\partial\mu_g/\partial P)_{\widetilde{T}} = 0,$$
$$v_g = K\widetilde{T}^{3/2}, \qquad P = \text{constant},$$

where K is a constant.

Table A4. Specific volume of steam, v_g.
Values of $R_g = v_g P/\tilde{T}$
Units 10^2 J kg^{-1} K^{-1}

$T(^\circ C)$	P (bar) 100	200	300	400	500	600	700	800	900	1000
100			0.83	1.10	1.37	1.63	1.90	2.16	2.42	2.68
200			0.72	0.95	1.18	1.40	1.63	1.85	2.07	2.29
300			0.70	0.91	1.12	1.33	1.53	1.73	1.93	2.12
400		2.96	1.26	1.13	1.28	1.46	1.63	1.80	1.98	2.15
500		3.82	3.37	2.91	2.51	2.29	2.23	2.26	2.34	2.45
600		4.16	3.93	3.71	3.50	3.32	3.18	3.10	3.06	3.06
700		4.34	4.21	4.08	3.97	3.87	3.78	3.72	3.67	3.63
800		4.44	4.37	4.29	4.23	4.17	4.12	4.09	4.06	4.05

Note. Values at: critical point, 1.083×10^2 J kg K^{-1}, $T \to \infty$, $P \to 0$, 4.615×10^2 J kg^{-1} K^{-1} .

Allowing for a weak molecular attraction, Sutherland showed

$$\mu_g = KT^{1/2}/(C + 1/\tilde{T}) \equiv a\tilde{T}^{1/2}/(1 + b/\tilde{T})$$

where K, C or a, b are constants. Various other empirical modifications have been proposed. Unfortunately these expressions are rather poor representations. They are however better than nothing.

Over the range $T = 100–1000°C$, $P = 1$ bar, the Sutherland form gives a fair fit to μ_g, in units 10^{-6} N m s^{-2}, ranging from 12.1 to 4.75, with $a = 1.8$, $b = 540$ with mean error of 1.6.

For saturated steam, over the range $T = 0–350°C$, $P = P_s$, for μ_g in units 10^{-6} N s m^{-2} ranging from 8.1 to 24.2, the simple Maxwell form has a mean error of 3.0; the Sutherland form 1.2. In fact over the range $0–300°C$ a better fit is simply $\partial \mu_g/\partial T \approx$ constant. A truncated power series gives a good fit: mean error 0.1 for order-4 series, over $T = 0–350°C$.

In summary:

(i)   Saturated steam: $\mu_g \approx (8 + 0.038\tilde{T})10^{-6}$ N m s^{-2}
(ii) Superheated steam: $\mu_g \approx 1.8\tilde{T}^{1/2}/(1 + 540/\tilde{T})10^{-6}$ N m s^{-2}

Note on the contributions of p_v to the enthalpies h_f and h_g of saturated water and steam

The proportion of the enthalpy which arises from pv is rather small. As shown in Fig. A4, for saturated steam the proportion is roughly constant, ranging from 3.4% to 7.1%; for saturated water it is very small at low temperatures and is everywhere less than 3.4%. In practice, except for the most careful measurements, the distinction between internal energy, u and enthalpy, $h = u + pv$ can be ignored.

Representation of functions near the critical point by interpolation

Near the critical point, usually somewhat above 350°C, the variation of some thermodynamic functions is extreme (see for example Fig. A5) and the possible accuracy of truncated polynomials of a few terms is limited. Interpolation, in spite of its rather cumbersome nature, is to be preferred. Consider a table of values (x_i, y_i), with $i = 0$, m, from which the interpolated value y for given x is required. Clearly we wish to have m as small as possible for a given accuracy.

The evaluation can use the method of differences. Alternatively, presume some local relationships such as $y = y(x, p, q)$, where p, q are two parameters to be found from the values at $j, j + 1$, and where

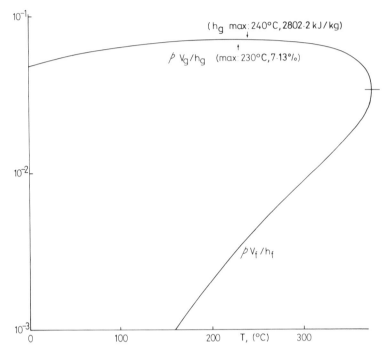

Fig. A4. Relative contribution of pv to specific enthalpy, h, as a function of temperature for saturated water-substance.

$x_j < x < x_{j+1}$. I have found, however, that a simple and often very accurate method is simply to "stretch" x. Transform all values to new quantities $x' = x^r$, $y' = y$, and obtain the required y by linear interpolation from (x'_j, y'_j), (x'_{j+1}, y'_{j+1}). For example, over the range $T = 350$–$374.15°C$, the function $10^3 (v_{fc} - v_f)$ $(m^3\ kg^{-1})$ ranges from 1.43 to 0, and with seven tabular values at 350(4)374 has a mean error of 0.0002 and a maximum error of 0.001, with $r = 0.153$. The errors are not critically dependent on r which can be chosen to provide acceptable mean error and maximum error. The best values of these occur at very different values of r, and it is usually simplest to choose that which gives the smallest maximum error. This device has been used here in the few calculations involving values near the critical point.

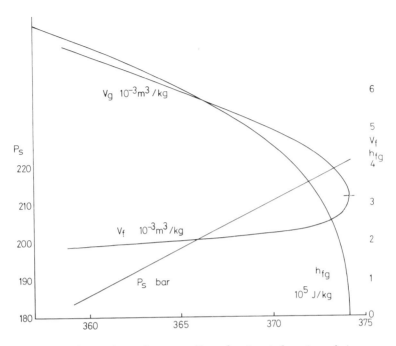

Fig. A5. Some thermodynamic properties of saturated water-substance as a function of temperature near the critical point.

Interpolation for T(P_s) near T_c

Adequate accuracy is obtained from $P_s = 170(10)220$ bar with stretching parameter $r = 0.14$. The mean error is $0.002°\mathrm{C}$ and the maximum error is $0.008°\mathrm{C}$.

A2. THE BOILING POINT

The boiling point as a function of depth, BPD

A useful reference situation is that of a static volume of liquid water just at the boiling point throughout its depth. The relation, $P(z)$, the BPD can be calculated for example from:

$$\rho_f = \rho_f(T_s), \qquad \mathrm{d}P/\mathrm{d}z = \rho f g$$

given $P = P_0$ or $T = T_0$ at $z = 0$, with the functions ρ_f, T_s given by tables, polynomials or a simple formula. This form is convenient and accurate since ρ_f is a slowly varying function of T_s except near the critical point.

Table A5. *Boiling point with depth.*

Z(m)	P(bar)	T(°C)	Z(m)	P(bar)	T(°C)
0	1.01	100.0	100	9.9	180
1	1.11	102.3	200	18.5	208
2	1.20	104.7	300	26.8	228
3	1.29	106.9	400	34.9	242
4	1.39	109.0	500	42.8	254
5	1.48	110.9	600	50.5	265
6	1.57	112.8	700	58.1	273
7	1.67	114.5	800	65.5	281
8	1.76	116.2	900	72.8	288
9	1.85	117.8			
			1000	80	295
10	1.94	119.2	1500	114	321
20	2.86	131.8	2000	145	339
30	3.77	141.4	2500	174	354
40	4.66	149.2	3000	200	366
50	5.56	155.9	3460	221	374
60	6.44	161.7			
70	7.33	166.8			
80	8.20	171.5			
90	9.08	175.8			

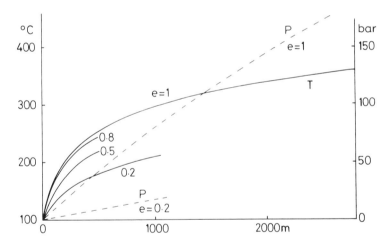

Fig. A6. Boiling point with depth relation (BPD). (i) Pressure, P, and temperature, T, for an entirely liquid column, $e = 1$. (ii) Temperature, T, for a column
of uniform liquid fraction $e = 0.2, 0.5, 0.8$, and also pressure, P, for $e = 0.2$.

The accompanying Table A5 and Fig. A6 have been obtained using
Simpson's method of integration with the 4-order polynomials given
above up to $T_s = 350°C$ and then tabular values to the critical point.
The quantities are accurate to about 1%. For $T_0 = 100°C$, the critical

point is reached (in the calculation) at 3458 m. The BPD changes most rapidly at the surface, with surface gradient $2.3\,\text{K}\,\text{m}^{-1}$ falling steadily to about $0.02\,\text{K}\,\text{m}^{-1}$ at the critical point.

Note: Use of $g = 9.8\,m\,s^{-2}$. This value is used here. The nominal acceleration of gravity over the geoid as summarized in the expression for the reference ellipsoid of the earth (IGRF, 1967) ranges from 9.780 to $9.832\,\text{m}\,\text{s}^{-2}$ from $0°$ to $90°$ latitude, with areal mean value of $9.81\,\text{m}\,\text{s}^{-2}$ and the value $9.8\,\text{m}\,\text{s}^{-2}$ near $40°$ latitude.

In calculations of conditions in the surface zone of a hydrothermal system the local variation of g with height should strictly be allowed for but usually the effects of vertical variation of g will be swamped by those of the surface pressure.

BPD in the atmosphere

For a variety of purposes, for example the study of hot springs at various altitudes, it is necessary to have a measurement of the local boiling point. If this is not available it can be estimated from the nominal pressure of the international standard atmosphere. Some values are given in Table A6.

The pressure of the standard atmosphere at sea level is taken as 1.01325 bar. The proportional change in pressure with altitude for the standard atmosphere is $0.12\,\text{bar}\,\text{km}^{-1}$. Thus, a standard atmospheric pressure of precisely 1 bar corresponds to an altitude of 110 m.

Table A6. *International standard atmosphere.*
Density at nominal sea level, $\rho_0 = 1.2250\,\text{kg}\,\text{m}^{-3}$ ($15°C$)

$Z(m)$	P(bar)	$T(K)$	ρ/ρ_0	BP($°C$)
−1000	1.1393	294.7	1.0996	
−500	1.0748	291.4	1.0489	
0	1.01325	288.15	1.0000	100.0
500	0.9546	284.9	0.9529	98.5
1000	0.8988	281.7	0.9075	97.0
1500	0.8456	278.4	0.8638	95.6
2000	0.7950	275.2	0.8217	94.1
2500	0.7469	271.9	0.7812	
3000	0.7012	268.7	0.7423	91.2
3500	0.6578	265.4	0.7048	
4000	0.6166	262.2	0.6689	88.3
4500	0.5775	258.9	0.6343	
5000	0.5405	255.7	0.6012	85.4

The boiling point gradient is nearly constant in the lower atmoshere, being $\partial T/\partial z \approx 2.96$ K km^{-1}. There is little error in taking this as 3 K km^{-1}.

BPD for various altitudes of the surface

The effect of altitude on the surface boiling point is quite pronounced. This effect continues into the ground but below a depth of 100 m or so is insignificant on the BPD. For example, for the two cases of $T_0 = 100°$C and $T_0 = 85°$C (corresponding to about 5 km above sea level) the values are

0 m:	1.0 bar, 100°C;	0.6 bar, 85°C
50 m:	5.2 bar, 153°C;	5.6 bar, 156°C
100 m:	9.6 bar, 178°C;	10.0 bar, 180°C
150 m:	14.3 bar, 196°C;	13.9 bar, 195°C
200 m:	18.5 bar, 208°C;	18.1 bar, 207°C

BPD in a partially drained surface zone

If, by exploitation or through natural processes, a previously liquid surface zone is partically drained the saturation, e, the volumetric fraction of liquid, may be such that $0 < e < 1$. It has been suggested that even with pronounced draining there will be a residual fraction of trapped liquid water with $e \approx 0.2$ or so. It is therefore of interest to have the reference function BPD(e). Some curves for various values of e are included in Fig. A6.

The computation is the same as that for $e = 1$ with the density obtained from

$$\rho = \frac{e}{v_f} + \frac{(1-e)}{v_g}$$

The values shown were obtained with v_g from the 4-order polynomial of $v_g P_s/T_s$. The effect on both $P(z)$, $T(z)$ is moderate for $e \gtrsim 0.5$, but is pronounced for smaller values of e.

Notation

Only symbols used throughout the text are listed here

a	radius, of source	s	specific entropy; stream-wide coordinate
b	uncased depth		
c	specific heat	t	time
d	diameter of pipe; reservoir depth	u	blocking ratio
		v	specific volume $(1/\rho)$; crustal level difference (top)
e	saturation ratio (volumetric fraction of liquid)		
		w	reservoir pressure change (equivalent water surface depth); vertical velocity; crustal level difference (bottom)
f	dimensionless friction factor; heat flux		
g	acceleration due to gravity, $9.8\,\text{m s}^{-2}$		
h	pressure in head of liquid water of density, $\rho_0 = 1\,\text{ton m}^{-3}$; specific enthalpy; cased depth	x	working or local variable
		y	level or head; mass fraction of soluble gas in gas—vapour mixture; local parameter
i	integer label		
j	integer label	z	vertical coordinate
k	permeability	A	area; Rayleigh number: A_m; porous medium Rayleigh number
l	length		
m	mass rate; mass	C	capacitance
n	number; soluble gas mass concentration	D	discharge (total artificial) equivalent volumetric rate; diameter of bore
p	pressure (local)		
q	velocity, velocity amplitude; volumetric velocity	E	(specific) enthalpy
		F	natural power input to reservoir; dimensionless permeability ratios
r	specific input resistance; radial coordinate; radius		

491

G	gravity constant, $6.67 \times 10^{-11}\,\mathrm{N\,m^2\,kg^{-2}}$	γ	cubical expansion coefficient
H	length of main; thickness	δ	length-scale of boundary layer region
K	thermal conductivity		
L	latent heat (h_{fg})	ϵ	porosity
M	mass rate (or discharge)	ζ	ratio, local only
N	Nusselt number	η	efficiency (of turbine)
P	pressure; change of (field) pressure	θ	temperature
		κ	thermal diffusivity
Q	volumetric discharge rate (volume calculated at standard density, $\rho_0 = 1\,\mathrm{ton\,m^{-3}}$)	λ	mass fraction of (soluble) gas
		μ	viscosity
R	resistance	ν	kinematic viscosity
T	temperature; thermal time-scale	ξ	steam mass proportion
		π	3.142
U	total energy	ρ	density
V	volume	σ	solute concentration, finite strength
W	work per unit mass		
Y	gross drawdown	τ	time-scale
Z	steam zone thickness	ϕ	angle
α	local parameter (velocity); two-phase flow factor; gas solubility factor, $\sim 4 \times 10^{-4}\,\mathrm{bar^{-1}}$ for CO_2	χ	value, local only
		ψ	stream function
		Δ	scale of microstructure
		Σ	total solute per sec
β	local parameter (temperature rate); two-phased flow factor (weight-term); $\rho T/P$	Φ	thermal volume
		w	unit of flow resistance, $1\,\mathrm{w} = 1\,\mathrm{S\,m^{-2}}$

Suffixes

0	initial value; reservoir; value with other variable(s) zero or small	1	intermediate value; in ground; outer radius

2	final value; at base of bore; inner radius	s	steam (water vapour); saturation value
c	soluble gas	$*$	property of rock; boiling onset; condensor value
f	fluid or liquid		
fg	change from liquid to gas	$'$	gradient; actual value (instead of scaled value): bore value
g	gas or vapour		
m	at well head: of porous medium	$-$	average

Others

| BPD | boiled point as a function of depth | WHT | well-head temperature |
| WHP | well-head pressure | SVP | saturation vapour pressure |

References

*UN*0 = Proceeding of UN conference on new sources of energy, Rome, 1961
*UN*1 = Geothermics (1970), Special Issue 2. (Proceedings of UN symposium on geothermal resources, Pisa 1970)
*UN*2 = Proceedings of UN symposium on geothermal resources, San Francisco, 1975.

Alldredge, L. R. (1977). Deep mantle conductivity, *J. Geophys. Res.* **82**, 5427–5431.

Allen E. T. and Zeis, E. G. (1923). A chemical study of the fumaroles of the Katmai region. *Natl. Geogr. Soc. Contrib. Tech. Papers*, Katmai Series 2, 75–155.

Allen E. T. and Day A. L. (1935). "Hot springs of the Yellowstone National Park." Carnegie Institute, Washington, Publ. 466.

Allis, R. G. (1978). "Thermal history of the Karapiti area, Wairakei." *NZ DSIR Geophs. Div. Report.*

Anderson, R. N. and Langseth, M. G. (1977). The mechanism of heat transfer through the floor of the Indian Ocean. *J. Geophys. Res.* **82**, 3391–3409.

Appleman, H. (1953). The formation of exhaust condensation trails by jet aircraft. *Bull. Amer. Met. Soc.* **34**, 14–20.

Athavale, R. N. and Sharma, P. V. (1975). Paleomagnetic results on early Tertiary lava flows from West Greenland and their bearing on the evolution of the Baffin Bay–Labrador Sea Region. *Can. J. Earth Sci.* **12**, 1–18.

Baker, S. R. and Friedman, G. M. (1973). Sedimentation in an arctic marine environment: Baffin Bay between Greenland and the Canadian Archipeligo. *Pap. Geol. Surv. Can.* **71–23**, 471–498.

Ballance, P. F. (1976). Evolution of the upper Cenozoic magmatic arc and plate boundary in northern New Zealand. *Earth Planet. Sci. Lett.* **28**, 356–370.

Banwell, C. J. (1957). Origin and flow of heat. *NZ DSIR Bull.* **123**, 9–71.

Banwell, C. J. (1963). Oxygen and hydrogen isotopes in New Zealand thermal areas. In Tongiori (1963) pp. 95–138.

Barnes, H. L. (ed.) (1979). "Geochemistry of Hydrothermal Ore deposits." Wiley, New York.

Barth, T. F. W. (1950). "Volcanic geology, hot springs, and geysers in Iceland." Carnegie Institute, Washington, Publ. 587.

Baumgart, I. L. and Healy, J. (1956). Recent volcanicity at Taupo, New Zealand. *Proc. 8th Pacif. Sci. Congr.* **2**, 113–25.

Benseman, R. F. (1959). The calorimetry of steaming ground in thermal areas. *J. Geophys. Res.* **64**, 123–6, 1057–62.

Benseman, R. F. (1965). The components of a Geyser. *NZ J. Sci.* **8**, 24–44.

495

Bodvarsson, G. (1948). On thermal activity in Iceland. Geothermal Dept. State Electricity Authority. Reykjavik, Iceland.

Bodvarsson, G. (1949). Drilling for heat in Iceland. *Oil and Gas J.* 47, 191—9.

Bodvarsson, G. (1950). Geophysical methods in prospecting for hot water in Iceland (in Danish). *Timarit Verkfraed. Islands* 35, 49—59.

Bodvarsson, G. (1954). Terrestrial heat balance in Iceland. *Timarit Verkfraed. Islands* 39, 69—76.

Bodvarsson, G (1961). Physical characteristics of natural heat resources in Iceland. *UN0*, paper G, p. 6.

Bolton, R. S. (1969). Wairakei geothermal power project: report on the partial shut down of the field, 21 December 1967 to 2 April 1968. M.O.W. Report 92/14/20/2.

Bolton, R. S. (1970). The behaviour of the Wairakei geothermal field during exploitation. *UN1*, pp. 1426—1449.

Boussinesq, J. (1903). "Theorie Analytique de la Chaleur". Gauthier-Villars, Paris.

Burgassi, R. (1961). Prospecting of geothermal fields. *UN0*, paper G, p. 65.

Burgassi, R., Battini, F. and Mouton, J. (1961). Geothermal prospecting for endogenous energy. *UN0*, paper G, p. 61.

Carslaw, H. S. and Jaeger, J. C. (1959). "Conduction of Heat in Solids." Clarendon Press, Oxford.

Chandrasekhar, S. (1961). "Hydrodynamic and Hydromagnetic Stability." Clarendon Press, Oxford.

Chapman, D. S. and Pollack, H. N. (1975). Global heat flow: a new look, *Earth Planet. Sci. Lett.* 28, 23—32.

Chierici, A. (1961). Planning of a geothermal power plant. *UN0*, paper G, p. 62.

Clarke, D. B. and Pederson, A. K. (1976). The tertiary volcanic province of West Greenland. *In* "The Geology of Greenland" (Escher, A. and Watt, W. S., eds.), pp. 364—385. Geological Survey of Greenland, Copenhagen.

Clayton, R. N. (1963). High-temperature isotope thermometry. Nuclear Geophysics, National Academy of Sciences — National Research Council Publications 1975.

Craig, H. (1961). Standard for reporting concentrations of deuterium and oxygen-18 in natural waters. *Science.* 133, 1833.

Craig, H. (1963). The isotopic geochemistry of water and carbon in geothermal areas. *In* Tongiorgi (1963), pp. 17—53.

Craig, H., Boato, G. and White D. E. (1956). Isotopic geochemistry of thermal waters. Proc. Second Conf. on Nuclear Processes in Geologic Settings. Nat. Acad. Sci.-Nat. Res. Council Pub. 400, p. 29.

Crane, K. and Normark, W. R. (1977). Hydrothermal activity and crestal structure of the East Pacific Rise at $21°N$. *J. Geophys. Res.* 82, 5336—5348.

Crane, K. (1977). Hydrothermal activity and near axis structure at mid-ocean spreading centres. Ph. D. Thesis, University of California at San Diego.

Darcy, H. P. G. (1856). "Les Fontaines Publiques de la Ville de Dijon." Victor Dalmont, Paris.

Dawson, G. B. and Dickinson, D. J. (1970). Heat flow studies in thermal areas of the North Island of New Zealand. *NZ DSIR Geophys. Div. Report* 59, 1—21.

Dawson, G. B. and Rayner, H. H. (1968). Discussion of the final 600 ft Wenner resistivity map of the Broadlands field. *NZ DSIR Geophys. Div. Report.* 53, 16—19.

Day, A. L. (1939). The hot-spring problem. *Bull. Geol. Soc. Amer.* **50**, 317.

Denham, L. R. (1974). Offshore geology of northern West Greenland (69°– 75°N), *Rapp. Grønlands Geol. Unders.* **63**, 24.

Dickinson, D. J. (1968). The natural heat output of the Broadlands geothermal area, 1967. *NZ DSIR Geophys. Div. Report.* **49**, 23–27.

Donaldson, I. G. (1958). The temperature, pressure and velocity distribution of water, steam or water-steam mixtures in a homogeneous permeable medium. *N.Z. Phys. Engng. Lab. Rep.* R. 287.

Donaldson, I. G. (1974). Underground waters of the lower Hutt Valley – a model study. *J. Hydrol. Geol. (NZ)* **13**, 81–97.

Drever, H. I. (1958). Geological results of four expeditions to Ubekendt Ejland, West Greenland, Arctic 11, pp. 198–210.

Einarsson, T. (1942). Uber das Wesen der Heissen Quellen Islands. *Rit. Visind. Isl.* **26**, 1–89.

Elder, J. W. (1966). Hydrothermal systems. *Bull.* 169 *NZ Dep. Scient. Ind. Res.*

Elder, J. W. (1968). The unstable thermal interface. *J. Fluid Mech.* **32**, 69–96.

Elder, J. W. (1969). The temporal development of a model of high Rayleigh number convection. *J. Fluid Mech.* **35**, 417–37.

Elder, J. W. (1976). "The Bowels of the Earth." University Press, Oxford.

Elder, J. W. (1978). Magma traps. *PAGEOPH* **117**, 1–33.

Elder, J. W. and Kerr, R. P. (1954). *NZ App. Maths. Div. Rep.* (Unpublished).

Ellis, A. J. and Golding, R. M. (1963). The solubility of carbon dioxide above 100°C in water and in sodium chloride solutions. *Amer. J. Sci.* **261**, 47–60.

Ellis, A. J. and Mahon, W. A. J. (1977). "Chemistry and Geothermal Systems. Academic Press, London and New York.

Ellis, A. J. and Wilson, S. H. (1955). The heat from the Wairakei-Taupo thermal region calculated from the chloride output. *NZ J. Sci. Tech.* B. **36**, 622–31.

Ellis, A. J. and Wilson, S. H. (1960). The geochemistry of alkali metal ions in the Wairakei hydrothermal system. *NZ J. Geol. Geophys.* **3**, 593–617.

Escher, A. and Pulvertaft, T. C. R. (1976). Rinkian mobile belt of West Greenland. *In* "The Geology of Greenland." (Escher, A. and Watt, W. S., eds.), pp. 104–119. Geological Survey of Greenland, Copenhagen.

Fergusson, G. J. and Knox, F. B. (1959). The possibilities of natural radiocarbon as a ground water tracer in thermal areas. *NZ J. Sci.* **2**, 431.

Fridleifsson, I. B. (1976). Lithology and structure of geothermal reservoir rocks in Iceland. *UN2*, p. 371–376.

Gass, I. (ed.) (1977). "Volcanic processes in ore genesis." Geological Society London, Spec. Publication 7.

Gerard, V. B. and Lawrie, J. A. (1955). Aeromagnetic surveys in New Zealand 1949–1952. Geophysics Memoir 3, NZDSIR.

Glover, R. B. (1970). Interpretation of gas compositions from the Wairakei field over 10 years. *UN2*, p. 1355–1366.

Grange, L. I. (ed.) (1955). Geothermal steam for power in New Zealand. *NZ DSIR Bull.* **117**.

Grange, L. I. (1937). The geology of the Rotorua-Taupo subdivision, Rotorua and Kaimanawa Divisions. *NZ DSIR Bull.* **37**.

Grange, L. I. (ed.) (1955). Geothermal steam for power in New Zealand. *NZ DSIR Bull.* **117**.

Grant, M. A. (1977). Broadlands – a gas dominated geothermal field. *Geothermics* **6**, 9–29.

Gregg, D. R. (1958). Natural heat flow from the thermal areas of the Taupo sheet district, *NZ J. Geol. Geophys.* **1**, 65–75.

Grindley, G. W. (1965). The geology, structure and exploitation of the Wairakei geothermal field, Taupo, New Zealand, *NZ DSIR Bull.* **75**.

Guilbert, J. M. and Lowell, J. D. (1974). Variations in zoning patterns in porphyry copper deposits. *Can. Min. Metall. Bull.* **67**, 99–109.

Gustafson, L. B. and Hunt, J. P. (1975). The porphyry copper deposit at El Salvador, Chile. *Econ. Geol.* **70**, 857–913.

Hald, N. (1973). Preliminary results of the mapping of the Tertiary basalts in western Nûgssuaq. *Rapp. Grønlands Geol. Unders*, **53**, 11–19.

Hald, N. and Pedersen, A. K. (1975). Lithostratigraphy of the early Tertiary volcanic rocks of central West Greenland. *Rapp. Grønlands Geol. Unders.* **69**, 17–23.

Harris, P. G., Kennedy, W. Q. and Scarfe, C. M. (1970). Volcanism versus plutonism – the effect of chemical composition. *Geological Journal* Special Issue, **2**, 187–200.

Haskell, N. A. (1937). The viscosity of the asthenosphere. *Amer. J. Sci.* **33**, 22–28.

Heirtzler, J. R., Dickson, G. O., Herron, E. M., Pitman, W. C. and Le Pichon, Z. (1968). Marine magnetic anomalies, geomagnetic field reversals, and motions of the ocean floors and continents. *J. Geophys. Res.* **73**, 2119–2136.

Henderson, F. M. (1950). Analysis of some problems in heat transfer through rock. *NZ Dom. Phys. Lab. Rep.* 8/7/112.

Henderson, G. (1973). The geological setting of the West Greenland basin in the Baffin Bay region. *Pap. Geol. Surv. Can.* 71–23, pp. 521–544.

Henderson, G., Rosenkrantz, A. and Schiener, E. J. S. (1976). Cretaceous–Tertiary sedimentary rocks of West Greenland. *In* "The Geology of Greenland." (Escher, A. and Watt, W. S., eds.), pp. 340–3620. Geological Survey of Greenland, Copenhagen.

Henley, R. W. and McNabb, A. (1978). Magmatic vapour plumes and groundwater interaction in porphyry copper emplacement. *Economic Geology* **73**, 1–20.

Hitchcock, G. W. and Bixley, P. F. (1972). Observations of the effect of a three-year shutdown at Broadlands geothermal field, New Zealand. *UN2*.

Hochstetter, F. von. (1864). "Geologie von Neu Seeland". Novara Exped., pt. 1, p. 274.

Holmes, A. (1965). "Principles of Physical Geology." Nelson, London.

Horai, K. and Simmons, G. (1969). Spherical harmonic analysis of terrestrial heat flow. *Earth Planet. Sci. Lett..* **6**, 386–394.

Howard, L. N. (1964). Convection at high Rayleigh number. "Proc. Int. Congress Applied Mechanics. Munich" (ed. H. Görtler), pp. 1109–1115. Berlin, Springer-Verlag.

Hughes, T. (1972). Thermal convection in polar ice sheets related to various empirical flow laws of ice. *Geophys. J. R. Astron. Soc.* **27**, 215–99.

Hunt, T. M. (1977). Recharge of water in Wairakei geothermal field determined from repeat gravity measurements. *NZ J. Geol. Geophys.* **20**, 303–317.

Hunt, T. M. and Hicks, S. R. (1975). Repeat gravity measurements at Broadlands geothermal field, 1967–1974. *NZ DSIR Geophys. Div. Rep.* **113**.

Jacchia, L. G. (1967). Upper atmosphere temperature. *In* "*International Dictionary of Geophysics.*" Pergamon Press, Oxford.

Jagger, T. A. (1940). Magmatic gases. *Amer. J. Sci.* **238**, 313—353.

Jessop, A. M., Hobart, M. A. and Sclater, J. G. (1975). The world heat flow data collection. Geothermal Series, Earth Physics Branch, Ottawa, Canada.

Jürgensen, T. and Mikkelsen, N. (1974). Coccoliths from volcanic sediments (Danian) in Nugssuaq, West Greenland. *Bull. Geol. Soc. Denmark* **23**, 225—229.

Kear, D. (1957). Erosional stages of volcanic cones as indicators of age. *NZ J. Sci. Tech. B* **38**, 671—682.

Kear, D. (1959). Stratigraphy of New Zealand's Cenozoic volcanism northwest of the volcanic belt. *NZ J. Geol. Geophys.* **2**, 578—589.

Keen, C. E., Keen, M. J., Ross, D. I. and Lack, M. (1974). Baffin Bay: a small ocean basin formed by sea floor spreading. *Bull. Amer. Ass. Petrol. Geol.* **58**, 1089—1108.

Kennedy, G. C. (1950). A portion of the system silica—water. *Amer. J. Sci.* **45**, 629—653.

Lal, D. and Peters, B. (1962). Cosmic ray produced isotopes and their application to problems in geophysics. *In* "Progress in Elementary Particle and Cosmic Ray Physics, VI" (Wilson, J. G. and Wouthuysen, S. A., eds). North-Holland, Amsterdam.

Lapwood, E. R. (1948). Convection of a fluid in a porous medium. *Proc. Camb. Phil. Soc.* **44**, 508—521.

Lee, W. H. K. (1970). On the global variations of terrestrial heat flow. *Phys. Earth Planet. Interiors* **2**, 332—341.

Lee, W. H. K. and MacDonald, G. J. F. (1963). The global variation of terrestrial heat flow. *J. Geophys. Res.* **68**, 6481—6492.

Lee, W. H. K. and Uyeda, S. (1965). "Review of Heat Flow Data, in Terrestrial Heat Flow," pp. 87—180. Geophysical Monograph Series No. 8. American Geophysical Union, Washington DC.

Le Pichon, X., Hyndman, R. D. and Pautot, G. (1971). Geophysical study of the opening of the Labrador Sea. *J. Geophys. Res.* **76**, 4724—4743.

Lindgren, W. (1933). "Mineral Deposits," 4th edn. McGraw-Hill, New York.

Lloyd, E. F. (1959). The hot springs and hydrothermal eruptions of Waiotapu. *NZ J. Geol. Geophys.* **2**, 141—76.

Malahoff, A. (1969). Gravity anomalies over volcanic regions. *In* "The earth's crust and upper mantle." (Hart, P. ed.), pp. 364—379.. Geophysical Monograph No. 13. American Geophysical Union.

Malinin, S. D. (1974). Thermodynamics of the H_2O—CO_2 system. *Geochem. Internat.* **7**, 1060—1070.

Malkus, J. S. (1960). Recent developments in studies of penetrative convection and an application to hurrican cumulonimbus towers. *In* "Cumulus Dynamics." (Anderson, C. E., ed.), pp. 65—84. Pergamon Press, Oxford.

McCree, K. J. (1957). The Great Wairakei Geyser: the state of its activity in 1951. *NZ DSIR Bull.* **123**, 97—104.

Menard, H. W. (1964). "Marine Geology of the Pacific." McGraw-Hill, New York.

Modriniak, N. and Studt, F. E. (1959). Geological structure and volcanism of the Taupo-Tarawera district. *NZ J. Geol. Geophys.* **2**, 654—684.

Molnar, P., Atwater, T., Mammerickx, J. and Smith, S. M. (1975). Magnetic anomalies, bathymetry and tectonic evolution of the South Pacific since the Late Cretaceous. *Geophys. J. R. Astron. Soc.* **40**, 383—420.

Münther, V. (1973). Results from a geological reconnaissance around Svartenhuk Halvø, West Greenland. *Rapp. Grønlands Geol. Unders.* **50**, 26.

Murase, T., Jushiro, I. and Fujii, T. (1977). Electrical conductivity of partially molten peridotite. Annual Report Geophysical Laboratory, Washington, 1976—1977, p. 416—419.

Muskat, M. (1937). "The Flow of Homogeneous Fluids Through Porous Media." McGraw-Hill, New York.

Nencetti, R. (1961). Methods and appratus used for well mouth measurements. UN0, paper G, p. 75

Norton, D. and Cathles, L. M. (1979). Thermal aspects of ore deposition. In "Geochemistry of hydrothermal ore deposits" (Barnes, H. L., ed.), pp. 611—629. Wiley, New York.

NZMWD Report (1977). Broadlands geothermal field investigation report.

Park, I., Clarke, D. B., Johnson, J. and Keen, M. J. (1971). Seaward extension of the West Greenland tertiary volcanic province. Earth Planet. Sci. Lett. 10, 235—238.

Pedersen, A. K. (1975). New mapping in North-Eastern Disko 1972. Rapp Grønlands Geol. Unders. 69, 25—32.

Penta, F. (1954). Studies of exhalative hydrothermal phenomena. Ann. Geofisica. 7, 317—408. [In Italian]

Petracco, C. and Squarci, P. (1970). Hydrological balance of Larderello geothermal region. UN1, pp. 521—530.

Pritchett, J. W., Rice, L. F. and Garg, S. K. (1978). "Reservoir engineering data: Wairakei geothermal field, New Zealand." Vols. 1 and 2. Systems, Science and Software, La Jolla, California.

Pulvertaft, T. C. R. and Clarke, D. B. (1969). New mapping on Svartenhuk peninsula. Rapp Grønlands Geol. Unders. 11, 15—17.

Ramberg, H. (1963). Experimental study of gravity tectonics by means of centrifuged models. Bull. Geol. Soc. Uppsala 42, 1—97.

Rayleigh, Lord (1916). On convection currents in a horizontal layer of fluid when the higher temperature is on the under side. Phil. Mag. 32, 529—46.

Rogers, F. T., Schilberg, L. E. and Morrison, H. L. (1951). Convection currents in porous media. J. Appl. Phys. 22, 1476—1479.

Robertson, E. I. and Dawson, G. B. (1964). Geothermal heat flow through the soil at Wairakei. NZ J. Geol. Geophys. 7, 134—143.

Rose, A. W. (1970). Zonal relations of wallrock alteration and sulfide distribution in porphyry copper deposits. Econ. Geol. 65, 920—936.

Rouse, H., Yih, C. S. and Humphreys, H. W. (1952). Gravitational convection from a boundary source. Tellus 4, 201—210.

Roy, J. L. (1973). Latitude maps of the eastern North American—Western European paleoblock. Pap. Geol. Surv. Can. 71—23, pp. 3—22.

Rubey, W. W. (1951). Geologic history of sea water — an attempt to state the problem. Geol. Soc. Amer. Bull. 62, 1111—1147.

Saffman, P. G. and Taylor, G. I. (1958). The penetration of a fluid into a porous medium or Hele—Shaw cell containing a more viscous liquid. Proc. R. Soc. A. 245, 312—329.

Saunders, P. M. and Sheppard, P. A. (1967). Atomspheric water vapour. In "International Dictionary of Geophysics", pp. 113—116. Pergamon Press, Oxford.

Sclater, J. G. and Francheteau, J. (1970). The implications of terrestrial heat flow observations on current tectonic and geochemical models of the crust and upper mantle of the earth. Geophys. J. R. Astron. Soc. 20, 509—542.

Sestini, G. (1970). Superheating of geothermal steam. *UN*1, pp. 622–648.

Shankland, T. J. and Waff, H. S. (1977). Partial melting and electrical conductivity anomalies in the upper mantle. *J. Geophys. Res.* 82, 5409–5417.

Sheppard, P. A. (1967). Atmospheric turbulence. *In* "International Dictionary of Geophysics", p. 111–113. Pergamon Press, Oxford.

Sherzer, W. H. 1933 An interpretation of Bunsen's geyser theory. *J. Geol.* 41, 512.

Skinner, B. J., White, D. E., Rose, H. J. and Mays, R. E. (1967). Sulphides associated with the Salton Sea geothermal brines. *Econ. Geol.* 62, 316–330.

Smith, J. H. (1958). Production and utilization of geothermal steam. *NZ Engng.* 13, 354–75.

Somerscales, E. F. C. and Gazda, I. W. (1969). Thermal convection in high Prandtl number liquids at high Rayleigh numbers. *Int. J. Heat Mass Transfer* 12, 1491–1511.

Sourirajan, S. and Kennedy, G. C. (1962). The system H_2O–NaCl at elevated temperatures and pressures. *Amer. J. Sci.* 260, 115–141.

Steiner, A. (1958). Petrogenetic implications of the 1954 Ngauruhoe lava and its xenoliths. *NZ J. Geol. Geophys.* 1, 325–363.

Stocker, R. L. and Gordon, R. B. (1975). Velocity and internal friction in partial melts. *J. Geophys. Res.* 80, 4828–4836.

Studt, F. E. (1957). Wairakei hydrothermal system and the influence of ground water. *NZ J. Sci. Tech. B* 38, 596–622.

Studt, F. E. (1958). The Wairakei hydrothermal field under exploitation. *NZ J. Geol. Geophys.* 1, 703–23.

Studt, F. E. and Thompson, G. E. K. (1966). Geothermal heat flow in the North Island of New Zealand. *NZ J. Geol. Geophys.* 12, 673–683.

Sutton, F. M. (1976). Pressure temperature curves for a two-phase mixture of water and carbon dioxide. *NZ J. Sci.* 19, 297–301.

Sutton, F. M. and McNabb, A. (1977). Boiling curves at Broadlands geothermal field. *NZ J. Sci.* 20, 333–337.

Thompson, G. E. K. (1957). Some physical measurements in the Wairakei–Taupo area. *NZ DSIR Bull.* 123, 81–95.

Thompson, G. E. K. (1960). Shallow temperature surveying in the Wairakei–Taupo area. *NZ J. Geol. Geophys.* 3, 553–62.

Thompson, G. E. K., Banwell, C. J., Dawson, G. B. and Dickinson, D. J. (1961). Prospecting of hydrothermal areas by surface thermal surveys. *UN*0, paper G, p. 54.

Thorarinson, S. (1970). The Lakagigar eruption of 1783. *Bull. Volcanol.* 33, 910–927.

Thorkelsson, T. (1928). On thermal activity in Reykjanes, Iceland. *Reykjajik Pretsmidjan Gutenberg*, p. 34.

Tongiorgi, E. (ed.) (1963). "Nuclear Geology on Geothermal Areas". Lab. geol. Nucleare, Pisa.

Turner, J. S. (1973). "Buoyancy Effects in Fluids." Cambridge University Press, London.

U.K. Steam tables in SI units (1970). Edward Arnold, London.

Uyeda, S., Nomura, K. and Watanabe, T. (1976). Buried thermistor probe method — a proposal for regional heat flow studies. *In* "Volcanoes and Tectonosphere" (Aoki, H. and Iizuka, S., eds.), pp. 301–308. Tokai University Press, Tokyo.

Villas, R. N. N. (1975). Fracture analysis, hydrodynamic properties and mineral abundance in the altered igneous wallrocks of the Mayflower Mine, Park City District, Utah. Ph.D. Thesis, University of Utah.

Warner, J. and Squires, R. (1958). Liquid water content and the adiabatic model of cumulus development. *Tellus* 10, 390—394 (see also pp. 372—380).

Watanabe, R., Epp, D., Uyeda, S., Langseth, M. and Yasui, M. (1970). Heat flow in the Phillipine Sea. *Tectonophysics* 10, 205—224.

White, D. E. (1957). Thermal waters of volcanic origin. *Geol. Soc. Amer. Bull.* 68, 1637—57 (and 1659—82).

Whiteford C. M. (1976). Magnetic map of the central volcanic region. *NZ DSIR Geophys. Div. Rep.* 101.

Williams, D. L., Von Herzen, R. P., Sclater, J. G. and Anderson, R. N. (1974). The Galapagos spreading centre: lithospheric cooling and hydrothermal circulation. *Geophys. J. R. Astron. Soc.* 38, 587—608.

Williams, D. L. (1974). Heat loss and hydrothermal circulation due to sea-floor spreading. Ph.D. Thesis, Woods Hole Oceanographic Institution, Mass.

Wilson, S. H. (1939). Measurement of amount of steam escaping from areas of volcanic or solfataric activity. *Nature* 143, 802—803.

Wilson, S. H. (1955). Geothermal steam for power in New Zealand. *NZ DSIR Bull.* 117, 27—42.

Wilson, S. H. (1959). Physical and chemical investigations. *NZ DSIR Bull.* 127, 32—50.

Wilson, S. H. (1960). Physical and chemical investigation of Ketetahi hot springs. *NZ DSIR Geol. Surv. Bull.* 40, 124—144.

Wilson, S. H. (1966). Minerals in geothermal waters. *Chemistry and Industry in New Zealand* 2, 169—181.

Wooding, R. A. (1957). Steady state free thermal convection of liquid in a saturated permeable medium. *J. Fluid Mech.* 2, 273—385.

Wooding, R. A. (1958). An experiment on free thermal convection of water in a saturated permeable material. *J. Fluid Mech.* 3, 582—600.

Wooding, R. A. (1959). The stability of a viscous liquid in a vertical tube containing porous material. *Proc. R. Soc. A* 252, 120—134.

Wooding, R. A. (1960a). Instability of a viscous liquid of variable density in a vertical Hele—Shaw cell. *J. Fluid Mech.* 7, 501—515.

Wooding, R. A. (1960b). Rayleigh instability of a thermal boundary layer in flow through a porous medium. *J. Fluid Mech.* 9, 183—92.

Wooding, R. A. (1962). Free convection of fluid in a vertical tube filled with porous material. *J. Fluid Mech.* 13, 129—44.

Wooding, R. A. (1963). Convection in a saturated porous medium at large Rayleigh number or Peclet number. *J. Fluid Mech.* 15, 527—44.

Woollard, G. P. (1969). Standardization of gravity measurements, and regional variations in gravity. *In* "The Earth's Crust and Upper Mantle" (Hart, P. J., ed.), pp. 283—292, 320—341. Geophysical monograph American Geophysical Union.

Wyllie, P. J. (1971). "The Dynamic Earth". Wiley, New York.

Van Wylen, G. J. and Sonntag, R. E. (1976). "Fundamentals of classical thermodynamics" (2nd ed., SI version). Wiley, New York.

Index